Springer-Verlag Wien GmbH

# CISM COURSES AND LECTURES



The series presents lecture notes, monographs, edited works and proceedings in the field of Mechanics, Engineering, Computer Science and Applied Mathematics.
Purpose of the series is to make known in the international scientific and technical community results obtained in some of the activities organized by CISM, the International Centre for Mechanical Sciences.

INTERNATIONAL CENTRE FOR MECHANICAL SCIENCES

COURSES AND LECTURES - No. 464

# MECHANICS OF MICROSTRUCTURED MATERIALS

EDITED BY

HELMUT J. BÖHM
VIENNA UNIVERSITY OF TECHNOLOGY

Springer-Verlag Wien GmbH

The publication of this volume was co-sponsored and co-financed by the UNESCO Venice Office - Regional Bureau for Science in Europe (ROSTE) and its content corresponds to a CISM Advanced Course supported by the same UNESCO Regional Bureau.

This volume contains 180 illustrations


Originally published by CISM, Udine in 2004.
SPIN 11370475

In order to make this volume available as economically and as rapidly as possible the authors' typescripts have been reproduced in their original forms. This method unfortunately has its typographical limitations but it is hoped that they in no way distract the reader.

ISBN 978-3-211-24154-7 ISBN 978-3-7091-2776-6 (eBook)
DOI 10.1007/978-3-7091-2776-6

# PREFACE

*The last decades have witnessed a strong thrust towards improving existing structural materials and to develop new ones, the long-term aim being a capability to tailor materials for specific applications. This environment — in combination with a remarkable increase in computing power — has been instrumental in giving rise to marked improvements in modeling capabilities and experimental methods aimed at obtaining a better understanding of the thermomechanical behavior of engineering materials. One field that has benefitted strongly from these trends is continuum mechanical methods for studying inhomogeneous materials.*

*The CISM Course on "Mechanics of Microstructured Materials", which was held in Udine from July 7 to July 11, 2003, accordingly aimed at presenting selected state-of-the-art theoretical tools in this field of research in close integration with related experimental developments.*

*Like the course, the present book is structured to consist of a short overview of modeling approaches in continuum micromechanics together with contributions on a number of research topics that are of current interest in the study of microstructured materials. Methods for modeling damage of and cracks in microstructured materials are a core issue, with the topics covered including descriptions of the fracture of brittle reinforcements in composites, thermomechanical cohesive zone models, and simulations of creep rupture in polycrystalline materials. A second focus lies in experimental and statistical methods for the description and classification of microstructures and statistically based methods for assessing the stress and strain fields. Advanced descriptions for the elastoplastic behavior of metals are a third point of emphasis, with contributions on crystal plasticity and discrete dislocation models. Finally, studies are provided that concentrate on modeling strategies for the thermomechanical behavior of specific microstructured materials, such as discontinuously reinforced composites and graded carbon–carbon composites. Together these contributions are intended to provide a framework for introducing the reader to current trends and different points of view in studying the mechanical behavior of microstructured materials.*

*Finally, it is my pleasure to thank the lecturers at the CISM course and contributors to this book, Javier LLorca, Peter McHugh, Ryszard Pyrz, Thomas Siegmund and Erik van der Giessen. I would also like to express my appreciation of the active participation by the 37 attendees of the course, who came from 15 different countries. Last, but not least, the efficient support by the rectors and the staff of CISM is most gratefully acknowledged.*

*Helmut J. Böhm*

## CONTENTS

# A Short Introduction to Continuum Micromechanics

Helmut J. Böhm

Institute of Lightweight Design and Structural Biomechanics,
Vienna University of Technology, Vienna, Austria

**Abstract** Basic issues in continuum mechanical modeling of microstructured materials are discussed and a number of physically based modeling approaches are presented, among them mean field and bounding methods as well as unit cell and embedding models. In addition, important aspects of multi-scale modeling strategies are addressed and a short introduction to the treatment of damage at the constituent level within micromechanical models is given.

## 1 Basic Considerations

In the following some basic issues in continuum mechanical modeling of microstructured materials are discussed. The emphasis is put on application related aspects of the field of continuum micromechanics of inhomogeneous materials, the topics being presented mainly from an "engineering point of view". For more formal treatments the reader is referred to the books by Mura (1987), Aboudi (1991), Nemat-Nasser and Hori (1993), Suquet (1997a), Markov and Preziosi (2000), Bornert et al. (2001), Jeulin and Ostoja-Starzewski (2001), Torquato (2001), and Milton (2002). Short overviews were given e.g. by Hashin (1983) and Zaoui (2002).

For simplicity modeling approaches are presented within the framework of two-phase materials consisting of reinforcements embedded in a contiguous matrix, i.e. composites in the strict sense. The concepts involved, however, can be extended to microstructured materials containing a higher number of constituents or showing other microtopologies (e.g. polycrystals) as well as to porous and, to a considerable extent, to cellular materials. Materially and geometrically linear behavior is assumed unless explicitly stated otherwise.

### 1.1 Length Scales

As implied by their name, microstructured materials are inhomogeneous at some small length scale, where different constituents (or phases) can be distinguished. The phase arrangements at the microscale may show matrix–inclusion (composites), interwoven (e.g. some duplex steels), or granular (polycrystals) microtopologies.

The microgeometries of most practically relevant materials are too complex for a deterministic description in full detail, so that the geometrical arrangement of phases has to be treated statistically and described by appropriate distribution or correlation functions, compare Torquato (2001), Jeulin and Ostoja-Starzewski (2001) and Pyrz (2004). Accordingly, statistically based microstructure–property relations for inhomogeneous materials are an important issue in micromechanical studies (Pyrz, 2004; Siegmund et al., 2004).

In many materials of interest the microstructure is statistically homogeneous (or stationary), i.e. the statistical descriptors of the geometrical arrangement do not depend on the position they are evaluated at. For such systems it makes sense to define volume averaged properties, which are then independent of the size and position of the volume element considered, provided it is sufficiently large. A volume element that contains all the necessary information for the statistical description of a given microstructure is called a reference volume element (RVE). Statistically homogeneous microstructures may be statistically isotropic, the statistical descriptors being rotationally invariant, or statistically anisotropic, in which case there are preferred directions in the shapes, orientations or positions of the constituents.

In the simplest case a microstructured material is associated with two characteristic length scales, viz. the length scale of components or samples, called the macroscale, and the much smaller length scale at which the material is inhomogeneous, referred to as the microscale. At the macroscale, the behavior of a microstructured material can be described by that of a suitably chosen equivalent homogeneous material, in which the stresses and strains vary "slowly" according to the macroscopic geometry and loading conditions. In the absence of marked macroscopic gradients in composition and loads, the slow variations are not evident at the microscale, where the heterogeneity of the material gives rise to small-scale ("fast") fluctuations of the local stresses and strains. Accordingly, the stress and strain tensors, $\boldsymbol{\sigma}$ and $\boldsymbol{\varepsilon}$, as well as other variables may be split into slow (marked by overbars) and fast (marked by primes) contributions to give

$$\boldsymbol{\sigma}(\mathbf{x}) = \bar{\boldsymbol{\sigma}} + \boldsymbol{\sigma}'(\mathbf{x}) \qquad \text{and} \qquad \boldsymbol{\varepsilon}(\mathbf{x}) = \bar{\boldsymbol{\varepsilon}} + \boldsymbol{\varepsilon}'(\mathbf{x}) \quad . \tag{1.1}$$

For inhomogeneous materials that show sufficient separation between the length scales the relation

$$\int_{\Omega} \boldsymbol{\sigma}^*(\mathbf{x})\ \boldsymbol{\varepsilon}^*(\mathbf{x}) d\Omega = \langle \boldsymbol{\sigma}^*\ \boldsymbol{\varepsilon}^* \rangle = \langle \boldsymbol{\sigma}^* \rangle \langle \boldsymbol{\varepsilon}^* \rangle \tag{1.2}$$

can be shown to hold for general statically admissible stress fields $\boldsymbol{\sigma}^*$ and kinematically admissible strain fields $\boldsymbol{\varepsilon}^*$ (Hill, 1967). Here $\Omega$ denotes an appropriate control volume that is at least as large as an RVE. Equation (1.2) is known as Hill's macrohomogeneity condition or the Mandel–Hill condition. For the special cases of homogeneous stress and strain boundary conditions (for which the above relation can be shown to hold by transforming the volume integrals into surface integrals, to which the fluctuations $\boldsymbol{\sigma}'$ and $\boldsymbol{\varepsilon}'$ give zero contributions) it is referred to as Hill's lemma.

The macrohomogeneity condition implies that the volume averaged strain energy density of an inhomogeneous material can be obtained from the volume averages of the stresses and strains, provided the micro- and macroscales are sufficiently different. As a consequence of such relations the notion of an equivalent homogeneous material, i.e. a homogeneous material that is energetically equivalent to a given microstructured material, is viable.

In cases where the micro- and macroscales do not differ to a sufficient degree, special homogenization methods have to be used such as gradient enhanced techniques that give rise to nonlocal descriptions for the homogenized material behavior, see e.g. Kouznetsova et al. (2002).

## 1.2 Homogenization and Localization

The central aim of micromechanical descriptions of microstructured materials is the "bridging of the length scales".

On the one hand this involves deducing the macroscopic mechanical behavior (or effective response) from the known mechanical properties and microscopic arrangements of the constituents. This scale transition from the smaller to the larger length scale is referred to as homogenization and can be carried out by volume averaging, i.e. the homogenized strains and stresses take the form

$$\bar{\varepsilon} = \frac{1}{\Omega}\int_{\Omega} \varepsilon(\mathbf{x})\, d\Omega = \frac{1}{2\Omega}\int_{\Gamma} \big[\mathbf{u}(\mathbf{x}) \otimes \mathbf{n}_{\Gamma}(\mathbf{x}) + \mathbf{n}_{\Gamma}(\mathbf{x}) \otimes \mathbf{u}(\mathbf{x})\big]\, d\Gamma \tag{1.3}$$

$$\bar{\boldsymbol{\sigma}} = \frac{1}{\Omega}\int_{\Omega} \boldsymbol{\sigma}(\mathbf{x})\, d\Omega = \frac{1}{\Omega}\int_{\Gamma} \mathbf{t}(\mathbf{x}) \otimes \mathbf{x}\, d\Gamma \quad . \tag{1.4}$$

Here $\Gamma$ is the surface of the control volume $\Omega$, $\mathbf{n}_{\Gamma}$ stands for the associated surface normal unit vector, $\mathbf{u}$ denotes the displacement field, $\mathbf{t}$ is the traction vector at the surface, and $\otimes$ represents the dyadic product of vectors. Note that the volume averaged strains and stresses are fully determined by the surface displacements and tractions, respectively, provided the interfaces between the constituents are perfect and no cracks are present. In case the latter conditions are not fulfilled appropriate corrections terms must be introduced into the surface integral formulation for $\bar{\varepsilon}$ given in eqn.(1.3), see Nemat-Nasser and Hori (1993). For a discussion of averaging at large strains see e.g. Nemat-Nasser (1999). Important applications of homogenization analyses, especially in nonlinear regimes, are materials characterization, i.e. the simulation of simple mechanical tests leading e.g. to uniaxial stress vs. strain diagrams, and micromechanically based constitutive models, which are capable of providing estimates of the material response for any sequence of loading conditions.

On the other hand, the local stress and strain states at the microscale under given macroscopic loading conditions are of considerable interest, especially if the macroscopic damage and failure behavior of a microstructured material is to be understood and described in terms of microscopic damage mechanisms. Such a transition from the larger to the smaller length scale is referred to as localization. In the isothermal case it can be formally denoted as

$$\varepsilon(\mathbf{x}) = \mathbf{A}(\mathbf{x})\, \bar{\varepsilon} \tag{1.5}$$

$$\boldsymbol{\sigma}(\mathbf{x}) = \mathbf{B}(\mathbf{x})\, \bar{\boldsymbol{\sigma}} \tag{1.6}$$

for the strain and stress fields, where $\mathbf{A}(\mathbf{x})$ and $\mathbf{B}(\mathbf{x})$ are known as the strain and stress concentration tensors (or localization tensors), respectively.

Due to the complexity of real microstructured materials approximations must be resorted to in homogenization and localization analyses. In the following two important groups of modeling strategies serving this purpose are discussed.

# 2 Mean Field Estimates and Bounding Methods

One strategy for modeling the mechanical and thermal expansion behavior of microstructured materials consists in approximating the actual stress and strain fields by phase-wise constant fields, the statistics of the microgeometry being accounted for in terms of simple descriptors such as the phase volume fractions and the overall symmetry. This modeling philosophy leads to mean field estimates, Hashin–Shtrikman estimates, and Hashin–Shtrikman-type bounds, which in their basic forms provide analytical results for the linear behavior of inhomogeneous materials. Most of these approaches can be extended to the nonlinear range (the elastoplastic behavior being of special interest in the present context), where solutions typically are obtained on the basis of elastic reference composites and numerical methods may have to be used.

## 2.1 Basic Relations

Assuming thermoelastic behavior for both phases, i.e.

$$\bar{\boldsymbol{\sigma}}^{(\mathrm{p})} = \mathbf{E}^{(\mathrm{p})}\bar{\boldsymbol{\varepsilon}}^{(\mathrm{p})} + \mathbf{e}\Delta T \qquad \text{and} \qquad \bar{\boldsymbol{\varepsilon}}^{(\mathrm{p})} = \mathbf{C}^{(\mathrm{p})}\bar{\boldsymbol{\sigma}}^{(\mathrm{p})} + \boldsymbol{\alpha}\Delta T \quad , \tag{2.1}$$

the overall behavior of a microstructured two-phase material is also thermoelastic and can be described as

$$\bar{\boldsymbol{\sigma}} = \mathbf{E}^{*}\bar{\boldsymbol{\varepsilon}} + \mathbf{e}^{*}\Delta T \qquad \text{and} \qquad \bar{\boldsymbol{\varepsilon}} = \mathbf{C}^{*}\bar{\boldsymbol{\sigma}} + \boldsymbol{\alpha}^{*}\Delta T \quad . \tag{2.2}$$

Here $\mathbf{E}$ and $\mathbf{C}$ denote elastic and compliance tensors with $\mathbf{C} = \mathbf{E}^{-1}$, $\boldsymbol{\alpha}$ and $\mathbf{e}$ denote thermal expansion and specific thermal stress tensors with $\mathbf{e} = -\mathbf{E}\boldsymbol{\alpha}$, and $\Delta T$ is a spatially uniform temperature difference with respect to some stress-free reference temperature.

Phase-wise constant strain and stress fields are obtained by phase averaging as

$$\bar{\boldsymbol{\varepsilon}}^{(\mathrm{p})} = \frac{1}{\Omega^{(\mathrm{p})}} \int_{\Omega^{(\mathrm{p})}} \boldsymbol{\varepsilon}(\mathbf{x})\, d\Omega \qquad \text{and} \qquad \bar{\boldsymbol{\sigma}}^{(\mathrm{p})} = \frac{1}{\Omega^{(\mathrm{p})}} \int_{\Omega^{(\mathrm{p})}} \boldsymbol{\sigma}(\mathbf{x})\, d\Omega \quad , \tag{2.3}$$

where $\Omega^{(\mathrm{p})} \subset \Omega$ denotes the volume occupied by phase (p). The localization relations then take the form

$$\bar{\boldsymbol{\varepsilon}}^{(\mathrm{p})} = \bar{\mathbf{A}}^{(\mathrm{p})}\,\bar{\boldsymbol{\varepsilon}} + \bar{\mathbf{a}}^{(\mathrm{p})}\Delta T \tag{2.4}$$

$$\bar{\boldsymbol{\sigma}}^{(\mathrm{p})} = \bar{\mathbf{B}}^{(\mathrm{p})}\,\bar{\boldsymbol{\sigma}} + \bar{\mathbf{b}}^{(\mathrm{p})}\Delta T \quad , \tag{2.5}$$

$\bar{\mathbf{a}}$ and $\bar{\mathbf{b}}$ being known as the (phase averaged) thermal strain and stress concentration tensors. Note that, in contrast to eqns.(1.5) and (1.6), the phase averaged elastic concentration tensors $\bar{\mathbf{A}}^{(\mathrm{p})}$ and $\bar{\mathbf{B}}^{(\mathrm{p})}$ are no longer functions of the position $\mathbf{x}$.

Denoting the phase volume fractions as $\xi^{(\mathrm{p})} = \Omega^{(\mathrm{p})}/\Omega$, the relations

$$\bar{\boldsymbol{\varepsilon}} = \xi^{(\mathrm{r})}\bar{\boldsymbol{\varepsilon}}^{(\mathrm{r})} + \xi^{(\mathrm{m})}\bar{\boldsymbol{\varepsilon}}^{(\mathrm{m})} \qquad \text{and} \qquad \bar{\boldsymbol{\sigma}} = \xi^{(\mathrm{r})}\bar{\boldsymbol{\sigma}}^{(\mathrm{r})} + \xi^{(\mathrm{m})}\bar{\boldsymbol{\sigma}}^{(\mathrm{m})} \tag{2.6}$$

follow immediately from the definitions of the macroscopic and phase averaged fields and the expressions

$$\xi^{(\mathrm{r})}\bar{\mathbf{A}}^{(\mathrm{r})} + \xi^{(\mathrm{m})}\bar{\mathbf{A}}^{(\mathrm{m})} = \mathbf{I} \qquad \xi^{(\mathrm{r})}\bar{\mathbf{B}}^{(\mathrm{r})} + \xi^{(\mathrm{m})}\bar{\mathbf{B}}^{(\mathrm{m})} = \mathbf{I} \tag{2.7}$$

$$\xi^{(\mathrm{r})}\bar{\mathbf{a}}^{(\mathrm{r})} + \xi^{(\mathrm{m})}\bar{\mathbf{a}}^{(\mathrm{m})} = \mathbf{0} \qquad \xi^{(\mathrm{r})}\bar{\mathbf{b}}^{(\mathrm{r})} + \xi^{(\mathrm{m})}\bar{\mathbf{b}}^{(\mathrm{m})} = \mathbf{0} \quad , \tag{2.8}$$

which link the concentration tensors of reinforcements (r) and matrix (m), can be obtained from eqns.(2.1) to (2.5). Here **I** stands for the unit tensor and **0** for the null tensor. Furthermore, the effective thermoelastic tensors of a microstructured material can be expressed in terms of the concentration tensors and the material properties of the constituents as

$$\begin{aligned}
\mathbf{E}^* &= \xi^{(r)}\mathbf{E}^{(r)}\bar{\mathbf{A}}^{(r)} + \xi^{(m)}\mathbf{E}^{(m)}\bar{\mathbf{A}}^{(m)} = \mathbf{E}^{(m)} + \xi^{(r)}(\mathbf{E}^{(r)} - \mathbf{E}^{(m)})\bar{\mathbf{A}}^{(r)} &\quad (2.9)\\
\mathbf{C}^* &= \xi^{(r)}\mathbf{C}^{(r)}\bar{\mathbf{B}}^{(r)} + \xi^{(m)}\mathbf{C}^{(m)}\bar{\mathbf{B}}^{(m)} = \mathbf{C}^{(m)} + \xi^{(r)}(\mathbf{C}^{(r)} - \mathbf{C}^{(m)})\bar{\mathbf{B}}^{(r)} &\quad (2.10)\\
\mathbf{e}^* &= \xi^{(r)}[\mathbf{E}^{(r)}\bar{\mathbf{A}}^{(r)} + \mathbf{e}^{(r)}] + \xi^{(m)}[\mathbf{E}^{(m)}\bar{\mathbf{A}}^{(m)} + \mathbf{e}^{(m)}] &\quad (2.11)\\
\boldsymbol{\alpha}^* &= \xi^{(r)}[\mathbf{C}^{(r)}\bar{\mathbf{B}}^{(r)} + \boldsymbol{\alpha}^{(r)}] + \xi^{(m)}[\mathbf{C}^{(m)}\bar{\mathbf{B}}^{(m)} + \boldsymbol{\alpha}^{(m)}] \quad . &\quad (2.12)
\end{aligned}$$

The elastic stress and strain concentration tensors of a given phase are linked by the relations

$$\bar{\mathbf{A}}^{(p)} = \mathbf{C}^{(p)}\bar{\mathbf{B}}^{(p)}\mathbf{E}^* \qquad \text{and} \qquad \bar{\mathbf{B}}^{(p)} = \mathbf{E}^{(p)}\bar{\mathbf{A}}^{(p)}\mathbf{C}^* \quad , \tag{2.13}$$

from which the effective elastic tensors may be eliminated by using eqns.(2.10) and (2.9). In addition, the thermal strain and stress concentration tensors are connected to the elastic concentration tensors by expressions such as

$$\bar{\mathbf{a}}^{(p)} = \mathbf{C}^{(p)}\{[\mathbf{A}^{(p)}]^T - \mathbf{I}\}\mathbf{e}^{(p)} \quad \text{and} \quad \bar{\mathbf{b}}^{(p)} = \mathbf{E}^{(p)}\{[\mathbf{B}^{(p)}]^T - \mathbf{I}\}\boldsymbol{\alpha}^{(p)} \quad , \tag{2.14}$$

where the superscript $T$ denotes transposition. Alternative relations were given e.g. by Benveniste and Dvorak (1990).

Equations (2.7) to (2.14), on the one hand, imply that within the mean field framework both homogenization and localization problems are essentially solved once the concentration tensors are known. On the other hand, they show that all concentration tensors of a two-phase material are available once any of the elastic concentration tensors is known. Alternatively, the concentration tensors can be deduced when the both the phase-level and the overall elastic behavior are known.

## 2.2 Eshelby's Model for Dilute Reinforcement Volume Fractions

From section 2.1 it is evident that the derivation of appropriate concentration tensors is of major importance in continuum micromechanics of inhomogeneous materials. A group of results obtained by Eshelby (1957) forms the basis of many such expressions.

The immediate subject of Eshelby's studies were the stress and strain distributions in infinite elastic media containing a subregion, referred to as a homogeneous inclusion, that spontaneously changes its shape and/or size. His results show that a uniform stress-free strain (or eigenstrain) acting on an inclusion gives rise to a uniform in-situ ("constrained") strain state, provided the inclusion is of ellipsoidal shape. This constrained strain, $\boldsymbol{\varepsilon}_c$, can be related to the stress-free strain, $\boldsymbol{\varepsilon}_t$, by the expression

$$\boldsymbol{\varepsilon}_c = \mathbf{S}\boldsymbol{\varepsilon}_t \quad , \tag{2.15}$$

**S** being known as the Eshelby tensor.

The above result on homogeneous inclusions forms the basis for describing the behavior of an ellipsoidal inhomogeneity (with elastic properties $\mathbf{E}^{(r)}$) embedded in an infinite matrix (with elastic properties $\mathbf{E}^{(m)}$). When such a system is subjected to a far field strain $\varepsilon^{a}$ the elastic mismatch between the constituents gives rise to stress and strain fields in the inhomogeneity and in the surrounding matrix. Identical stress and strain states, however, can also be induced in a homogeneous inclusion by subjecting it to an appropriately chosen "equivalent" eigenstrain, $\varepsilon_{\tau}$. For this homogeneous system eqn.(2.15) holds, so that the conditions of equal stress and strain fields in the (actual) inhomogeneity and in the (fictitious) equivalent inclusion can be denoted as

$$\begin{aligned}\bar{\sigma}^{(r)} &= \mathbf{E}^{(r)}\bar{\varepsilon}^{(r)} = \mathbf{E}^{(m)}(\bar{\varepsilon}^{(r)} - \varepsilon_{\tau}) \\ \bar{\varepsilon}^{(r)} &= \varepsilon^{a} + \varepsilon_{c} = \varepsilon^{a} + \mathbf{S}\varepsilon_{\tau} \quad . \end{aligned} \tag{2.16}$$

From these relations the equivalent eigenstrain $\varepsilon_{\tau}$ can be determined and the strain in the inhomogeneity $\bar{\varepsilon}^{(r)}$ can consequently be expressed in terms of the far field strain $\bar{\varepsilon} = \varepsilon^{a}$, which allows the reinforcement strain and stress concentration tensors of the system to be extracted as

$$\bar{\mathbf{A}}^{(r)}_{\text{dil}} = [\mathbf{I} + \mathbf{S}\mathbf{C}^{(m)}(\mathbf{E}^{(r)} - \mathbf{E}^{(m)})]^{-1} \tag{2.17}$$

$$\bar{\mathbf{B}}^{(r)}_{\text{dil}} = [\mathbf{I} + (\mathbf{I} - \mathbf{S})\mathbf{E}^{(m)}(\mathbf{C}^{(r)} - \mathbf{C}^{(m)})]^{-1} \quad . \tag{2.18}$$

Here the subscript "dil" is used to indicate that the concentration tensors pertain to dilute, i.e. very low, reinforcement volume fractions for which the stress and strain fields in a given inhomogeneity are not perturbed by those of its neighbors. Equations (2.17) and (2.18) were derived by Hill (1965b) and elaborated by Benveniste (1987); different but equivalent expressions were given by a number of authors, among them Wakashima et al. (1988) and Ju and Sun (1999).

The Eshelby tensor $\mathbf{S}$ depends only on the material properties of the matrix and on the shape of the inhomogeneities, which can be described by an aspect ratio $a$ for the practically important case when they are spheroids, i.e. ellipsoids of rotation. Expressions for the components of the Eshelby tensor pertaining to spheroidal inclusions in an isotropic matrix are available in many works, among them (Tandon and Weng, 1984), (Mura, 1987), and (Clyne and Withers, 1993). These formulae are very simple for continuous fibers ($a \to \infty$), spherical inclusions ($a = 1$), and thin circular discs ($a \to 0$). Analytical expressions for the Eshelby tensor of spheroidal inclusions in an appropriately oriented transversely isotropic matrix were published e.g. by Withers (1989). For cases of lower material symmetry the Eshelby tensor can be evaluated numerically, compare (Gavazzi and Lagoudas, 1990).

Finally, it is worth noting that the inhomogeneous stress and strain fields outside ellipsoidal inclusions or inhomogeneities as well as the stress and strain jumps at the interface to the matrix can also be described analytically. This may be done in terms of a position dependent exterior field Eshelby tensor, see (Eshelby, 1959), (Mura, 1987), and (Ju and Sun, 1999). When eqns.(2.7) and (2.17) are combined to obtain mean field matrix strain concentration tensors for the dilute case, the variations in the matrix strain field are, of course, averaged out.

## 2.3 Estimates for Nondilute Reinforcement Volume Fractions

In composites with reinforcement volume fractions of more than a few percent interaction effects between individual reinforcements, on the hand, give rise to inhomogeneous stress and strain fields within each inhomogeneity ("intra-particle fluctuations"). On the other hand, they cause the levels of the average stresses and strains in individual inhomogeneities to differ ("inter-particle fluctuations"). Within the mean field framework such reinforcement interaction effects as well as the concomitant perturbations of the stress and strain fields in the matrix are accounted for via suitable approximations.

**Mori–Tanaka Approaches.** At dilute reinforcement volume fractions inhomogeneities can be viewed as being acted upon by the macroscopic ("applied") stresses and strains, compare eqn.(2.16). One strategy for introducing interaction effects in nondilute cases amounts to having the inhomogeneities "feel" suitable effective homogeneous matrix strain and stress fields instead, which account for all perturbations from other reinforcements. Following Benveniste (1987) this idea leads to the ansatz

$$\begin{aligned}\bar{\boldsymbol{\varepsilon}}^{(\mathrm{r})} &= \bar{\mathbf{A}}^{(\mathrm{r})}_{\mathrm{dil}}\bar{\boldsymbol{\varepsilon}}^{(\mathrm{m})} &= \bar{\mathbf{A}}^{(\mathrm{r})}_{\mathrm{dil}}\bar{\mathbf{A}}^{(\mathrm{r})}_{\mathrm{M}}\bar{\boldsymbol{\varepsilon}} \\ \bar{\boldsymbol{\sigma}}^{(\mathrm{r})} &= \bar{\mathbf{B}}^{(\mathrm{r})}_{\mathrm{dil}}\bar{\boldsymbol{\sigma}}^{(\mathrm{m})} &= \bar{\mathbf{B}}^{(\mathrm{r})}_{\mathrm{dil}}\bar{\mathbf{B}}^{(\mathrm{r})}_{\mathrm{M}}\bar{\boldsymbol{\sigma}} \quad ,\end{aligned} \tag{2.19}$$

which may be viewed as the central assumption of the so called effective field approximations or Mori–Tanaka estimates. By combining eqns. (2.6) and (2.19) the Mori–Tanaka matrix concentration tensors can be expressed in terms of dilute concentration tensors as

$$\bar{\mathbf{A}}^{(\mathrm{m})}_{\mathrm{M}} = [\xi^{(\mathrm{m})}\mathbf{I} + \xi^{(\mathrm{r})}\bar{\mathbf{A}}^{(\mathrm{r})}_{\mathrm{dil}}]^{-1} \qquad \text{and} \qquad \bar{\mathbf{B}}^{(\mathrm{m})}_{\mathrm{M}} = [\xi^{(\mathrm{m})}\mathbf{I} + \xi^{(\mathrm{r})}\bar{\mathbf{B}}^{(\mathrm{r})}_{\mathrm{dil}}]^{-1} \quad , \tag{2.20}$$

with the corresponding relations for the reinforcement concentration tensors taking the form

$$\bar{\mathbf{A}}^{(\mathrm{r})}_{\mathrm{M}} = \bar{\mathbf{A}}^{(\mathrm{r})}_{\mathrm{dil}}\,[\xi^{(\mathrm{m})}\mathbf{I} + \xi^{(\mathrm{r})}\bar{\mathbf{A}}^{(\mathrm{r})}_{\mathrm{dil}}]^{-1} \qquad \text{and} \qquad \bar{\mathbf{B}}^{(\mathrm{r})}_{\mathrm{M}} = \bar{\mathbf{B}}^{(\mathrm{r})}_{\mathrm{dil}}\,[\xi^{(\mathrm{m})}\mathbf{I} + \xi^{(\mathrm{r})}\bar{\mathbf{B}}^{(\mathrm{r})}_{\mathrm{dil}}]^{-1} \quad . \tag{2.21}$$

Using the above concentration tensors the overall elastic and thermoelastic tensors can be evaluated via the relations given in section 2.1.

A number of authors gave different but essentially equivalent Mori–Tanaka-type expressions for the phase concentration tensors of two-phase materials, among them Pedersen (1983), Wakashima et al. (1988), Taya et al. (1991) and Clyne and Withers (1993). Tandon and Weng (1984) formulated a Mori–Tanaka method that directly yields estimates for the overall elasticity tensor in the form

$$\mathbf{E}^{*}_{\mathrm{M}} = \mathbf{E}^{(\mathrm{m})}\Big\{\mathbf{I} - \xi^{(\mathrm{r})}\big[(\mathbf{E}^{(\mathrm{r})} - \mathbf{E}^{(\mathrm{m})})\big(\mathbf{S} - \xi^{(\mathrm{r})}(\mathbf{S} - \mathbf{I})\big) + \mathbf{E}^{(\mathrm{m})}\big]^{-1}(\mathbf{E}^{(\mathrm{r})} - \mathbf{E}^{(\mathrm{m})})\Big\}^{-1} \tag{2.22}$$

which can be modified for describing porous materials by letting $\mathbf{E}^{(\mathrm{r})} \to \mathbf{0}$ which gives

$$\mathbf{E}^{*}_{\mathrm{M,por}} = \mathbf{E}^{(\mathrm{m})}\Big[\mathbf{I} + \frac{\xi^{(\mathrm{r})}}{\xi^{(\mathrm{m})}}(\mathbf{I} - \mathbf{S})^{-1}\Big]^{-1} \quad . \tag{2.23}$$

Mori–Tanaka-type theories cover the full range of phase volume fractions, $0 \le \xi^{(\mathrm{r})} \le 1$, and correspond to microstructures consisting of an ellipsoidal distribution of aligned

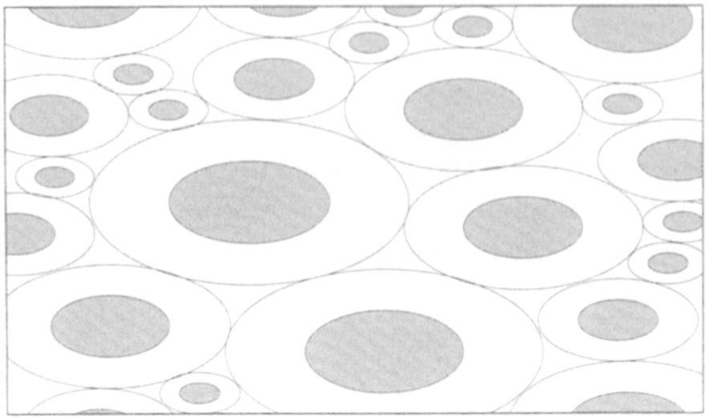

**Figure 1.** Sketch of an ellipsoidal spatial arrangement of aligned ellipsoidal reinforcements ($a$=2.0) of the type underlying Mori–Tanaka estimates.

ellipsoidal reinforcements (Ponte Castañeda and Willis, 1995), both sets of ellipsoids having the same aspect ratio and orientation, compare Figure 1. Accordingly, Mori–Tanaka methods are well suited for describing materials with aligned matrix–inclusion microtopologies; for extensions to the nonaligned case see e.g. (Böhm, 2004). For the typical case of stiffer reinforcements embedded in a more compliant matrix, Mori–Tanaka estimates for the effective elastic moduli are always on the low side, compare section 2.4. Mori–Tanaka predictions for the overall behavior of porous materials (which may be viewed as extremely compliant inhomogeneities embedded in a matrix of finite stiffness), however, are too stiff.

Even though the actual effective properties of composites may be underestimated by Mori–Tanaka methods to a noticeable degree in situations of elevated elastic contrast and at reinforcement volume fractions beyond, say, $\xi^{(\mathrm{r})} = 0.3$ (compare Figure 2 as well as Table 1), effective field methods are very useful micromechanical tools. On the one hand, their tendency towards underestimating the elastic stiffnesses of ideal composites is not necessarily a major liability in view of the presence of flaws in most real composites and, on the other hand, being explicit methods, they allow to carry out homogenization and localization analyses at minimum computational costs.

**Self-Consistent Approximations.** An alternative strategy for obtaining mean field estimates for the thermoelastic behavior of non-dilute inhomogeneous materials is based on the idea of placing each inhomogeneity in the effective material. Because the homogenized response is not known a priori, such effective medium theories are implicit methods that employ iterative self-consistent algorithms for evaluating the effective behavior.

Taking eqn.(2.9) as starting point and following Hill (1965b) the effective elastic tensor for the so-called classical (or two-phase) self-consistent scheme can be written as

$$\mathbf{E}^{*}_{\mathrm{SCS}} = \mathbf{E}^{(\mathrm{m})} + \xi^{(\mathrm{r})}[\mathbf{E}^{(\mathrm{r})} - \mathbf{E}^{(\mathrm{m})}]\bar{\mathbf{A}}^{(\mathrm{r},*)}_{\mathrm{dil}} \tag{2.24}$$

Here $\bar{\mathbf{A}}^{(r,*)}_{\mathrm{dil}}$ denotes the strain concentration tensor of an inhomogeneity embedded in the effective medium, which can be obtained by appropriately modifying eqn.(2.17). The self-consistent iteration scheme then takes the form

$$\begin{aligned} \mathbf{E}_{n+1} &= \mathbf{E}^{(\mathrm{m})} + \xi^{(\mathrm{r})}[\mathbf{E}^{(\mathrm{r})} - \mathbf{E}^{(\mathrm{m})}][\mathbf{I} + \mathbf{S}_n\mathbf{C}_n(\mathbf{E}^{(\mathrm{r})} - \mathbf{E}_n)]^{-1} \\ \mathbf{C}_{n+1} &= \left[\mathbf{E}_{n+1}\right]^{-1} \quad , \end{aligned} \tag{2.25}$$

where $\mathbf{E}_n$ is the $n$-th approximation to $\mathbf{E}_{\mathrm{SCS}}$. The Eshelby tensor used in this expression, $\mathbf{S}_n$, has to be evaluated with respect to the elastic properties of the current approximation to behavior of the effective material, $\mathbf{E}_n$.

Generally, classical self-consistent schemes are best suited for describing the overall elastic properties of two-phase materials that do not show a matrix–inclusion microtopology in at least part of their range of volume fractions (e.g. polycrystals, for which multi-phase analogs of eqn.(2.25) are often used, or materials with two interwoven phases, both of which percolate within a band of volume fractions around $\xi^{(\mathrm{r})}$=0.5). They may also be used to advantage for studying composite materials with a polycrystalline matrix in which the sizes of the reinforcements and the matrix grains are similar (Corvasce et al., 1991), the latter being modeled explicitly within a multi-phase scheme.

**Hashin–Shtrikman Estimates.** The stress field in an inhomogeneous material can be expressed in terms of that in a homogeneous comparison (or reference) medium as

$$\boldsymbol{\sigma}(\mathbf{x}) = \mathbf{E}^0\boldsymbol{\varepsilon}(\mathbf{x}) + \boldsymbol{\tau}(\mathbf{x}) \quad , \tag{2.26}$$

where $\boldsymbol{\tau}(\mathbf{x})$ is known as the polarization tensor and $\mathbf{E}^0$ is the elastic tensor of the comparison material. Phase-wise constant polarization tensors

$$\bar{\boldsymbol{\tau}}^{(\mathrm{p})} = (\mathbf{E}^{(\mathrm{p})} - \mathbf{E}^0)\bar{\boldsymbol{\varepsilon}}^{(\mathrm{p})} \tag{2.27}$$

can be obtained by phase averaging.

Equation (2.27) can be used as the starting point for deriving general mean field strain concentration tensors of the type

$$\bar{\mathbf{A}}^{(\mathrm{p})}_{\mathrm{HS}} = (\mathbf{L}^0 + \mathbf{E}^{(\mathrm{p})})^{-1}\left[\xi^{(\mathrm{r})}(\mathbf{L}^0 + \mathbf{E}^{(\mathrm{r})})^{-1} + \xi^{(\mathrm{m})}(\mathbf{L}^0 + \mathbf{E}^{(\mathrm{m})})^{-1}\right]^{-1} \quad . \tag{2.28}$$

This expression leads to estimates of the overall elastic stiffness of microstructured two-phase materials in the form

$$\mathbf{E}^*_{\mathrm{HS}} = \mathbf{E}^{(\mathrm{m})} + \xi^{(\mathrm{r})}(\mathbf{L}^0 - \mathbf{E}^{(\mathrm{m})})\left[\mathbf{I} + \xi^{(\mathrm{m})}(\mathbf{L}^0 + \mathbf{E}^{(\mathrm{m})})^{-1}(\mathbf{E}^{(\mathrm{r})} - \mathbf{E}^{(\mathrm{m})})\right]^{-1} \quad , \tag{2.29}$$

which are referred to as Hashin–Shtrikman elastic tensors. The tensor $\mathbf{L}^0$ appearing in eqns.(2.28) and (2.29) is known as the overall constraint tensor (Hill, 1965a) and is defined as

$$\mathbf{L}^0 = \mathbf{E}^0[(\mathbf{S}^0)^{-1} - \mathbf{I}] \quad . \tag{2.30}$$

Here the Eshelby tensor $\mathbf{S}^0$ must be evaluated with respect to the comparison material. $\mathbf{L}^0$ can be shown to depend on the two-point statistics of the microgeometrical arrangement, see (Bornert, 2001).

Standard mean field models can be obtained as special cases of Hashin–Shtrikman methods by appropriate choices of the comparison material. For example, using the matrix as the comparison material results in Mori–Tanaka methods, whereas the selection of the effective material as comparison material leads to classical self-consistent schemes.

The above Hashin–Shtrikman formalism can be extended in a number of ways, one of which consists in not only studying inhomogeneities but more complex geometrical entities, called patterns or motifs, embedded in a matrix, see (Bornert, 2001). Typically the stress and strain fields in such patterns are inhomogeneous even in the dilute case and numerical methods have to be resorted to in evaluating the corresponding overall constraint tensors. A simple example of such a pattern is a spherical particle or cylindrical fiber surrounded by a layer of matrix at an appropriate volume fraction, which, when embedded in the effective material, allows to recover the three-phase or generalized self-consistent scheme of Christensen and Lo (1979). The latter method is well suited for describing composites reinforced by spherical particles or aligned cylindrical fibers and leads to slightly stiffer overall moduli than Mori–Tanaka methods for composites consisting of stiff reinforcements embedded in a softer matrix.

### 2.4 Bounds

Rigorous bounds for the overall elastic properties of inhomogeneous materials can be obtained from variational (minimum energy) principles in combination with appropriate trial fields.

**Voigt and Reuss Bounds.** Standard expressions for the minimum potential energy and the minimum complementary energy in combination with uniform stress and strain trial functions lead to the simplest variational bounding expressions, the upper bounds of Voigt (1889) and the lower bounds of Reuss (1929). In their tensorial form (Hill, 1952) they are known as the Hill bounds and can be written as

$$\Big[\sum_{(p)} \xi^{(p)} \mathbf{C}^{(p)}\Big]^{-1} \leq \mathbf{E}^* \leq \sum_{(p)} \xi^{(p)} \mathbf{E}^{(p)} \quad . \tag{2.31}$$

These bounds, while universal and simple, contain little information on the microgeometry of an inhomogeneous material and are typically too slack to be of practical use. In contrast to the other estimates and bounds given in the present section, however, eqn.(2.31) also is strictly fulfilled for control volumes that are too small to be RVEs.

**Hashin–Shtrikman-Type Bounds.** Much tighter bounds can be obtained from a variational formulation due to Hashin and Shtrikman (1963) that employs polarization fields, see eqn.(2.26), to separate fast and slow variables. By optimizing phase-wise constant polarizations (and thus selecting optimal comparison materials) it can be shown that the use of the softer constituent as reference material leads to lower bounds for the effective behavior of elastic materials, whereas upper bounds are obtained when the stiffer constituent is chosen as reference medium. The original Hashin–Shtrikman bounds were derived for overall isotropic materials in the form of expressions for the effective shear moduli $G^*$ and bulk moduli $K^*$, from which bounds on the effective Young's moduli $E^*$ and the effective Poisson numbers $\nu^*$ can also be obtained, see Hashin (1983) and Zimmerman (1992). In addition, Hashin–Shtrikman bounds were given for the microgeometries in which the constituents are continuous and aligned in one direction (such as continuously reinforced unidirectional composites), see e.g. (Hashin, 1983).

Further developments led to the Walpole (1966) and Willis (1977) bounds, which pertain to thermoelastically anisotropic overall behavior due to anisotropic constituents, statistically anisotropic phase arrangements, and/or aligned nonspherical reinforcements, a typical example being the ellipsoidal microgeometry sketched in Figure 1. These bounds include the Hashin–Shtrikman bounds discussed above as special cases and they can also handle situations where a "softest" and "stiffest" constituent cannot be identified unequivocally (e.g. for constituents with $(K^{(\mathrm{r})} - K^{(\mathrm{m})})(G^{(\mathrm{r})} - G^{(\mathrm{m})}) < 0$) and a fictive reference material must be used. It is worth noting that in the absence of complications of the above kind, Mori–Tanaka estimates correspond to one of the extended Hashin–Shtrikman bounds for two-phase materials (Weng, 1990), the other being obtainable by a "color inversion" (i.e. exchanging the material properties of matrix and reinforcements).

Hashin–Shtrikman-type bounds are optimal in the sense that they can be shown to be the best bounds that can be obtained on the basis of two-point statistics. For porous materials only upper Hashin–Shtrikman bounds can be obtained, the lower bounds of the elastic moduli being zero.

**Improved Bounds.** When more complex trial functions are used in variational bounding, their optimization requires statistical information on the phase arrangement in the form of higher correlation functions. This way improved bounds can be generated that are significantly tighter than Hashin–Shtrikman-type expressions.

Three-point bounds for statistically isotropic two-phase materials can be formulated in such a way that the information on the phase arrangement statistics is contained in two three-point microstructural parameters, $\eta(\xi^{(\mathrm{r})})$ and $\zeta(\xi^{(\mathrm{r})})$. On the one hand, these parameters can be evaluated from actual samples by image analysis. On the other hand, analytical expressions or tabulated data in terms of the inclusion volume fraction $\xi^{(\mathrm{r})}$ are available for a number of generic microgeometries. Among the latter are statistically homogeneous isotropic materials containing identical, bidisperse and polydisperse impenetrable ("hard") spheres (describing matrix–inclusion composites), and interpenetrating spheres ("boolean models" that can describe many interwoven phase arrangements), as well as statistically homogeneous transversely isotropic materials that contain impenetrable or interpenetrating aligned cylinders.

References to some three-point bounds and to a number of expressions for $\eta$ and $\zeta$ applicable to certain two-phase composites can be found in section 2.5, where results from a number of mean field and bounding approaches are compared. For reviews of higher order bounds for elastic (as well as other) properties of inhomogeneous materials see (Torquato, 1991) and (Torquato, 2001). In the present context the second order estimates developed by Torquato (1998) are also worth mentioning, which use the same microstructural information as the three-point bounds and always lie between them. They give excellent analytical predictions for the overall thermoelastic response of inhomogeneous materials that show the appropriate microstructures and are free of flaws.

## 2.5 Comparison of Predictions for the Effective Thermoelastic Behavior

In order to give some idea of typical results obtained from mean field and bounding approaches, in this section some predictions for the overall thermoelastic behavior of an epoxy matrix reinforced by glass spheres are presented and discussed. The thermoelastic material parameters of the constituents of this two-phase composite are listed in Table 1, where the elastic contrast between particles and matrix, $E^{(r)}/E^{(m)}$, can be seen to be slightly in excess of 20.

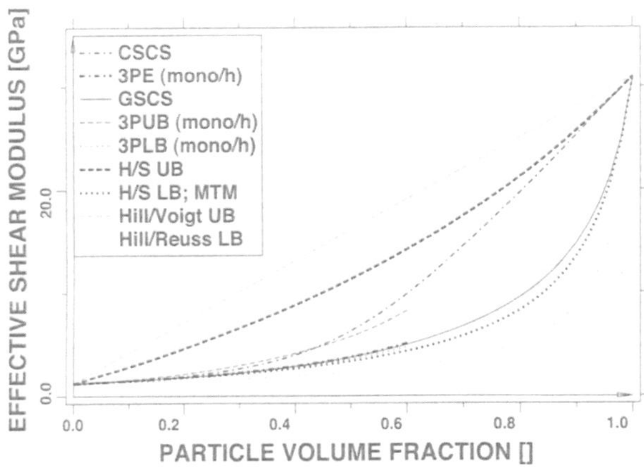


**Figure 2.** Bounds and estimates for the effective shear moduli of particle reinforced composites consisting of an epoxy matrix reinforced by glass particles.

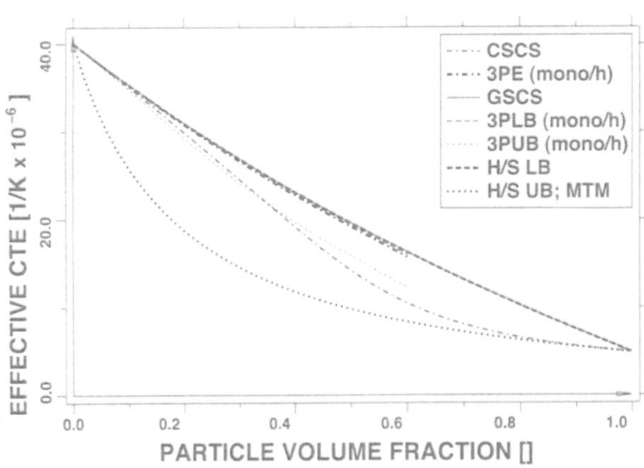


**Figure 3.** Bounds and estimates for the effective CTEs of particle reinforced composites consisting of an epoxy matrix reinforced by glass particles.

Figure 2 shows predictions for the effective shear modulus as functions of the particle volume fraction. The Voigt and Reuss bounds can be seen to be very slack, whereas the Hashin–Shtrikman (H/S) bounds are considerably tighter. The three-point bounds (3P) depicted correspond to the case of hard spheres of equal size (mono/h), the formalisms developed by Beran and Molyneux (1966), Milton (1981), and Phan-Thien and Milton (1983) being used with expressions for the statistical parameters $\eta$ and $\zeta$ reported by Miller and Torquato (1990) and Torquato et al. (1987). They are much tighter than the Hashin–Shtrikman bounds, especially for $\xi^{(r)} < 0.4$, the upper bound in particular being markedly reduced. Due to the limited volume fractions attainable with spheres of equal size the bounds are available only up to $\xi^{(r)}{\approx}0.6$. The Mori–Tanaka estimates (MTM) coincide with the lower Hashin–Shtrikman bounds and the results of both the generalized self-consistent scheme (GSCS) and Torquato's second order estimates (3PE; evaluated for monodisperse hard spheres) are somewhat stiffer than those of the MTM. All the estimates pertaining to matrix–inclusion microtopologies can be seen to be quite similar and much closer to the lower than to the upper bounds; this is a very typical behavior for composites. The results of the classical self-consistent scheme (CSCS), in contrast, are qualitatively different, being close to the lower H/S bounds for low values of $\xi^{(r)}$ and close to the upper H/S bounds as $\xi^{(r)}$ approaches unity. At intermediate volume fractions the predictions of the CSCS show a characteristic sigmoid shape.

Similar trends can be seen in the results for the thermoelastic moduli of a particle reinforced glass–epoxy composite at a volume fraction of $\xi^{(r)}{=}0.389$ listed in Table 1. The sensitivity of the three-point bounds and second order estimates to the statistics of the microgeometry is clearly evident: For both methods the effective moduli evaluated for polydispersely sized hard spheres (poly/h; three-point parameters following Thovert et al. (1990)) noticeably exceed those pertaining to uniformly sized spheres (mono/h).

A comparison of predictions for the effective coefficients of thermal expansion, $\alpha^*$, of glass particle reinforced epoxy is provided in Figure 3. For generating these curves, a relation proposed by Levin (1967) and Schapery (1968) was used that allows to obtain bounds and estimates for the effective CTE of statistically isotropic materials from bounds and estimates for the bulk modulus. Lower bounds on $K^*$ give rise to upper bounds of $\alpha^*$ and vice versa. Because they are based on the same estimates for the effective bulk modulus, the effective CTEs shown for the GSCS and the Mori–Tanaka-scheme coincide with the upper Levin/Hashin–Shtrikman bounds.

## 2.6 Inelastic Behavior

Descriptions for viscoelastic microstructured materials are closely related to those for elastic composites. Relaxation moduli and creep compliances can be obtained by applying micromechanical methods in the Laplace transform domain, where the problems are analogous to elastic ones for the same microgeometries. Standard micromechanical methods can be used in conjunction with complex moduli for steady state vibration analyses of viscoelastic composites. For correspondence principles between descriptions for elastic and viscoelastic inhomogeneous materials see e.g. Hashin (1983). The extension of mean field estimates and bounding methods to elastoplastic, viscoelastoplastic, and damaging inhomogeneous materials, however, has proven to be challenging.

**Estimates for the Elastoplastic Behavior of Microstructured Materials.** The aim of mean field models of microstructured materials with at least one elastoplastic constituent consists in providing accurate estimates for the material response for any load state and load history at reasonable computational cost. The main difficulties in attaining this goal lie in the typically strong intra-phase fluctuations of the stress and strain fields in elastoplastic inhomogeneous materials and in the hereditary nature of plasticity. The responses of the constituents can thus vary markedly at the microscale, with each point following its own trajectory in stress space. Accordingly, a 2-phase elastoplastic composite effectively behaves as a multiphase material and phase averages are less useful descriptors than in the linear elastic case.

Mean field models of elastoplastic inhomogeneous materials typically are based on solving sequences or sets of linearized problems in terms of linear inhomogeneous reference media. Accordingly, choices have to be made with respect to the linearization procedure, the linear homogenization model, and the phase-wise equivalent stresses and strains to be used in evaluating the elastoplastic constituent material behavior, see Zaoui (2001).

Several lines of development of mean field and Hashin–Shtrikman-type approaches for modeling elastoplastic inhomogeneous materials can be found in the literature. The most commonly used have been secant plasticity concepts, see e.g. Tandon and Weng (1988) or Dunn and Ledbetter (1997), and incremental plasticity models, see e.g. Hill (1965a) or Karayaka and Sehitoglu (1993). In addition, tangent concepts (Molinari et al., 1987) and affine formulations (Masson et al., 2000) were proposed for viscoplastic inhomogeneous materials; the latter also were extended to the elastoplastic case. Analogous methods can be used to describe inhomogeoneous materials at least one constituent of which is subject to damage.

Secant models aim at directly arriving at solutions in terms of the overall response for a given load state. They are based on the deformation theory of plasticity and can be formulated in terms of potentials, which allows for a concise mathematical presentation (Bornert and Suquet, 2001). Secant models are limited to monotonic loading and radial (or approximately radial) trajectories of the constituents in stress space, which precludes their use as micromechanically based constitutive models in multi-scale analyses. Advanced secant formulations, however, are highly suitable for materials characterization, where they give excellent results.

Incremental mean field models are based on formulations using phase averaged strain and stress rate tensors, $d\bar{\boldsymbol{\varepsilon}}_{\mathrm{t}}^{(\mathrm{p})}$ and $d\bar{\boldsymbol{\sigma}}_{\mathrm{t}}^{(\mathrm{p})}$. Equations (2.4) and (2.5) then take the form

$$\begin{aligned} d\bar{\boldsymbol{\varepsilon}}_{\mathrm{t}}^{(\mathrm{p})} &= \bar{\mathbf{A}}_{\mathrm{t}}^{(\mathrm{p})}\, d\bar{\boldsymbol{\varepsilon}} + \bar{\mathbf{a}}_{\mathrm{t}}^{(\mathrm{p})} dT \\ d\bar{\boldsymbol{\sigma}}_{\mathrm{t}}^{(\mathrm{p})} &= \bar{\mathbf{B}}_{\mathrm{t}}^{(\mathrm{p})}\, d\bar{\boldsymbol{\sigma}} + \bar{\mathbf{b}}_{\mathrm{t}}^{(\mathrm{p})} dT \quad , \end{aligned} \tag{2.32}$$

where $\bar{\mathbf{A}}_{\mathrm{t}}^{(\mathrm{p})}$, $\bar{\mathbf{B}}_{\mathrm{t}}^{(\mathrm{p})}$, $\bar{\mathbf{a}}_{\mathrm{t}}^{(\mathrm{p})}$, and $\bar{\mathbf{b}}_{\mathrm{t}}^{(\mathrm{p})}$ are instantaneous concentration tensors that must be evaluated by some appropriate micromechanical model on the basis of the instantaneous constituent behavior. The overall instantaneous stiffness tensor $\mathbf{E}_{\mathrm{t}}^{*}$ can then be obtained from equivalents of eqn. (2.9) such that it is a continuum tangent operator with $d\bar{\boldsymbol{\sigma}} = \mathbf{E}_{\mathrm{t}}^{*} d\bar{\boldsymbol{\varepsilon}}$. Incremental approaches are not subject to limitations in terms of the (mechanical) load cases they can handle and they can be extended to thermal loading (Pettermann, 1997). They have, however, typically shown a marked tendency towards being excessively stiff in the post-yield regime for matrix dominated deformation modes. The reason for this behavior is thought to be an accumulation of errors over the incremental procedure, so that recent algorithmic developments have been specifically aimed at this problem area (Doghri and Ouaar, 2003).

The intraphase variations of the stress and strain fields within the assumption of phase-wise uniformity tend to have a considerable influence on the predicted behavior. A typical case in point are the phase averaged von Mises equivalent stresses, which govern the onset of yielding in many plasticity models and which may be severely underestimated if they are evaluated from the phase averaged stress components. Improvements in this respect have been obtained by evaluating the phase averages of the von Mises equivalent stresses on the basis of energy considerations (Qiu and Weng, 1992) and by evaluating higher statistical moments of the microfields, see Buryachenko (1996) and Ju and Sun (2001). Such refinements have become an integral part of "modified secant models" (Ponte Castañeda and Suquet, 1998) but have seen little use in incremental approaches.

A specific type of geometrical nonlinearity can occur in porous elastoplastic materials, where the geometries of the voids may evolve in the course of loading histories. Problems of this type can be studied by Hashin–Shtrikman-models in combination with appropriate void shape and void orientation evolution conditions, see e.g. Kailasam et al. (2000).

**Bounds for the Nonlinear Behavior.** Analogs to the Hill bounds for nonlinear inhomogeneous materials were introduced by Bishop and Hill (1951). For polycrystals the nonlinear equivalents to Voigt and Reuss expressions are usually referred to as Taylor and Sachs bounds, respectively.

In analogy to mean field estimates for elastoplastic material behavior, nonlinear bounds are typically obtained by evaluating a sequence of linear bounds. An important development was the derivation by Ponte Castañeda (1992) of a variational principle that allows upper bounds on the effective nonlinear behavior of inhomogeneous materials to be generated on the basis of upper bounds for the elastic response. Essentially, the variational principle guarantees the best choice for the "comparison composite" at any given loading state. The Ponte Castañeda bounds were shown to be closely related to improved secant plasticity mean field methods.

The study of bounds as well as the development of improved estimates for the overall mechanical behavior of nonlinear inhomogeneous materials have been active fields of research during the last decade, see the recent overviews by Suquet (1997b), Ponte Castañeda and Suquet (1998), Willis (2000), and Bornert and Suquet (2001).

## 3 Models Based on Resolved Microstructures

Whereas the modeling approaches discussed in section 2 make homogenization and localization tractable by using piecewise constant approximations for the microscopic stress and strain fields of microstructured materials, another group of strategies employs simplified microgeometries for which, however, the microfields are spatially resolved to a high degree. The latter group of methods can provide information on the intraphase and interphase fluctuations of the microscopic stresses and strains and are well suited for studying nonlinear inhomogeneous materials. In the following discussion the main emphasis is on the small strain elastic and elastoplastic behavior; for homogenization methods geared specifically towards finite strain problems see e.g. Miehe et al. (1999) or Terada et al. (2003).

The three main representatives of modeling strategies involving resolved microstructures are periodic microfield approaches, in which periodic "model composites" are studied, embedding methods, in which a geometrically fully resolved "core" is embedded in homogenized material, and windowing approaches, in which heterogeneous control volumes are subjected to homogeneous stress and strain boundary conditions to generate bounds on the overall behavior. In Figure 4 typical geometries used with the above approaches are sketched.

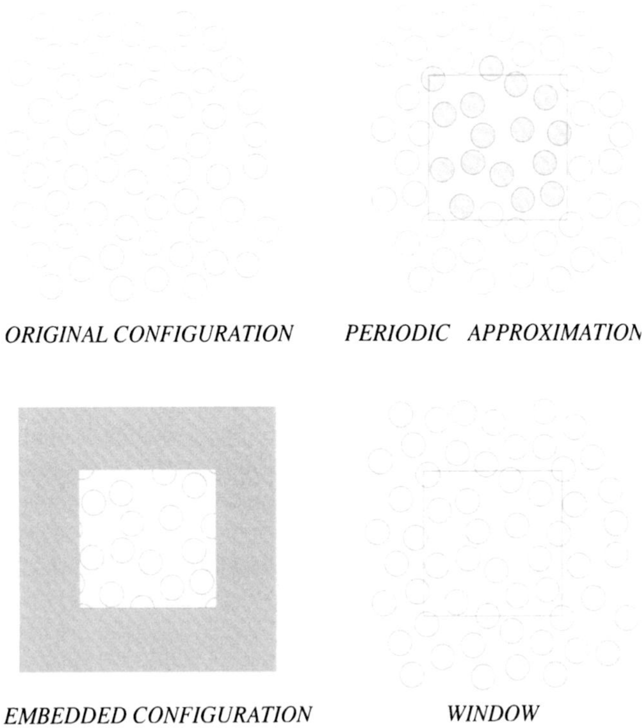


**Figure 4.** Sketch of a matrix–inclusion microstructure and of the microgeometries used by periodic microfield, embedded cell, and windowing approaches for modeling it.

### 3.1 General Remarks

**Model Geometries.** There are two, often complementary, philosophies for modeling inhomogeneous materials via discrete microgeometries. One of them is based on studying generic phase arrangements, which may range from simple periodic arrays of fibers or particles to highly complex microgeometries involving a considerable number of reinforcements, the positions of which are chosen to approximate some phase arrangement statistics. Phase arrangements of the latter type (multi-fiber or multi-particle geometries) can be generated by appropriate statistically based algorithms such as random insertion methods and Monte Carlo algorithms, which give rise to quasi-random microstructures, compare e.g. Gusev (1997), Yang et al. (1997), Zeman and Šejnoha (2001), Böhm et al. (2002), and LLorca (2004), as well as numerical annealing procedures, which allow the generation of statistically reconstructed microgeometries, see e.g. Rintoul and Torquato (1997), Roberts and Garboczi (1999), Bochenek and Pyrz (2002), and Pyrz (2004). Computer generated microgeometries tend to employ idealized reinforcement shapes, equiaxed particles embedded in a matrix, for example, being represented by spheres and non-equiaxed inhomogeneities by spheroids or cylinders. Where necessary appropriate orientation distributions of fibers or platelets can be incorporated into the models.

Alternatively, microgeometries may be chosen to follow as closely as possible the phase arrangement of a given sample of the material to be modeled, obtained from metallographic sections, serial sections, or tomographic data, see e.g. Fischmeister and Karlsson (1977), Terada and Kikuchi (1996), and Li et al. (1999). Descriptions of this type are often termed "real structure" models.

For models of either type, the question immediately arises what level of geometrical complexity of the model (and consequently what size of volume element) is required for describing microstructured materials at some desired level of accuracy. At present it can only be answered for elastic statistically isotropic composites with matrix–inclusion microtopology and sphere-like particles. For such materials Drugan and Willis (1996) gave estimates for the sizes of volume elements in terms of the particle diameter that are required to stay below a given relative error in the elastic moduli. The resulting cell sizes are surprisingly small: for an accuracy of 1% in terms of the overall moduli the required characteristic length of the control volume is approximately 5 particle diameters for any volume fraction. These estimates also hold reasonably well for periodic arrangements. Alternatively, adequate sizes of model geometries for elastic studies may be estimated from experimentally obtained correlation lengths (Bulsara et al., 1999), by requiring to have at least two statistically independent inhomogeneities in the volume element (Zeman and Šejnoha, 2001), or by using windowing approaches to bound the overall response from above and below (Huet, 1990). In addition, the adequacy of the size of a volume element may be judged on the basis of deviations from the required symmetry of the overall response. For nonlinear material behavior a number of studies, e.g. by Böhm and Han (2001), Jiang et al. (2001), and Zohdi and Wriggers (2001), have indicated that larger volume elements may be required than in the elastic case. Generally the accuracy of estimates obtained from a given volume element depends on the material property considered (Kanit et al., 2003). At present there appears to be no reliable method for directly assessing the representativeness of the microscopic stress and strain distributions obtained from a given volume element.

Microgeometries such as matrices reinforced by continuous aligned fibers or polycrystals consisting of long columnar grains can be described very well by two-dimensional models of the generalized plane strain type unless damage such as fiber fracture introduces strain gradients in the axial direction. Some care may, however, be required in handling axial shear, compare e.g. Pettermann and Suresh (2000). For other microstructures, such as composites with discontinuous reinforcements and polycrystals with approximately equiaxed grains, three-dimensional models are clearly superior to planar ones in describing arrangement effects, see e.g. the discussions by Iung and Grange (1995), Böhm and Han (2001), and Shen and Lissenden (2002). Properly used, however, planar models can nevertheless provide valuable insights into the behavior of microstructured materials. Accordingly, they are often used in complex analyses, such as studies involving microscopic damage on account of their much lower computational costs.

**Numerical Methods.** The majority of published micromechanical analyses of discrete microstructures have employed standard numerical engineering methods for resolving the microfields, such as Finite Difference methods, spring lattice models, the Boundary Element method, and the Finite Element method (FEM). In addition, algorithms have been reported that are specialized for handling periodic problems, among them Fast Fourier Transforms (Moulinec and Suquet, 1994) and Discrete Fourier Transforms (Müller, 1996), or that are specific to micromechanics, such as equivalent inclusion methods (Axelsen and Pyrz, 1995; Fond et al., 2001), the Transformation Field Analysis (Dvorak, 1992) with its extensions (Michel and Suquet, 2003), as well as the Method of Cells (Aboudi, 1989) and its developments (Aboudi et al., 2003). At present, the FEM is the most commonly used numerical scheme for evaluating the microfields of discrete microgeometries, especially in the nonlinear range, due to its geometrical flexibility and its capability of supporting a wide range of constitutive models for the phases, which range from isotropic elasticity to crystal plasticity (McHugh, 2004) and continuum damage models. Finite Element algorithms have also formed the basis of discrete dislocation plasticity models of composites with metallic matrices (van der Giessen, 2004b).

Applications of the FEM to micromechanical studies tend to fall into four main groups, compare Figure 5. Most commonly, phase arrangements are discretized by an often high number of standard continuum elements, the mesh being designed in such a way that element boundaries (and, where appropriate, special interface elements) are positioned at all interfaces between constituents. Such an approach has the advantage that almost any microgeometry can be handled and that readily available commercial Finite Element packages may be used. On the downside, the meshing of complex multi-reinforcement microgeometries can be challenging.

Alternatively, a smaller number of special hybrid elements may be used that are formulated to model the deformation, stress, and strain fields in an inhomogeneous region consisting of a single inclusion or void plus the surrounding matrix on the basis of an appropriate analytical theory. The most highly developed approach of this type at present is the Voronoi Finite Element Method (Ghosh et al., 1996), in which the mesh for the hybrid elements is obtained by Voronoi tesselations based on the reinforcement positions.

Especially when the phase arrangements to be studied are based on digital images of actual microgeometries, a third approach for discretizing microgeometries of interest. It consists of using a regular square or hexahedral mesh that has the same resolution as the digital image, each element being assigned to one of the constituents by operations such as thresholding of the grey values of the corresponding pixel or voxel, respectively. Such meshes have the advantage of allowing a straightforward automatic model generation from appropriate experimental data sets and of avoiding ambiguities in smoothing the

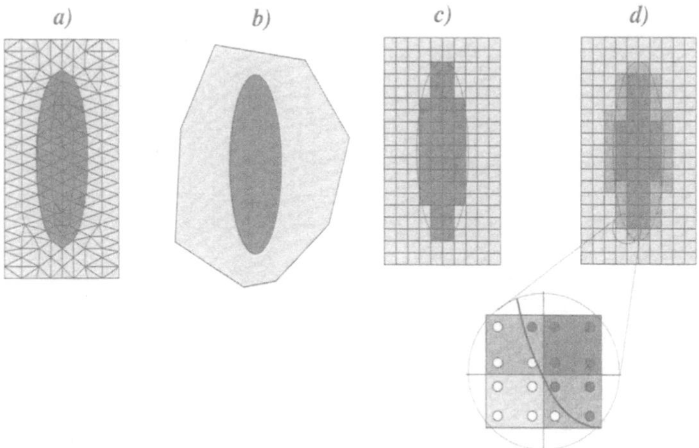


**Figure 5.** Sketch of FEM approaches used in micromechanics: a) discretization by standard elements, b) special hybrid elements, c) pixel/voxel discretization, d) "multiphase elements"

digital data, but they lead to ragged phase boundaries that may degrade accuracy.

A fourth approach also uses regular FE meshes, but assigns phase properties at the integration point level of standard elements ("multiphase elements"), compare e.g. Schmauder et al. (1996) or Quilici and Cailletaud (1999). Such procedures may be viewed as trading off ragged boundaries at the element edges for smeared-out microfields within those elements that contain a phase boundary, where stress or strain discontinuities within elements cannot be adequately resolved by standard FE shape functions. With respect to the element stiffnesses the latter concern can be much reduced by overintegrating elements containing phase boundaries, which leads to good approximations of integrals involving non-smooth displacements by the numerical quadrature, see Zohdi and Wriggers (2001). The rate of convergence of the microfields, however, remains slow. This disadvantage can be eliminated by using "extended" FE methods, in which the mesh does not have to conform to the phase boundaries, the latter being accounted for by appropriately enriching the Finite Element approximations (Moës et al., 2003).

A recent development for studying microgeometries involving a large number of inhomogeneities by Finite Element methods involves programs specially geared towards solving micromechanical problems. Such codes can, for example, be based on matrix-free iterative solvers such as Conjugate Gradient methods, for which Voigt-type approximations provide excellent starting solutions. Fast solvers of this type open the possibility of solving inverse problems, e.g. for finding optimal particle shapes for given load cases and damage modes, compare (Zohdi, 2003).

### 3.2 Periodic Microfields and Unit Cells

The basic idea behind periodic microfield approaches consists in approximating real microstructured materials by periodic model materials, which can be studied by evaluating the behavior of appropriate unit cells. The literature on such methods is fairly extensive and well developed mathematical theories are available on scale transitions in periodic structures, compare the discussion in (Michel et al., 2001).

Periodic microfield approaches have proven to be highly flexible tools, but they are subject to some intrinsic limitations: They cannot handle free surfaces in any direction for which the model is periodic, they act as filters in term of wave lengths in dynamic analyses (Suzuki and Yu, 1998) and buckling studies (Vonach, 2001), and any microscopic damage predicted by them must by definition come in periodic patterns.

**Boundary Conditions.** In order to allow the generation of proper periodic microfields, a unit cell must be subjected to appropriate boundary conditions (B.C.s) such that valid tilings result both for the undeformed geometry and for all deformed states pertinent to the investigation, i.e. gaps and overlaps between neighboring unit cells as well as unphysical constraints on the deformations must be avoided. The major types of boundary conditions used to enforce such behavior of unit cells are periodicity, symmetry, and point symmetry (antisymmetry) boundary conditions. Their most important features are discussed in the following for planar geometries, the extension to three-dimensional cases being trivial. More formal treatments of boundary conditions for unit cells were given e.g. by Anthoine (1995) and Michel et al. (1999).

The most general boundary conditions for unit cells are periodicity B.C.s, which can handle any possible deformation state of the cell and, consequently, of the inhomogeneous material to be modeled. Because such unit cells tile the computational space by translation, neighboring cells (and, consequently, opposite faces of a given cell) must fit into each other like parts of a jigsaw puzzle in both undeformed and deformed states. For the case of quadrilateral two-dimensional unit cells this can be achieved by pairing opposite faces and linking the corresponding degrees of freedom within each pair of faces. Using the conventions for naming the vertices and faces of unit cells shown in Figure 6 (left), the resulting equations for the periodically varying microscopic displacement vectors at the boundaries, $\mathbf{u}'$, can be symbolically denoted as

$$\mathbf{u}'_{\mathrm{N}}(\tilde{x}) = \mathbf{u}'_{\mathrm{S}}(\tilde{x}) + \mathbf{u}'_{\mathrm{NW}} \qquad \mathbf{u}'_{\mathrm{E}}(\tilde{y}) = \mathbf{u}'_{\mathrm{W}}(\tilde{y}) + \mathbf{u}'_{\mathrm{SE}} \qquad \mathbf{u}'_{\mathrm{NE}} = \mathbf{u}'_{\mathrm{NW}} + \mathbf{u}'_{\mathrm{SE}} \quad , \tag{3.1}$$

i.e. the displacements at the "slave faces", N and E, are determined by those of the "master faces", S and W, and of the "master nodes", SE and NW. In case the stresses are discretized in addition to the displacements (as in many hybrid Finite Element algorithms) they also have to follow periodicity conditions; the surface traction vectors are antiperiodic. Boundary conditions of the type described by eqn. (3.1) can be implemented as linear constraint equations between individual degrees of freedom. Periodicity B.C.s generally are the least restrictive option for multi-reinforcement unit cells generated by statistically based algorithms or from experimentally obtained geometries.

For rectangular and hexahedral unit cells the faces of which coincide with symmetry planes of the phase arrangement and for which this property is retained for all deformed states that are to be studied, periodicity B.C.s simplify to symmetry boundary conditions. For the arrangement shown in Figure 6 (center) they can be denoted as

$$u'_{\mathrm{E}}(\tilde{y}) = u'_{\mathrm{SE}} \qquad v'_{\mathrm{N}}(\tilde{x}) = v'_{\mathrm{NW}} \qquad u'_{\mathrm{W}}(\tilde{y}) = 0 \qquad v'_{\mathrm{S}}(\tilde{x}) = 0 \quad , \tag{3.2}$$

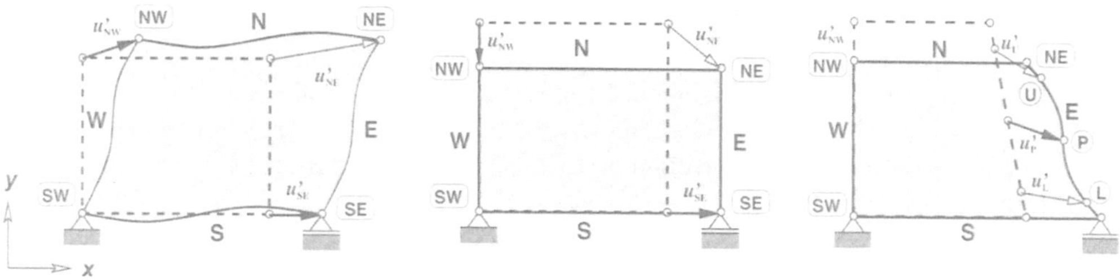


**Figure 6.** Sketch of periodicity (left), symmetry (center), and point symmetry (right) boundary conditions for two-dimensional unit cells.

where $u'$ and $v'$ are the components of $\mathbf{u}'$. The roles of slaves and masters are the same as for eqn.(3.1). Symmetry boundary conditions are fairly easy to use and tend to give rise to small unit cells for simply periodic phase arrangements, but the load cases that can be handled are limited to uniform thermal loads, mechanical loads that act in directions normal to one or more pairs of faces, and combinations of the above. This makes unit cells with symmetry B.C.s suitable for many materials characterization tasks.

Point symmetry (antisymmetry) boundary conditions are even more limited in terms of the microgeometries that they can be used on, because they require the presence of centers of symmetry ("pivot points") on at least one face of the unit cell. They do not, however, give rise to major restrictions in the load cases that can be handled. In Figure 6 (right) a unit cell is sketched the E side of which is subjected to antisymmetry boundary conditions, the pivot point being P. For this cell the B.C.s

$$\mathbf{u}'_\mathrm{U}(\tilde{s}) + \mathbf{u}'_\mathrm{L}(\tilde{s}) = 2\mathbf{u}'_\mathrm{P} \qquad v'_\mathrm{N}(\tilde{x}) = v'_\mathrm{NW} = 2v'_\mathrm{P} \qquad v'_\mathrm{S}(\tilde{x}) = 0 \qquad u'_\mathrm{W}(\tilde{y}) = 0 \tag{3.3}$$

must be fulfilled, where $\mathbf{u}'_\mathrm{U}(\tilde{s})$ and $\mathbf{u}'_\mathrm{L}(\tilde{s})$ are the deformation vectors of pairs of points U and L that are positioned symmetrically on face E (where a local coordinate system $\tilde{s}$ centered on the pivot is used) and deform symmetrically with respect to the pivot P.

The displacement components $u'_{SE}$, $v'_{NW}$, $u'_\mathrm{P}$, and $v'_\mathrm{P}$ of the master nodes SE, NW and P can be identified as the macroscopic contributions to the unit cells' displacements, see eqns. (3.1), (3.2) and (3.3) and Figure 6.

**Asymptotic Homogenization.** The most versatile and elegant strategy for linking the macroscale and microscale in unit cell analyses is based on a mathematical framework known as asymptotic homogenization or homogenization theory, see e.g. Suquet (1987). It introduces separate macroscopic coordinates $\mathbf{Z}$ and microscopic coordinates $\mathbf{z} = \mathbf{Z}/\epsilon$ into expressions such as eqn. (1.1), where $\epsilon \ll 1$ is a scaling parameter defined as the ratio between the characteristic lengths at the micro- and macroscales. When displacements, strains and stresses are expanded with respect to $\epsilon$ and introduced into the equilibrium equations, stress–strain relations, and strain–displacement relations, a set of hierarchical differential equations is obtained. This leads to a boundary value problem at the unit cell level and to concentration expressions linking the macro- and microscopic fields. These equations, in turn, can be discretized e.g. by Finite Element algorithms that allow effective homogenized elastic or tangential tensors to be evaluated, for details see e.g. Ghosh et al. (1996) or Chung et al. (2001). For this task special analysis codes are typically required. Homogenization theory can be used as the basis for solving macroscopic problems without explicitly specifying homogenized constitutive laws, so that simultaneous two-scale analyses to be carried out, see e.g. Ghosh et al. (1996) or Terada et al. (2003).

**Macroscopic Degrees of Freedom.** When asymptotic homogenization is not used, it is good practice to apply far field stresses and strains to a unit cell via concentrated nodal forces or prescribed displacements at the master nodes, an approach termed "the method of macroscopic degrees of freedom" by Michel et al. (1999). For load controlled analyses the forces to be applied to the master nodes SE and NW, $\mathbf{p}_{\mathrm{SE}}$ and $\mathbf{p}_{\mathrm{NW}}$, of a planar unit cell with periodicity boundary conditions are given by the surface integrals

$$\mathbf{p}_{\mathrm{SE}} = \int_{\Gamma_{\mathrm{E}}} \mathbf{t}^{\mathrm{a}}_{\mathrm{E}}(\mathbf{x})\, d\Gamma \qquad \text{and} \qquad \mathbf{p}_{\mathrm{NW}} = \int_{\Gamma_{\mathrm{N}}} \mathbf{t}^{\mathrm{a}}_{\mathrm{N}}(\mathbf{x})\, d\Gamma \quad , \tag{3.4}$$

the nomenclature and configuration of Figure 6 being used. Here $\mathbf{t}^{\mathrm{a}}(\mathbf{x}) = \boldsymbol{\sigma}^{\mathrm{a}}\mathbf{n}_{\Gamma}(\mathbf{x})$ stands for the surface traction vector corresponding to the applied macroscopic stress field $\boldsymbol{\sigma}_{\mathrm{a}}$ at point $\mathbf{x}$ on the cell's surface $\Gamma_{\mathrm{UC}}$ (note that in contrast to the actual distribution of tractions on $\Gamma_{\mathrm{UC}}$, denoted as $\mathbf{t}(\mathbf{x})$ in eqn.(1.4), $\mathbf{t}^{\mathrm{a}}(\mathbf{x})$ can be directly evaluated from the macroscopic stress). Equation (3.4) can be generalized to require that each master node is loaded by a force corresponding to the surface integral of the applied traction vectors over the face slaved to it via an equivalent of eqn.(3.1), compare e.g. Smit et al. (1998). Three-dimensional problems and symmetry as well as antisymmetry boundary conditions as described by eqns. (3.2) and (3.3) can be handled in an analogous way. For geometrically nonlinear analyses eqn. (3.4) must be applied to the current configuration.

For strain controlled analyses using the method of macroscopic degrees of freedom appropriate displacements are specified for the master nodes, which are evaluated by integrating the homogeneous macroscopic strain to be applied.

**Evaluation of Averaged Fields.** In small deformation analyses, the macroscopic stresses strains in a unit cell can be evaluated by volume averaging according to eqns. (1.3) and (1.4). When effective values or phase averages are to be generated of variables that are nonlinear functions of the stress or strain components (e.g. equivalent stresses or strains, stress triaxialities and principals), the variable must be evaluated first (e.g. at the integration point level in FE codes) and volume averaging performed afterwards.

In many Finite Element programs it is possible to approximately evaluate volume integrals of a function $f(\mathbf{x})$ as

$$\bar{f} = \frac{1}{\Omega_{\mathrm{e}}} \int_{\Omega} f(\mathbf{x}) d\Omega \approx \frac{1}{\Omega} \sum_{k=1}^{N_{\mathrm{i}}} f_k \Omega_k \quad , \tag{3.5}$$

where $f_k$ is the function value at the $k$-th integration point to which an "integration point volume" $\Omega_k$ is associated in the sense of an integration weight. $N_{\mathrm{i}}$ is the number of integration points within the integration volume $\Omega_e$. Equation (3.5) can be used to obtain overall averages, phase averages or averages within individual reinforcements; it fully accounts for both intra-phase and inter-phase fluctuations of the microfields. Higher statistical moments of functions $f(\mathbf{x})$ can be evaluated in analogy.

**Periodic Microgeometries and Unit Cells.** The simplest periodic microgeometries are hexagonal or square arrays of aligned fibers or grains and cubic arrays of particles. Such arrangements clearly are highly idealized compared to the geometries of real materials. Nevertheless they can capture many features of the latters' nonlinear behavior at least in a qualitative way and their relatively low computational requirements allow much leeway in terms of complex material models for constituents and interfaces. Some of these "simple periodic arrangements" are noticeably more anisotropic in their elastic and inelastic responses than the materials they aim to model. Typical cases for this are square and simple cubic arrays of reinforcements or grains, which can give rather soft or very stiff responses depending on the loading direction. The overall hardening behavior of elastoplastic inhomogeneous materials tends to be rather sensitive to symmetry effects. For example, due to inter-phase fluctuations of plastic strains elastic isotropy of a microgeometry does not guarantee elastoplastic anisotropy.

A number of issues pertaining to unit cell models that use simple periodic arrangements of discontinuous reinforcements are discussed elsewhere in this volume (LLorca, 2004; Böhm, 2004) for the cases of particles and aligned short fibers.

The realism of models involving periodic microgeometries can be markedly improved by using unit cells that contain a considerable number of fibers, particles or grains as mentioned in section 3.1. Phase arrangements to be used with multi-reinforcement unit cells

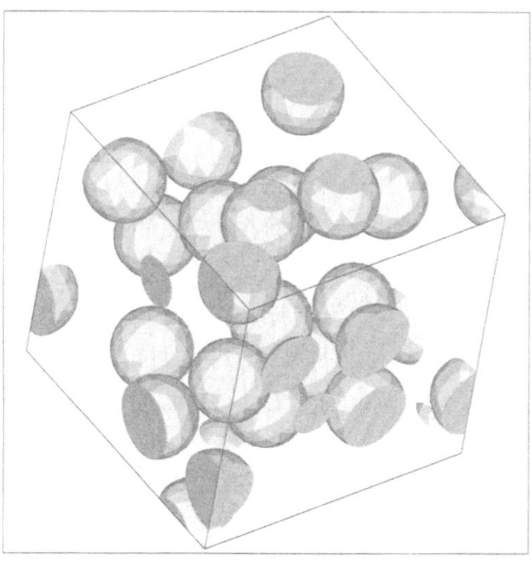

**Figure 7.** Multi-particle unit cell for a particle reinforced MMC ($\xi^{(r)} = 0.2$ nominal) containing 20 spherical particles in a quasi-random arrangement suitable for using periodicity B.C.s.

obviously must be periodic (or must show appropriate planes of symmetry), which implies special provisions in algorithms for generating such microgeometries or the necessity of suitably modifying experimentally obtained microgeometries. The latter requirement evidently influences the arrangement statistics of "small" multi-reinforcement unit cells, but is of decreasing importance as the size of the cells increases.

For unit cell models that combine periodicity or antisymmetry boundary conditions with numerical models employing unstructured meshes, it is necessary that not only the phase distributions, but also the discretizations of the geometry are compatible (e.g. the same surface meshes must be used on opposite faces of unit cells). This adds to the challenges in meshing complex unit cell microgeometries, for which in most cases triangular or tetrahedral elements have been used.

The first reports of planar multi-fiber unit cells for continuously reinforced composites date from the early 1990s, see e.g. Brockenbrough and Suresh (1990), whereas three-dimensional multi-particle unit cells for nondilute particle reinforced composites were introduced considerably later (Gusev, 1997). Recently, also multi-fiber unit cell analyses for nonaligned short fiber reinforced composites were reported, see Böhm et al. (2002) and Hine et al. (2002). Figure 7 shows a typical multi-particle unit cell containing 20 spherical particles that are randomly positioned within the constraint of maintaining a periodic phase arrangement. Because such unit cell geometries may be interpreted as realizations of the underlying arrangement statistics, it is usually advisable to evaluate a number of such microgeometries and ensemble average their responses for generating results on the overall behavior.

In Table 1 results of three-dimensional multi-particle unit cell analyses (3D MPUC) carried out with the commercial micromechanics code PALMYRA (MatSim Ges.m.b.H., Zurich, Switzerland), which allows to generate, mesh and evaluate multi-reinforcement unit cells, are compared with mean field estimates and bounds for the thermoelastic behavior of a glass particle reinforced epoxy. The unit cells used for generating the above predictions contained 80 randomly positioned spherical particles of fixed size (mono) or with a size distribution (poly). The results given for $E$, $\nu$, $G$ and $\alpha$ are averages over a number of loading directions.

Generally very good agreement between advanced analytical estimates and the unit cell results is evident, with the latter following the three-point bounds for all moduli listed, and similar trends are predicted for the influence of the particle size distributions by both approaches. Excellent agreement between the analytical bounds and estimates on the one hand and multi-particle unit cell predictions on the other hand was also reported by Segurado et al. (2002) for the nonlinear behavior of particle reinforced composites.

In general, periodic microfield approaches are well suited for doing materials characterization in the nonlinear range, for following complex macroscopic load paths, and for studying inter-phase and intra-phase variations of stresses and strains.

### 3.3 Embedding Models

Embedding models typically are built up from a core ("local heterogeneous region"), in which the microstructure is well resolved geometrically, that is embedded in an outer region that serves mainly for transmitting far field loads, compare Figure 8. This model-

**Table 1.** Overall thermoelastic properties of a particle reinforced glass/epoxy composite (spherical reinforcements, $\xi^{(r)}$=0.389) as predicted by the Hashin–Shtrikman (H/S) and three-point (3P) bounds, by the Mori–Tanaka method (MTM), by the generalized self-consistent scheme (GSCS), by Torquato's second order estimates (3PE), as well as by three-dimensional multi-particle unit cells (MPUC) using monodispersely and polydispersely sized particles.

| | $E$ [GPa] | $\nu$ [ ] | $G$ [GPa] | $K$ [GPa] | $\alpha$ [$K^{-1}\times 10^{-6}$] |
|---|---|---|---|---|---|
| glass particles | 73.1 | 0.18 | 30.1 | 38.1 | 4.90 |
| epoxy matrix | 3.16 | 0.35 | 1.17 | 3.51 | 40.0 |
| H/S LB | 6.80 | 0.024 | 2.59 | 6.11 | 12.1 |
| H/S UB | 20.9 | 0.404 | 8.53 | 12.7 | 23.5 |
| 3PLB (mono/h) | 7.05 | 0.236 | 2.69 | 6.16 | 20.1 |
| 3PUB (mono/h) | 10.0 | 0.334 | 3.95 | 7.22 | 23.6 |
| 3PLB (poly/h) | 7.17 | 0.195 | 2.74 | 6.26 | 17.2 |
| 3PUB (poly/h) | 12.1 | 0.356 | 4.79 | 8.56 | 23.0 |
| MTM | 6.80 | 0.315 | 2.59 | 6.11 | 23.5 |
| GSCS | 7.23 | 0.303 | 2.77 | 6.11 | 23.5 |
| 3PE (mono/h) | 7.29 | 0.304 | 2.80 | 6.22 | 23.2 |
| 3PE (poly/h) | 7.52 | 0.304 | 2.88 | 6.39 | 22.6 |
| 3/D MPUC (mono) | 7.95 | 0.291 | 3.11 | 6.34 | 22.8 |
| 3/D MPUC (poly) | 8.06 | 0.288 | 3.11 | 6.36 | 22.7 |

ing strategy avoids the main restriction of unit cell models, viz. that the geometry and all microscopic fields are strictly periodic, which can be a major advantage when studying damage and failure at the microscale. The accuracy of embedding models, however, is predicated on the availability of suitable approximations for describing the behavior of the outer region. Also, some care is required with respect to spurious boundary layers that tend to occur at the "interfaces" between the core and the surrounding embedding medium. Note that mesh superposition or submodeling techniques, which use a coarse mesh over the whole model plus a geometrically independent much finer mesh in regions of interest, compare e.g. Takano et al. (2001), are conceptually closely related to embedding schemes.

Three basic types of embedding approaches have been proposed in the literature. One of them employs discrete phase arrangements in both the core region and in the surrounding material, the latter, however, being discretized by a much coarser FE mesh, see e.g. Sautter et al. (1993). In the second group of embedding methods the behavior of the embedding region is described via appropriate macroscopic material models. These typically take the form of semiempirical or micromechanically based constitutive laws that are prescribed a priori for the embedding zone and that must be chosen to correspond closely to the homogenized behavior of the core. This way, conceptually simple models are obtained that are very well suited for studying the stress and strain distributions in the vicinity of crack tips or at macroscopic interfaces in composites (Chimani et al.,

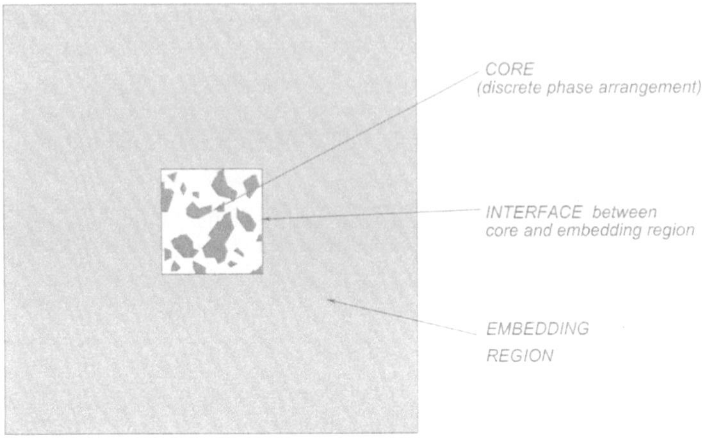


**Figure 8.** Schematic depiction of the arrangement of core, embedding region, and interface in an embedding model.

1997) and for modeling the growth of cracks in inhomogeneous materials, see e.g. Wulf et al. (1996) and van der Giessen (2004a).

The third type of embedding scheme uses the homogenized mechanical response of the core to determine the effective behavior of the surrounding medium, giving rise to models of the self-consistent type, which are mainly employed for materials characterization. The use of such approaches, of course, hinges on the availability of suitable parameterizable constitutive laws for the embedding material that can follow the core's homogenized behavior with sufficient accuracy for all load cases and for any loading history. This requirement typically is difficult to fulfill in a strict sense when at least one of the constituents shows elastoplastic material behavior. Models of this type were discussed e.g. by Bornert et al. (1994) and Dong and Schmauder (1996).

When embedding models are used for materials characterization it is typically sufficient to load them by appropriate homogeneous stresses or strains at the boundaries of the embedding region. For studying crack problems and related questions, however, it is often preferable to impose displacement boundary conditions corresponding to the far field behavior of suitable analytical solutions. Embedding models also allow complete samples to be considered when simulating experiments, e.g. by combining an embedding region that corresponds to the geometry of a given specimen and uses a smeared-out material description with a small fully resolved region for zeroing in on a crack tip and its immediate surroundings. Despite their conceptual simplicity, embedding models tend to be quite expensive in terms of computational resources, especially when nonlinear material behavior is to be accounted for.

### 3.4 Windowing Models

An alternative way of avoiding the restrictions faced by periodic microfield analyses in studying geometrically resolved microstructures is subsumed here as "windowing approaches". Such models, which can provide upper and lower bounds on the effective properties of inhomogeneous materials, were originally developed for dealing with volume elements that are known to be too small for being proper RVEs, see Huet (1990). Alternatively, they can be used for assessing the suitability of a given volume element by checking how far the bounds are apart, compare Ostoja-Starzewski (1998).

Windowing approaches are based on placing mesoscopic test windows at random positions in an inhomogeneous material and subjecting the resulting samples of the microstructure to homogeneous stress (uniform traction) and strain (uniform displacement) boundary conditions at their whole surface. Provided sufficiently large and "nonpathological" windows are chosen, predictions of this type give rise to lower and upper estimates, respectively, compare e.g. Nemat-Nasser and Hori (1993). Ensemble averages of such estimates in turn provide lower and upper bounds on the overall behavior of the material. By definition, when the test windows are sufficiently large for their contents to be proper reference volume elements the lower and upper estimates for and bounds on the overall properties must coincide.

By using a series of windows of increasing size a hierarchy of bounds can be generated for a given microstructure, which allows to assess the dependence of the predicted overall moduli on the size of the volume elements, compare e.g. Huet (1999). Alternatively, one of the bounds may used as an estimate for the overall response of an inhomogeneous material, see e.g. Zohdi and Wriggers (2001). Recently windowing approaches were extended into the elastoplastic regime (Jiang et al., 2001), where the bounding properties of the two types of boundary conditions can be proven within the context of deformation theory.

## 4 Multi-Scale Models

The micromechanical methods discussed in the previous two sections are each designed to handle one single scale transition between a lower and a higher length scale ("microscale" and "macroscale"), overall responses being obtained by homogenization and local fields by localization. Many inhomogeneous materials, however, show more than two clearly distinct characteristic length scales, typical examples being laminated composites, woven and braided composites reinforced by tows containing many parallel thin fibers, materials in which there are well defined clusters of particles, as well as many biomaterials. In such cases an obvious modeling strategy are hierarchical (or multi-scale) approaches that can handle a sequence of scale transitions, the material response at any given length scale being described on the basis of the homogenized behavior of the next lower one. Figure 9 schematically shows such a hierarchical model for a particle reinforced composite with a clustered mesostructure.

Any of the methods discussed in sections 2 and 3 (with the exception of bounding approaches at the lower length scales) can be used as "building blocks" in such a scheme. Among the continuum mechanical multi-scale descriptions of the mechanical behavior of inhomogeneous materials reported in the literature, some combine mean field methods at the higher length scale with mean field methods, see e.g. Hu et al. (1998) and Siegmund et al. (2004), or periodic microfield approaches, compare González and LLorca (2000) and LLorca (2004), at the lower length scale.

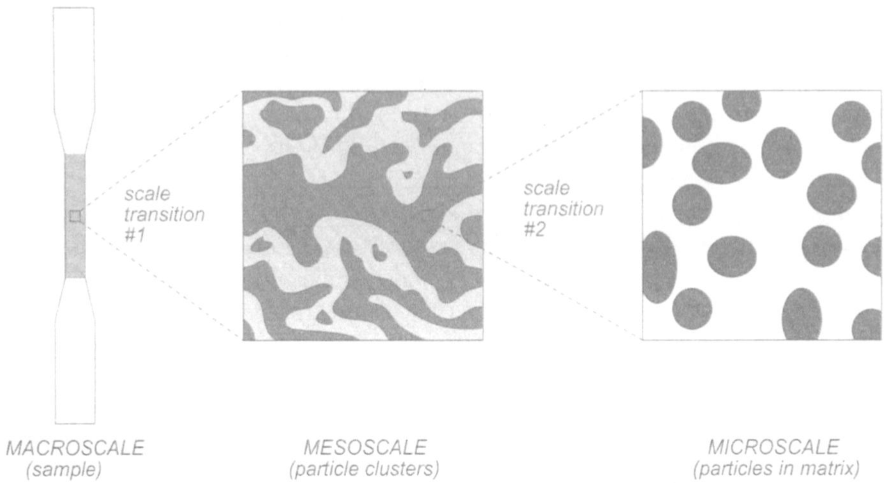


**Figure 9.** Schematic representation of a multi-scale approach for studying a material consisting of reinforcements in a matrix that shows mesoscopic regions of high (dark grey) and low (light grey) particle volume fractions. Two scale transitions (macro⟷meso and meso⟷micro) are used.

The most common strategy for multi-scale modeling, however, uses Finite Element based unit cell methods at the topmost length scale and handles the lower length scales via micromechanically based material models that are evaluated at each integration point. The numerical requirements of such approaches typically are quite manageable in the elastic range, but they can be very high indeed if elastoplastic material behavior is considered, where essentially a full micromechanical submodel has to be maintained and solved for at each integration point in order to account for the history dependence of the local responses. Despite the substantial computational costs involved, studies that employ two-dimensional multi-fiber cell models based on asymptotic homogenization for the "lower" scale transition have been reported within such a framework (Lee et al., 1999). This approach was developed into an FE-based framework in which the computational domain is adaptively split into subregions that are resolved at appropriate length scales (Ghosh et al., 2001). Non-critical regions are described by continuum elements with appropriate homogenized material properties, in regions of high macroscopic stress and strain gradients unit cell models are activated at the integration points, and embedded models of the fully resolved microstructure come into play near free edges or when damage is detected. Lower (but by no means negligible) computational costs can be achieved by using constitutive models based on mean field approaches, such as incremental Mori–Tanaka methods, see e.g. Pettermann et al. (1999) or Doghri and Ouaar (2003).

Multi-scale approaches are not limited to continuum mechanics. Especially the capability of the Finite Element method for handling highly complex material models at the integration point level can be used to build multi-scale descriptions that include atomistic models, see e.g. Shenoy et al. (1998).

# 5 Treatment of Damage within Micromechanical Models

An important aspect of many micromechanical models is the study of the damage and failure behavior of microstructured materials, the underlying idea being that macroscopic damage and failure are the consequences of microscopic damage processes in the constituents or in the interfaces between them. Accordingly, some aspects of such modeling issues are discussed in the following, the emphasis being put on resolved microstructure models employing the Finite Element method. It is worth noting at this point that other numerical techniques such as spring lattice networks and meshless techniques have been used with considerable success in modeling the progress of cracks in inhomogeneous materials.

If information is required only on the microstructural features or the overall loads at which microscopic damage will initiate, it is often sufficient to monitor appropriate "damage relevant fields" such as the the maximum principal stress in brittle materials, a ductile damage indicator (Fischer et al., 1995) for metals, or the traction vectors at interfaces. In a similar vein, shakedown theorems may be used on the microscale to assess the vulnerability of microstructured materials to low-cycle fatigue (Böhm, 1993). Such models can account for the presence of pre-existing flaws such as broken fibers and can often be implemented as postprocessing operations, but they are not designed to handle any degradation of stiffness due to damage.

If the microcracks or microvoids underlying a given damage mode are much smaller than the characteristic length scale of the microgeometry to be studied, the degradation of the stiffness of the constituents can be described by continuum damage models and related approaches (essentially this amounts to multi-scale approaches as discussed in section 4). The evolution of damage in such models typically will cause softening material behavior at the constituent and macroscopic levels, which tends to give rise to localization and may lead to the formation of shear bands on the microscale. A numerical difficulty typically encountered by continuum damage models in such situations, viz. a marked mesh dependence of the solutions (typically the localized region has a characteristic length scale of one or a few elements, regardless of the element size), can be resolved by appropriately regularizing the problem. This can be achieved by using continuum damage theories that are enriched by higher order gradients, see e.g. Geers et al. (2001), by introducing nonlocal averaging of an appropriate internal variable, compare Peerlings et al. (2001), both of which explicitly introduce an absolute length scale into the problem, or by employing time dependent formulations involving rate effects (viscoelasticity, viscoplasticity). A rather ad-hoc alternative consists in selecting a mesh size that leads to physically reasonable sizes of localization features.

Macrocracks as well as microcracks that have characteristic lengths comparable to the microscale of a given phase arrangement must be resolved explicitly in micromechanical studies. Typical modeling strategies for studying the propagation of discrete cracks within a Finite Element framework are based on generating new crack surfaces at element boundaries, e.g. by node release techniques or cohesive surface models, compare Elices et al. (2002), on using specially formulated finite elements with embedded discontinuities (Jirásek, 2000), or on smearing out the crack within some process zone ("fictitious crack techniques"), see e.g. Weihe and Kröplin (1995).

In composite materials with matrix–inclusion microtopology, damage may either occur in the reinforcements, in the matrix, or within the interface between them. Similarly, in polycrystals intergranular or intragranular failure may be observed. The dominant microscopic damage process is decided by the strength parameters of the constituents and interfaces, by the microgeometry, and by the loading conditions.

**Modeling of Reinforcement Damage and Failure.** Particulate and fibrous reinforcements in composites typically fail by brittle cleavage. Within the framework of micromechanical models using the FEM, the failure of such phases can be modeled by following the extension of a crack through the reinforcement or by instantaneously opening a crack along some predefined crack plane.

Even though the former type of approach in principle is capable of following complex crack paths, it has mostly been applied for simple microgeometries where the cracks in the reinforcements follow planes of symmetry. For such situations and for cases where cracks open along predefined cracks paths in more complex microgeometries, cohesive surface models may be used to advantage, see e.g. Tvergaard (1994) and LLorca (2004).

Within "instantaneous failure" models the activation of cracks in particles or fibers may either be controlled by deterministic (Rankine) criteria, e.g. critical normal maximum principal stresses, compare Berns et al. (1998) and Ghosh and Moorthy (1998), or by probabilistic criteria such as Weibull-type weakest link theories. The latter group of models, when applied at the level of individual particles, have the advantage of introducing an absolute length scale into the problem. In "fully probabilistic" implementations of this type a random process is provided that selects the particles that fail at a given load level, see e.g. Pandorf (2000). When such descriptions are used, modeling runs using the same microgeometry and loading conditions will typically give different failure sequences of the reinforcement, so that results are best evaluated in terms of averages over a number of runs (Eckschlager, 2002).

**Modeling of Matrix Damage and Failure.** Damage in ductile matrices typically occurs by the nucleation and growth of voids, by shear localization between broken reinforcements, or by grain boundary failure, compare LLorca (2004). Micromechanical models of microstructured materials have mainly concentrated on the first mentioned damage mechanism, one of the following two approaches being typically followed. On the hand, damage-type descriptions employing the ductile rupture model of Gurson (1977) and its developments (Tvergaard and Needleman, 1984) have been applied successfully to such problems, see e.g. Needleman et al. (1993) or LLorca (2004). On the other hand, element elimination models that are based on a ductile damage indicator (Fischer et al., 1995) or on other criteria have been used, compare e.g. Berns et al. (1998). The local versions of both groups of algorithms are liable to show a marked mesh dependence in their solution, which can be resolved by regularization as mentioned above.

Damage in composites with brittle elastic matrix behavior can be investigated via fracture mechanics based methods, see e.g. Ismar and Reinert (1997), or Weibull-type criteria may be employed (Brockmüller et al., 1995). Individual cracks in polymer matrices have been studied, for example, by node release techniques together with criteria for determining the crack direction. Due to the complexity of crack paths in inhomogeneous media frequent remeshing steps may be necessary to provide appropriate crack paths when such algorithms are used, compare Vejen and Pyrz (2002).

**Modeling of Interface Failure.** Probably the most commonly used description for the debonding of an inclusion from a matrix in numerical micromechanics have been cohesive surface models such as the one developed by Needleman (1987) and extended by Tvergaard (1990), compare also LLorca (2004). An alternative approach suitable for describing very weakly bonded reinforcements consists in using standard contact elements at the interface, the composite being held together in a "shrink fit" by the thermal stresses caused by cooling down from the processing temperature, see e.g. Sherwood and Quimby (1995). Other interfacial decohesion models for composites reported in the literature include strain softening interfacial springs (Li and Wisnom, 1994), node release techniques (Vejen and Pyrz, 2002), as well as fictitious crack models for interphases (i.e. layers of finite thickness between the phases proper), see e.g. McHugh and Connolly (1994).

Similar modeling strategies can be used to study the evolution of interfacial damage between individual grains in the intergranular failure of polycrystals. For example, the use of interfacial elements for modeling high temperature creep failure due to intergranular void nucleation is discussed by van der Giessen (2004a).

A special approach was developed specifically for accounting for the stiffness loss due to interfacial debonding within mean field models. It is based on replacing debonded reinforcements by perfectly bonded anisotropic "fictitious inclusions" of appropriately reduced stiffness. Within such a framework, the fraction of debonded reinforcements can e.g. be controlled via Weibull-type models, compare Zhao and Weng (1995).

## 6 Closing Remarks

The methods of continuum micromechanics of materials have enjoyed significant success over some four decades in describing the behavior of microstructured materials and in providing predictive tools for engineers and materials scientists. In applying such methods, however, some practical limitations must be kept in mind.

It is important to be aware of the fact that — with the exception of some embedding models that study macrocracks in inhomogeneous materials — all approaches discussed in the present contribution do not have an intrinsic absolute length scale. This implies that effects of the absolute size of microstructural features can only be studied in a physically meaningful way if a length scale is introduced via the constitutive behavior of the constituents or of the interfaces between them. Also, one should be aware of the possibility that mesh dependent behavior (e.g. of damage models, compare the remarks in section 5) can introduce spurious length scales into a model.

A major practical difficulty in the use of continuum micromechanics approaches has been to identify appropriate constitutive models and to obtain realistic material parameters for the constituents. Typically, available data pertain to the behavior of the bulk materials (as measured from homogeneous macroscopic samples), whereas what is actually required are parameters (and, in some cases, constitutive theories) describing the in-situ response of the phases at the microscale. For example, there is considerable evidence that classical continuum plasticity theories may not adequately describe the behavior of metallic materials in the presence of marked strain gradients, which readily occur in metal matrix composites.

Another point of practical relevance is the fact that the overall stiffness of microstructured materials tends to be much less sensitive to details of the phase geometry than are the distributions of the microfields and the overall strength properties. Together with the dearth of dependable material parameters this makes quantitative predictions of the strength of inhomogeneous materials by micromechanical methods a difficult challenge.

Finally it is worth noting that diffusion and transport problems (such as heat conduction) in microstructured materials can be solved by methods that are closely related to the ones presented here for modeling the mechanical responses.

## Bibliography

J. Aboudi. Micromechanical analysis of composites by the method of cells. *Appl. Mech. Rev.*, 42:193–221, 1989.

J. Aboudi. *Mechanics of Composite Materials.* Elsevier, Amsterdam, 1991.

J. Aboudi, M.J. Pindera, and S.M. Arnold. Higher-order theory for periodic multiphase materials with inelastic phases. *Int. J. Plast.*, 19:805–847, 2003.

A. Anthoine. Derivation of the in-plane elastic characteristics of masonry through homogenization theory. *Int. J. Sol. Struct.*, 32:137–163, 1995.

M.S. Axelsen and R. Pyrz. Correlation between fracture toughness and the microstructure morphology in transversely loaded unidirectional composites. In R. Pyrz, editor, *Microstructure–Property Interactions in Composite Materials*, pages 15–26, Dordrecht, 1995. Kluwer Academic Publishers.

Y. Benveniste. A new approach to the application of Mori–Tanaka's theory in composite materials. *Mech. Mater.*, 6:147–157, 1987.

Y. Benveniste and G.J. Dvorak. On a correspondence between mechanical and thermal effects in two-phase composites. In G.J. Weng, M. Taya, and H. Abé, editors, *Micromechanics and Inhomogeneity*, pages 65–82, New York, NY, 1990. Springer–Verlag.

M.J. Beran and J. Molyneux. Use of classical variational principles to determine bounds for the effective bulk modulus in heterogeneous media. *Quart. Appl. Math.*, 24:107–118, 1966.

H. Berns, A. Melander, D. Weichert, N. Asnafi, C. Broeckmann, and A. Gross-Weege. A new material for cold forging tools. *Comput. Mater. Sci.*, 11:166–188, 1998.

J.F.W. Bishop and R. Hill. A theory of the plastic distortion of a polycrystalline aggregate under combined stress. *Phil. Mag.*, 42:414–427, 1951.

B. Bochenek and R. Pyrz. Reconstruction methodology for planar and spatial random microstructures. In R. Pyrz, J. Schjødt-Thomsen, J.C. Rauhe, T. Thomsen, and L.R. Jensen, editors, *New Challenges in Mesomechanics*, pages 565–572, Aalborg, Denmark, 2002. Aalborg University.

H.J. Böhm. Numerical investigation of microplasticity effects in unidirectional longfiber reinforced metal matrix composites. *Modell. Simul. Mater. Sci. Engng.*, 1:649–671, 1993.

H.J. Böhm. Modeling the mechanical behavior of short fiber reinforced composites. *This volume*, pages 41–56, 2004.

H.J. Böhm, A. Eckschlager, and W. Han. Multi-inclusion unit cell models for metal matrix composites with randomly oriented discontinuous reinforcements. *Comput. Mater. Sci.*, 25:42–53, 2002.

H.J. Böhm and W. Han. Comparisons between three-dimensional and two-dimensional multi-particle unit cell models for particle reinforced metal matrix composites. *Modell. Simul. Mater. Sci. Engng.*, 9:47–65, 2001.

M. Bornert. Homogénéisation des milieux aléatoires: bornes et estimations. In M. Bornert, T. Bretheau, and P. Gilormini, editors, *Homogénéisation en mécanique des materiaux 1. Matériaux aléatoires élastiques et milieux périodiques*, pages 133–221, Paris, 2001. Editions Hermès.

M. Bornert, T. Bretheau, and P. Gilormini, editors. *Homogénéisation en mécanique des matériaux.* Editions Hermès, Paris, 2001.

M. Bornert, E. Hervé, C. Stolz, and A. Zaoui. Self consistent approaches and strain heterogeneities in two-phase elastoplastic materials. *Appl. Mech. Rev.*, 47:66–S76, 1994.

M. Bornert and P. Suquet. Propriétés non linéaires des composites: Approches par les potentiels. In M. Bornert, T. Bretheau, and P. Gilormini, editors, *Homogénéisation en mécanique des matériaux 2. Comportements non linéaires et problèmes ouverts*, pages 45–90, Paris, 2001. Editions Hermès.

J.R. Brockenbrough and S. Suresh. Plastic deformation of continuous fiber-reinforced metal-matrix composites: Effects of fiber shape and distribution. *Scr. metall. mater.*, 24:325–330, 1990.

K.M. Brockmüller, O. Bernhardi, and M. Maier. Determination of fracture stress and strain of highly oriented oriented short fibre-reinforced composites using a fracture mechanics-based iterative finite-element method. *J. Mater. Sci.*, 30:481–487, 1995.

V.N. Bulsara, R. Talreja, and J. Qu. Damage initiation under transverse loading of unidirectional composites with arbitrarily distributed fibers. *Compos. Sci. Technol.*, 59:673–682, 1999.

V.A. Buryachenko. The overall elastoplastic behavior of multiphase materials with isotropic components. *Acta Mech.*, 119:93–117, 1996.

C.M. Chimani, H.J. Böhm, and F.G. Rammerstorfer. On stress singularities at free edges of bimaterial junctions — A micromechanical study. *Scr. mater.*, 36:943–947, 1997.

R.M. Christensen and K.H. Lo. Solutions for effective shear properties in three phase sphere and cylinder models. *J. Mech. Phys. Sol.*, 27:315–330, 1979.

P.W. Chung, K.K. Tamma, and R.R. Namburu. Asymptotic expansion homogenization for heterogeneous media: Computational issues and applications. *Composites A*, 32A: 1291–1301, 2001.

B. Clyne and P.J. Withers. *An Introduction to Metal Matrix Composites.* Cambridge University Press, Cambridge, 1993.

F. Corvasce, P. Lipiński, and M. Berveiller. The effects of thermal, plastic and elastic stress concentrations on the overall behavior of metal matrix composites. In G.J. Dvorak, editor, *Inelastic Deformation of Composite Materials*, pages 389–408, New York, NY, 1991. Springer–Verlag.

I. Doghri and A. Ouaar. Homogenization of two-phase elasto-plastic composite materials and structures. *Int. J. Sol. Struct.*, 40:1681–1712, 2003.

M. Dong and S. Schmauder. Modeling of metal matrix composites by a self-consistent embedded cell model. *Acta mater.*, 44:2465–2478, 1996.

W.J. Drugan and J.R. Willis. A micromechanics-based nonlocal constitutive equation and estimates of representative volume element size for elastic composites. *J. Mech. Phys. Sol.*, 44:497–524, 1996.

M.L. Dunn and H. Ledbetter. Elastic–plastic behavior of textured short-fiber composites. *Acta mater.*, 45:3327–3340, 1997.

G.J. Dvorak. Transformation field analysis of inelastic composite materials. *Proc. Roy. Soc. London*, A437:311–327, 1992.

A. Eckschlager. *Simulation of Particle Failure in Particle Reinforced Ductile Matrix Composites.* PhD thesis, TU Wien, Vienna, Austria, 2002.

M. Elices, G.V. Guinea, J. Gomez, and J. Planas. The cohesive zone model: Advantages, limitations and challenges. *Engng. Fract. Mech.*, 69:137–163, 2002.

J.D. Eshelby. The determination of the elastic field of an ellipsoidal inclusion and related problems. *Proc. Roy. Soc. London*, A241:376–396, 1957.

J.D. Eshelby. The elastic field outside an ellipsoidal inclusion. *Proc. Roy. Soc. London*, A252:561–569, 1959.

F.D. Fischer, O. Kolednik, G.X. Shan, and F.G. Rammerstorfer. A note on calibration of ductile failure damage indicators. *Int. J. Fract.*, 73:345–357, 1995.

H.F. Fischmeister and B. Karlsson. Plastizitätseigenschaften grob-zweiphasiger Werkstoffe. *Z. Metallkd.*, 68:311–327, 1977.

C. Fond, A. Riccardi, R. Schirrer, and F. Montheillet. Mechanical interaction between spherical inhomogeneities: An assessment of a method based on the equivalent inclusion. *Eur. J. Mech. A/Solids*, 20:59–75, 2001.

A.C. Gavazzi and D.C. Lagoudas. On the numerical evaluation of Eshelby's tensor and its application to elastoplastic fibrous composites. *Comput. Mech.*, 7:12–19, 1990.

M.G.D. Geers, R.A.B. Engelen, and R.J.M. Ubachs. On the numerical modelling of ductile damage with an implicit gradient-enhanced formulation. *Rev. Eur. Elem. Fin.*, 10:173–191, 2001.

S. Ghosh, K.H. Lee, and S. Moorthy. Two scale analysis of heterogeneous elastic-plastic materials with asymptotic homogenization and Voronoi cell finite element model. *Comput. Meth. Appl. Mech. Engng.*, 132:63–116, 1996.

S. Ghosh, K.H. Lee, and P. Raghavan. A multi-level computational model for multi-scale analysis in composite and porous materials. *Int. J. Sol. Struct.*, 38:2335–2385, 2001.

S. Ghosh and S. Moorthy. Particle fracture simulation in non-uniform microstructures of metal-matrix composites. *Acta mater.*, 46:965–982, 1998.

C. González and J. LLorca. A self-consistent approach to the elasto-plastic behavior of two-phase materials including damage. *J. Mech. Phys. Sol.*, 48:675–692, 2000.

A.L. Gurson. Continuum theory of ductile rupture by void nucleation and growth: Part I — Yield criteria and flow rules for porous ductile media. *J. Engng. Mater. Technol.*, 99:2–15, 1977.

A.A. Gusev. Representative volume element size for elastic composites: A numerical study. *J. Mech. Phys. Sol.*, 45:1449–1459, 1997.

Z. Hashin. Analysis of composite materials — A survey. *J. Appl. Mech.*, 50:481–505, 1983.

Z. Hashin and S. Shtrikman. A variational approach to the theory of the elastic behavior of multiphase materials. *J. Mech. Phys. Sol.*, 11:127–140, 1963.

R. Hill. The elastic behavior of a crystalline aggregate. *Proc. Phys. Soc. London*, A65: 349–354, 1952.

R. Hill. Continuum micro-mechanics of elastic-plastic polycrystals. *J. Mech. Phys. Sol.*, 13:89–101, 1965a.

R. Hill. A self-consistent mechanics of composite materials. *J. Mech. Phys. Sol.*, 13: 213–222, 1965b.

R. Hill. The essential structure of constitutive laws for metal composites and polycrystals. *J. Mech. Phys. Sol.*, 15:79–95, 1967.

P.J. Hine, H.R. Lusti, and A.A. Gusev. The numerical simulation of the elastic and thermoelastic properties of short fibre composites. *Compos. Sci. Technol.*, 62:1445–1453, 2002.

G.K. Hu, G. Guo, and D. Baptiste. A micromechanical model of influence of particle fracture and particle cluster on mechanical properties of metal matrix composites. *Comput. Mater. Sci.*, 9:420–430, 1998.

C. Huet. Application of variational concepts to size effects in elastic heterogeneous bodies. *J. Mech. Phys. Sol.*, 38:813–841, 1990.

C. Huet. Coupled size and boundary-condition effects in viscoelastic heterogeneous and composite bodies. *Mech. Mater.*, 31:787–829, 1999.

H. Ismar and U. Reinert. Modelling and simulation of the macromechanical nonlinear behavior of fibre-reinforced ceramics on the basis of a micromechanical–statistical material description. *Acta Mech.*, 120:47–60, 1997.

T. Iung and M. Grange. Mechanical behavior of two-phase materials investigated by the finite element method: Necessity of three-dimensional modeling. *Mater. Sci. Engng.*, A201:L8–L11, 1995.

D. Jeulin and M. Ostoja-Starzewski, editors. *Mechanics of Random and Multiscale Microstructures*. Springer–Verlag, Vienna, 2001.

M. Jiang, M. Ostoja-Starzewski, and I. Jasiuk. Scale-dependent bounds on effective elastoplastic response of random composites. *J. Mech. Phys. Sol.*, 49:655–673, 2001.

M. Jirásek. Comparative study on finite elements with embedded discontinuities. *Comput. Meth. Appl. Mech. Engng.*, 188:307–330, 2000.

J.W. Ju and L.Z. Sun. A novel formulation for the exterior point Eshelby's tensor of an ellipsoidal inclusion. *J. Appl. Mech.*, 66:570–574, 1999.

J.W. Ju and L.Z. Sun. Effective elastoplastic behavior of metal matrix composites containing randomly located aligned spheroidal inhomogeneities. Part I: Micromechanics-based formulation. *Int. J. Sol. Struct.*, 38:183–201, 2001.

M. Kailasam, N. Aravas, and P. Ponte Castañeda. Porous metals with developing anisotropy: Constitutive models, computational issues and applications to deformation processing. *Comput. Model. Engng. Sci.*, 1:105–118, 2000.

T. Kanit, S. Forest, I. Gallier, V. Mounoury, and D. Jeulin. Determination of the size of the representative volume element for random composites: Statistical and numerical approach. *Int. J. Sol. Struct.*, 40:3647–3679, 2003.

M. Karayaka and H. Sehitoglu. Thermomechanical deformation modeling of Al2xxx-T4/SiC$_p$ composites. *Acta metall. mater.*, 41:175–189, 1993.

V.G. Kouznetsova, M.G.D. Geers, and W.A.M. Brekelmans. Multi-scale constitutive modeling of heterogeneous materials with a gradient-enhanced computational homogenization scheme. *Int. J. Num. Meth. Engng.*, 54:1235–1260, 2002.

K.H. Lee, S. Moorthy, and S. Ghosh. Multiple scale computational model for damage in composite materials. *Comput. Meth. Appl. Mech. Engng.*, 172:175–201, 1999.

V.M. Levin. On the coefficients of thermal expansion of heterogeneous materials. *Mech. Sol.*, 2:58–61, 1967.

D.S. Li and M.R. Wisnom. Unidirectional tensile stress–strain response of BP-SiC fiber reinforced Ti-6Al-4V. *J. Compos. Technol. Res.*, 16:225–233, 1994.

M. Li, S. Ghosh, O. Richmond, H. Weiland, and T.N. Rouns. Three dimensional characterization and modeling of particle reinforced metal matrix composites, Part I: Quantitative description of microstructural morphology. *Mater. Sci. Engng.*, A265: 153–173, 1999.

J. LLorca. Deformation and damage in particle reinforced composites: Experiments and models. *This volume*, pages 87–124, 2004.

K.Z. Markov and L. Preziosi. *Heterogeneous Media: Micromechanics Modeling Methods and Simulations.* Birkhäuser, Boston, MA, 2000.

R. Masson, M. Bornert, P. Suquet, and A. Zaoui. An affine formulation for the prediction of the effective properties of nonlinear composites and polycrystals. *J. Mech. Phys. Sol.*, 48:1203–1227, 2000.

P.E. McHugh. Introduction to crystal plasticity theory. *This volume*, pages 125–172, 2004.

P.E. McHugh and P. Connolly. Modelling the thermo-mechanical behavior of an Al alloy-SiC$_p$ composite. Effects of particle shape and microscale failure. *Comput. Mater. Sci.*, 3:199–206, 1994.

J.C. Michel, H. Moulinec, and P. Suquet. Effective properties of composite materials with periodic microstructure: A computational approach. *Comput. Meth. Appl. Mech. Engng.*, 172:109–143, 1999.

J.C. Michel, H. Moulinec, and P. Suquet. Composites à microstructure périodique. In M. Bornert, T. Bretheau, and P. Gilormini, editors, *Homogénéisation en mécanique des materiaux 1. Matériaux aléatoires élastiques et milieux périodiques*, pages 57–94, Paris, 2001. Editions Hermès.

J.C. Michel and P. Suquet. Nonuniform transformation field analysis. *Int. J. Sol. Struct.*, 40:6937–6955, 2003.

C. Miehe, J. Schröder, and J. Schotte. Computational homogenization analysis in finite plasticity. simulation of texture development in polycrystalline materials. *Comput. Meth. Appl. Mech. Engng.*, 171:387–418, 1999.

C.A. Miller and S. Torquato. Effective conductivity of hard sphere suspensions. *J. Appl. Phys.*, 68:5486–5493, 1990.

G.W. Milton. Bounds on the electromagnetic, elastic, and other properties of two-component composites. *Phys. Rev. Lett.*, 46:542–545, 1981.

G.W. Milton. *The Theory of Composites.* Cambridge University Press, Cambridge, 2002.

N. Moës, N. Cloirec, P. Cartraud, and J.F. Remacle. A computational approach to handle complex microstructure geometries. *Comput. Meth. Appl. Mech. Engng.*, 192: 3163–3177, 2003.

A. Molinari, G.R. Canova, and S. Ahzi. A self-consistent approach for large deformation viscoplasticity. *Acta metall.*, 35:2983–2984, 1987.

H. Moulinec and P. Suquet. A fast numerical method for computing the linear and nonlinear mechanical properties of composites. *C. R. Acad. Sci. Paris, série II*, 318: 1417–1423, 1994.

W.H. Müller. Mathematical versus experimental stress analysis of inhomogeneities in solids. *J. Phys. IV*, 6:1–139–C1–148, 1996.

T. Mura. *Micromechanics of Defects in Solids.* Martinus Nijhoff, Dordrecht, 1987.

A. Needleman. A continuum model for void nucleation by inclusion debonding. *J. Appl. Mech.*, 54:525–531, 1987.

A. Needleman, S.R. Nutt, S. Suresh, and V. Tvergaard. Matrix, reinforcement, and interfacial failure. In S. Suresh, A. Mortensen, and A. Needleman, editors, *Fundamentals of Metal Matrix Composites*, pages 233–250, Boston, MA, 1993. Butterworth–Heinemann.

S. Nemat-Nasser. Averaging theorems in finite deformation plasticity. *Mech. Mater.*, 31: 493–523, 1999.

S. Nemat-Nasser and M. Hori. *Micromechanics: Overall Properties of Heterogeneous Solids.* North–Holland, Amsterdam, 1993.

M. Ostoja-Starzewski. Random field models of heterogeneous materials. *Int. J. Sol. Struct.*, 35:2429–2455, 1998.

R. Pandorf. *Ein Beitrag zur FE-Simulation des Kriechens partikelverstärkter metallischer Werkstoffe.* Reihe **5**, Nr.585. VDI–Verlag, Düsseldorf, 2000.

O.B. Pedersen. Thermoelasticity and plasticity of composites — I. Mean field theory. *Acta metall.*, 31:1795–1808, 1983.

R.H.J. Peerlings, M.G.D. Geers, R. de Borst, and W.A.M. Brekelmans. A critical comparison of nonlocal and gradient-enhanced softening continua. *Int. J. Sol. Struct.*, 38: 7723–7746, 2001.

H.E. Pettermann. *Derivation and Finite Element Implementation of Constitutive Material Laws for Multiphase Composites Based on Mori–Tanaka Approaches.* Reihe **18**, Nr.217. VDI–Verlag, Düsseldorf, 1997.

H.E. Pettermann, A.F. Plankensteiner, H.J. Böhm, and F.G. Rammerstorfer. A thermo-elasto-plastic constitutive law for inhomogeneous materials based on an incremental Mori–Tanaka approach. *Comput. Struct.*, 71:197–214, 1999.

H.E. Pettermann and S. Suresh. A comprehensive unit cell model: A study of coupled effects in piezoelectric 1–3 composites. *Int. J. Sol. Struct.*, 37:5447–5464, 2000.

N. Phan-Thien and G.W. Milton. New third-order bounds on the effective moduli of $n$-phase composites. *Quart. Appl. Math.*, 41:59–74, 1983.

P. Ponte Castañeda. Bounds and estimates for the properties on nonlinear inhomogeneous systems. *Phil. Trans. Roy. Soc.*, A340:531–567, 1992.

P. Ponte Castañeda and P. Suquet. Nonlinear composites. In E. van der Giessen and T.Y. Wu, editors, *Advances in Applied Mechanics 34*, pages 171–302, New York, NY, 1998. Academic Press.

P. Ponte Castañeda and J.R. Willis. The effect of spatial distribution on the effective behavior of composite materials and cracked media. *J. Mech. Phys. Sol.*, 43:1919–1951, 1995.

R. Pyrz. Microstructural description of composites, statistical methods. *This volume*, pages 173–234, 2004.

Y.P. Qiu and G.J. Weng. A theory of plasticity for porous materials and particle-reinforced composites. *J. Appl. Mech.*, 59:261–268, 1992.

S. Quilici and G. Cailletaud. FE simulation of macro-, meso- and microscales in polycrystalline plasticity. *Comput. Mater. Sci.*, 16:383–390, 1999.

A. Reuss. Berechnung der Fließgrenze von Mischkristallen auf Grund der Plastizitätsbedingung für Einkristalle. *ZAMM*, 9:49–58, 1929.

M. Rintoul and S. Torquato. Reconstruction of the structure of dispersions. *J. Colloid Interf. Sci.*, 186:467–476, 1997.

A.P. Roberts and E.J. Garboczi. Elastic properties of a tungsten–silver composite by reconstruction and computation. *J. Mech. Phys. Sol.*, 47:2029–2055, 1999.

M. Sautter, C. Dietrich, M.H. Poech, S. Schmauder, and H.F. Fischmeister. Finite element modelling of a transverse-loaded fibre composite: Effects of section size and net density. *Comput. Mater. Sci.*, 1:225–233, 1993.

R.A. Schapery. Thermal expansion coefficients of composite materials based on energy principles. *J. Compos. Mater.*, 2:380–404, 1968.

S. Schmauder, J. Wulf, T. Steinkopff, and H. Fischmeister. Micromechanics of plasticity and damage in an Al/SiC metal matrix composite. In A. Pineau and A. Zaoui, editors, *Micromechanics of Plasticity and Damage of Multiphase Materials*, pages 255–262, Dordrecht, 1996. Kluwer.

J. Segurado, J. LLorca, and C. González. On the accuracy of mean-field approaches to simulate the plastic deformation of composites. *Scr. mater.*, 46:525–529, 2002.

H. Shen and C.J. Lissenden. 3D finite element analysis of particle-reinforced aluminum. *Mater. Sci. Engng.*, A338:271–281, 2002.

V.B. Shenoy, R. Miller, E.B. Tadmor, R. Phillips, and M. Ortiz. Quasicontinuum models of interfacial structures and deformation. *Phys. Rev. Lett.*, 80:742–745, 1998.

J.A. Sherwood and H.M. Quimby. Micromechanical modeling of damage growth in titanium based metal-matrix composites. *Comput. Struct.*, 56:505–54, 1995.

T. Siegmund, R. Cipra, J. Liakus, B. Wang, M. LaForest, and A. Fatz. Processing–microstructure–property relationships in short fiber reinforced carbon–carbon composite system. *This volume*, pages 235–258, 2004.

R.J.M. Smit, W.A.M. Brekelmans, and H.E.H. Meijer. Prediction of the mechanical behavior of non-linear heterogeneous systems by multi-level finite element modeling. *Comput. Meth. Appl. Mech. Engng.*, 155:181–192, 1998.

P. Suquet. Elements of homogenization for inelastic solid mechanics. In E. Sanchez-Palencia and A. Zaoui, editors, *Homogenization Techniques in Composite Media*, pages 194–278, Berlin, 1987. Springer–Verlag.

P. Suquet, editor. *Continuum Micromechanics*. Springer–Verlag, Vienna, 1997a.

P. Suquet. Effective properties of nonlinear composites. In P. Suquet, editor, *Continuum Micromechanics*, pages 197–264, Vienna, 1997b. Springer–Verlag.

T. Suzuki and P.K.L. Yu. Complex elastic wave band structures in three-dimensional periodic elastic media. *J. Mech. Phys. Sol.*, 46:115–138, 1998.

N. Takano, M. Zako, and T. Okazaki. Efficient modeling of microscopic heterogeneity and local crack in composite materials by finite element mesh superposition method. *JSME Int. J. Srs.A*, 44:602–609, 2001.

G.P. Tandon and G.J. Weng. The effect of aspect ratio of inclusions on the elastic properties of unidirectionally aligned composites. *Polym. Compos.*, 5:327–333, 1984.

G.P. Tandon and G.J. Weng. A theory of particle-reinforced plasticity. *J. Appl. Mech.*, 55:126–135, 1988.

M. Taya, W.D. Armstrong, M.L. Dunn, and T. Mori. Analytical study on dimensional changes in thermally cycled metal matrix composites. *Mater. Sci. Engng.*, A143: 143–154, 1991.

K. Terada and N. Kikuchi. Nonlinear homogenization method for practical applications. In S. Ghosh and M. Ostoja-Starzewski, editors, *Computational Methods in Micromechanics*, pages 1–16, New York, NY, 1996. ASME.

K. Terada, I. Saiki, K. Matsui, and Y. Yamakawa. Two-scale kinematics and linearization for simultaneous two-scale analysis of periodic heterogeneous solids at finite strain. *Comput. Meth. Appl. Mech. Engng.*, 192:3531–3563, 2003.

J.F. Thovert, I.C. Kim, S. Torquato, and A. Acrivos. Bounds on the effective properties of polydispersed suspensions of spheres: An evaluation of two relevant parameters. *J. Appl. Phys.*, 67:6088–6098, 1990.

S. Torquato. Random heterogeneous media: Microstructure and improved bounds on effective properties. *Appl. Mech. Rev.*, 44:37–75, 1991.

S. Torquato. Morphology and effective properties of disordered heterogeneous media. *Int. J. Sol. Struct.*, 35:2385–2406, 1998.

S. Torquato. *Random Heterogeneous Media*. Springer–Verlag. New York, NY, 2001.

S. Torquato, F. Lado, and P.A. Smith. Bulk properties of two-phase disordered media. IV. Mechanical properties of suspensions of penetrable spheres at nondilute concentrations. *J. Chem. Phys.*, 86:6388–6392, 1987.

V. Tvergaard. Analysis of tensile properties for a whisker-reinforced metal-matrix composite. *Acta metall. mater.*, 38:185–194, 1990.

V. Tvergaard. Fibre debonding and breakage in a whisker-reinforced metal. *Mater. Sci. Engng.*, A190:215–222, 1994.

V. Tvergaard and A. Needleman. Analysis of the cup-cone fracture in a round tensile bar. *Acta metall.*, 32:157–169, 1984.

E. van der Giessen. Creep rupture in polycrystalline materials. *This volume*, pages 283–306, 2004a.

E. van der Giessen. Discrete dislocation plasticity. *This volume*, pages 259–282, 2004b.

N. Vejen and R. Pyrz. Transverse crack growth in glass/epoxy composites with exactly positioned long fibers. Part II: Numerical. *Composites B*, 33B:279–290, 2002.

W. Voigt. Über die Beziehung zwischen den beiden Elasticitäts-Constanten isotroper Körper. *Ann. Phys.*, 38:573–587, 1889.

W. Vonach. *A General Solution to the Wrinkling Problem of Sandwiches.* Reihe **18**, Nr.268. VDI–Verlag, Düsseldorf, 2001.

K. Wakashima, H. Tsukamoto, and B.H. Choi. Elastic and thermoelastic properties of metal matrix composites with discontinuous fibers or particles: Theoretical guidelines toward materials tailoring. In *The Korea–Japan Metals Symposium on Composite Materials*, pages 102–115, Seoul, 1988. The Korean Institute of Metals.

L.J. Walpole. On bounds for the overall elastic moduli of inhomogeneous systems — I. *J. Mech. Phys. Sol.*, 14:151–162, 1966.

S. Weihe and B.H. Kröplin. The fictitious crack concept in the mechanics of composites. In D.R.J. Owen, E. Oñate, and E. Hinton, editors, *Computational Plasticity: Fundamentals and Applications*, pages 1215–1226, Swansea, 1995. Pineridge Press.

G.J. Weng. The theoretical connection between Mori–Tanaka theory and the Hashin–Shtrikman–Walpole bounds. *Int. J. Engng. Sci.*, 28:1111–1120, 1990.

J.R. Willis. Bounds and self-consistent estimates for the overall moduli of anisotropic composites. *J. Mech. Phys. Sol.*, 25:185–202, 1977.

J.R. Willis. The overall response of nonlinear composite media. *Eur. J. Mech. A/Solids*, 19:S165–S184, 2000.

P.J. Withers. The determination of the elastic field of an ellipsoidal inclusion in a transversely isotropic medium, and its relevance to composite materials. *Phil. Mag.*, A59: 759–781, 1989.

J. Wulf, T. Steinkopff, and H. Fischmeister. FE-simulation of crack paths in the real microstructure of an Al(6061)/SiC composite. *Acta mater.*, 44:1765–1779, 1996.

S. Yang, A. Tewari, and A. Gokhale. Modeling of non-uniform spatial arrangement of fibers in a ceramic matrix composite. *Acta mater.*, 45:3059–3069, 1997.

A. Zaoui. Plasticité: Approches en champ moyen. In M. Bornert, T. Bretheau, and P. Gilormini, editors, *Homogénéisation en mécanique des matériaux 2. Comportements non linéaires et problèmes ouverts*, pages 17–44, Paris, 2001. Editions Hermès.

A. Zaoui. Continuum micromechanics: Survey. *J. Engng. Mech.*, 128:808–816, 2002.

J. Zeman and M. Šejnoha. Numerical evaluation of effective elastic properties of graphite fiber tow impregnated by polymer matrix. *J. Mech. Phys. Sol.*, 49:69–90, 2001.

Y.H. Zhao and G.J. Weng. A theory of inclusion debonding and its influence on the stress–strain relations of a ductile matrix composite. *Int. J. Dam. Mech.*, 4:196–211, 1995.

R.W. Zimmerman. Hashin-Shtrikman bounds on the Poisson ratio of a composite material. *Mech. Res. Comm.*, 19:563–569, 1992.

T.I. Zohdi. Constrained inverse formulations in random material design. *Comput. Meth. Appl. Mech. Engng.*, 192:3179–3194, 2003.

T.I. Zohdi and P. Wriggers. A model for simulating the deterioration of structural-scale material responses of microheterogeneous solids. *Comput. Meth. Appl. Mech. Engng.*, 190:2803–2823, 2001.

# Modeling the Mechanical Behavior of Short Fiber Reinforced Composites

Helmut J. Böhm

Institute of Lightweight Design and Structural Biomechanics,
Vienna University of Technology, Vienna, Austria

**Abstract** A discussion of selected modeling approaches for studying the mechanical and thermal expansion behavior of short fiber reinforced composites is given, the emphasis being put on mean field and unit cell models. Aligned and randomly oriented reinforcement geometries as well as more general fiber orientations described via orientation distribution functions are addressed.

## 1 Short Fiber Reinforced Composites — Introduction

Short fiber reinforced composites are materials of considerable technological and economical importance. Especially for polymeric matrices reinforcing by short fibers can confer improvements in terms of stiffness and strength while allowing the manufacture of components with complex geometries by cost effective processes such as transfer molding, sheet molding, injection molding, etc., see e.g. De and White (1996). Reinforcing by short fibers has also been applied to many other composite systems, two examples being carbon–carbon composites produced by spray deposition, compare Siegmund et al. (2004), and copper matrix composites for thermal management applications produced by consolidating metal coated fibers (Korb et al., 1998).

Most production routes for short fiber reinforced composites give rise to materials in which the fibers are at least to some extent statistically oriented rather than perfectly aligned. At low to moderate fiber volume fractions the orientations of neighboring fibers may be viewed as essentially independent within the geometrical constraints of non-penetration. For fiber arrangements of this type, which are referred to as dispersed systems, the microgeometrical configuration of the composite can be statistically described in terms of fiber orientation distribution functions (ODFs) and fiber length distribution functions (LDFs), both of which can be determined experimentally, see e.g. (Clarke et al., 1995) or (Pyrz, 2004). The effective thermomechanical properties of short fiber reinforced composites with dispersed microstructures can be evaluated by averaging over the fiber orientation and length distributions (configurational averaging). At elevated fiber volume fractions, however, local domains of (nearly) aligned fibers are typically observed, which may give rise to a "grainy" mesostructure as found in many sheet molding compounds. Microgeometries of this type are referred to as aggregated systems (Eduljee and McCullough, 1993). If such a domain structure is a well developed, effective properties can be obtained by two-step procedures involving homogenization at the level of individual domains plus averaging over all domains. If the fibers are sufficiently long and flexible they may form part of a number of domains in aggregated systems. Conversely, fibers that are rather brittle tend to break in the course of processing, with the consequence that the in-situ fiber length distribution in the composite may differ significantly from the initial one of the fibers, see e.g. Narkis et al. (1991).

Obviously the complexity of their microgeometries can make the modeling of short fiber reinforced composites in terms of homogenization and localization procedures rather demanding. Accordingly, modeling techniques for these materials are not as highly developed as for composites reinforced by particles or continuous fibers.

# 2 Models for Composites Reinforced by Aligned Short Fibers

Because of their relatively simple microgeometries short fiber composites with aligned reinforcements have been the focus of considerable modeling efforts over the last decades. These have provided many contributions to the present understanding of the thermomechanical behavior of discontinuous fibers embedded in a matrix.

Composites reinforced by aligned short fibers show transversally isotropic overall behavior, the most important parameters for describing the microgeometries being the fiber volume fraction $\xi$ and the fiber aspect ratio $a$.

## 2.1 Basic and Semi-Empirical Models

**Shear Lag Models.** Shear lag models, which concentrate on studying the load transfer between a compliant matrix and stiff fibers, were among the first theoretical descriptions of short fiber reinforced composites and go back to studies by Cox (1952) of the mechanical behavior of paper. Shear lag models for composites with elastoplastic matrix are based on simplified displacement and stress fields in a fiber and the surrounding matrix under uniaxial tensile loading in the fiber direction as sketched in Figure 1.

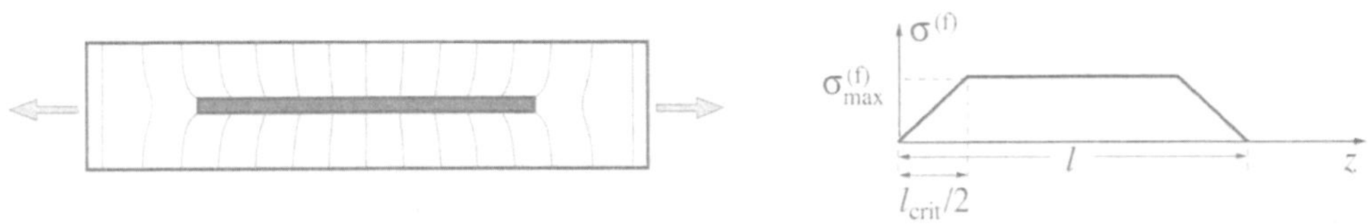


**Figure 1.** Sketch of a matrix deformation pattern of the type assumed by shear lag models (left) and the resulting distribution of the axial stresses in a fiber that exceeds the critical length (right).

Following Rosen (1965) the stress equilibrium in an infinitesimal fiber element allows the build-up of the axial stress in the fibers, $\sigma^{(f)}(z)$, starting from the fiber ends to be approximated as

$$\sigma^{(f)}(z) = \frac{2\tau_y z}{r} \quad . \tag{2.1}$$

Here $\tau_y$ stands for the yield shear stress of the matrix or the shear strength of the interface between fiber and matrix, $z$ is the axial position along the fiber, and $r$ is the fiber radius.

Such a linear build-up of stresses takes place at both ends of the fiber until its strength, $\sigma_{\max}^{(f)}$, is reached. Provided the fiber is sufficiently long, this results in a trapezoidal distribution of the fiber stress as shown in Figure 1 (right). This simple model directly leads to the concepts of a critical fiber length $l_{\text{crit}}$ and a critical fiber aspect ratio $a_{\text{crit}}$,

$$l_{\text{crit}} = r\frac{\sigma_{\max}^{(f)}}{\tau_y} \quad \text{and} \quad a_{\text{crit}} = \frac{\sigma_{\max}^{(f)}}{2\tau_y} \quad , \tag{2.2}$$

respectively, which must be attained by a fiber in order to make full use of its strength within a given composite system. The average stress in a fiber of length $l$ follows as

$$\bar{\sigma}^{(f)} = \frac{\tau_y l}{2r} \quad \text{for} \quad l < l_{\text{crit}} \qquad \text{and} \qquad \bar{\sigma}^{(f)} = \sigma_{\max}^{(f)}\left(1 - \frac{l_{\text{crit}}}{2l}\right) \quad \text{for} \quad l \geq l_{\text{crit}} \quad . \tag{2.3}$$

Modified shear lag models also account for the stress transfer via the fiber ends to give improved predictions for the axial variation of $\sigma^{(f)}$. When combined with "rule of mixture" expressions, they can provide estimates for the axial stiffness of composites reinforced by aligned short fibers (Clyne and Withers, 1993). Shear lag models, however, are limited in their capability of resolving fiber arrangement effects, compare Weißenbek and Rammerstorfer (1993). For a broader discussion of shear lag and related models see e.g. (Fukuda and Takao, 2000).

**Halpin–Tsai Relationships.** The basic rules of mixture, which give very satisfactory predictions for the axial stiffness (but not the other elastic moduli) of continuously reinforced composites, are of limited applicability to short fiber reinforced materials. A more general semi-empirical relationship that connects the effective properties of fiber reinforced composites with the elastic moduli of matrix (m) and fibers (f) was proposed by Halpin and Tsai (1969). It takes the form

$$P^* = P^{(m)}\left(\frac{1+\xi^{(f)}\zeta\,\eta_P}{1-\xi^{(f)}\eta_P}\right) \qquad \text{with} \qquad \eta_P = \frac{P^{(f)} - P^{(m)}}{P^{(f)} + \zeta P^{(m)}} \quad , \tag{2.4}$$

where $P$ stands for the property to be evaluated, $\eta_P$ is a dimensionless "reduced property", and $\zeta$ is a scaling parameter that may be interpreted as the "reinforcing efficiency" of the composite system. The reinforcement factor $\zeta$ is treated as an adjustable parameter that depends on the fiber geometry, the constituent properties, and the modulus $P$ to be considered; it must be fitted from experimental results. For $\zeta \to 0$ and $\zeta \to \infty$ Reuss and Voigt expressions, respectively, are recovered. For short fiber composites a typical choice for describing the longitudinal Young's modulus is $\zeta = 2a$, whereas for the other moduli $\zeta$ is of order unity (Tucker and Liang, 1999).

## 2.2 Mean Field Methods and Bounds

As discussed elsewhere in this volume (Böhm, 2004) mean field methods, especially effective field approaches of the Mori–Tanaka-type, and Hashin–Shtrikman estimates describe the thermoelastic behavior of matrix–inclusion composites with aligned ellipsoidal reinforcements of prescribed aspect ratios. Such microgeometries obviously are suitable for modeling composites reinforced by aligned short fibers. In a similar vein, secant and incremental methods based on mean field models can be used to generate estimates for the nonlinear thermomechanical responses of such materials, see e.g. Hu (1996), Pettermann et al. (1999), Lee and Simunovic (2000), or Ju and Sun (2001).

The Willis bounds (Willis, 1977), one of which coincides with Mori–Tanaka expressions in most cases, compare Weng (1992) or Ponte Castañeda and Willis (1995), pertain to the elastic behavior of ellipsoidal arrangements of aligned ellipsoids. They can thus be used to bound the effective elastic behavior of composites reinforced by aligned short fibers. No bounds for the effective elastoplastic behavior of such macroscopically transversely isotropic materials, however, appear to be available in the literature.

## 2.3 Unit Cell Models for Aligned Short Fiber Composites

Most unit cell models for composites reinforced by aligned short fibers have been based on relatively simple three-dimensional periodic microgeometries, the most important of which are hexagonal or square arrangements of cylindrical non-staggered short fibers, see Figure 2 (left), and square arrangements of staggered short fibers, see Figure 2 (right).

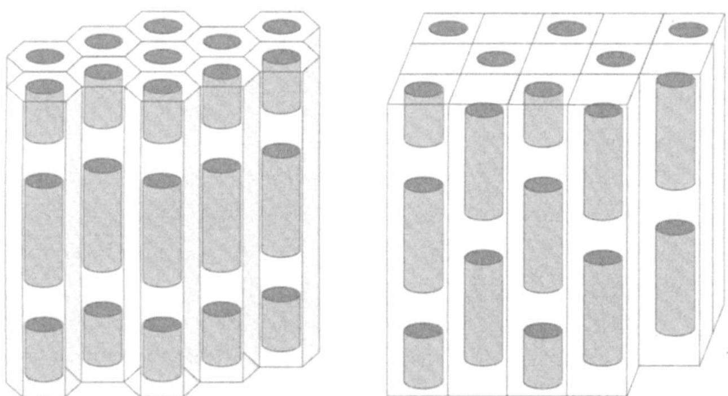

**Figure 2.** Sketch of fiber positions in non-staggered (left) and staggered (right) arrangements used for composites reinforced by aligned short fibers.

Three-dimensional unit cells that describe either of the above phase arrangements and employ periodicity boundary conditions can be used to evaluate the overall thermoelastic tensors of aligned short fiber composites. Such models are also capable of following any load path for inelastic constituent behavior. When symmetry boundary conditions are specified, smaller unit cells result that cannot handle all load cases but are highly suitable for materials characterization and for studying thermal residual stress states caused by the thermal expansion mismatch between fibers and matrix during processing, see e.g. Levy and Papazian (1991). Unit cells of the above types show orthotropic overall elastic behavior when based on square arrangements of cylinders or ellipsoids, whereas the hexagonal arrays are elastically transversally isotropic. For comparisons between the predictions of such unit cells and results from analytical models see e.g. Tucker and Liang (1999). There are no simple geometric arrangements of aligned short fibers that give transversally isotropic overall behavior in the inelastic range, the symmetry of hexagonal arrangements being broken by the low symmetry of the inelastic strains on the microscale for most load cases.

Three-dimensional unit cells of the above types are rather restrictive with respect to the fiber arrangements that can be studied. This limitation can be circumvented by using unit cells that contain a number of randomly positioned aligned reinforcements, compare Ingber and Papathanasiou (1997). In a recent study Gusev et al. (2002b) reported multi-fiber unit cells of this type that are generated by random insertion and Monte Carlo techniques and can handle a wide range of fiber length and fiber diameter distributions at fairly high fiber volume fractions.

For many materials characterization investigations a more economical alternative to the above three-dimensional unit cells is provided by axisymmetric cells that describe the behavior of non-staggered or staggered arrays of aligned short fibers in an approximate way. The basic idea behind these models consists in replacing square or hexagonal arrays of fibers by arrangements of axisymmetric composite cells of the same cross sectional area, fiber volume fraction, and fiber as well as cell aspect ratio. Figure 3 sketches such axisymmetric cells that correspond to a staggered fiber arrangement of the type given in Figure 2 (right). Axisymmetric cells that describe non-staggered microgeometries of the type shown in Figure 2 (left) can be developed in direct analogy, see e.g. Nutt and Needleman (1987).

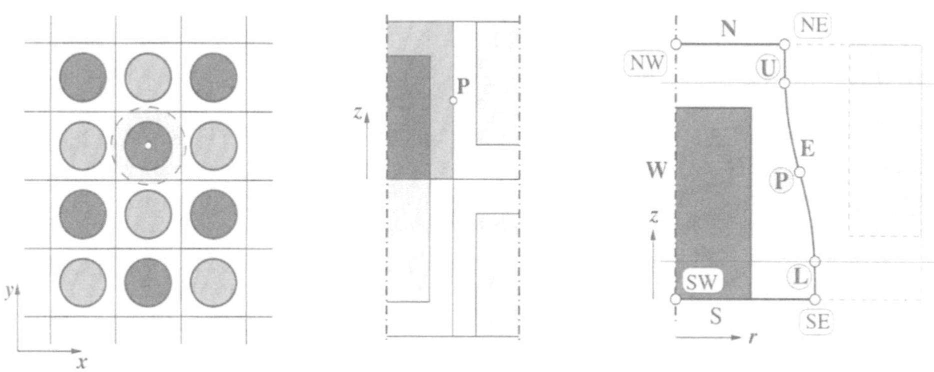


**Figure 3.** Approximation of a square arrangement of staggered fibers (left) by pairs of axisymmetric cells (center) and axisymmetric cell for staggered short fibers (right).

Axisymmetric cells are not unit cells in the strict sense, because they overlap and are not space filling. They also do not show the same transverse fiber spacing as the underlying three-dimensional arrays. Obviously, models based on axisymmetric cells are restricted to handling thermomechanical load cases that lead to axisymmetric deformations, so that they cannot be used to study the response to transverse uniaxial loading or any kind of shear loading. They have the advantage, however, of significantly reduced computational requirements.

In terms of the identifiers of faces and vertices given in Figure 3 (right), axisymmetric cell models employ symmetry boundary conditions for the top (N) and bottom (S) faces.

The B.C.s at the cylindrical surface (E) must be chosen to maintain the same cross sectional area along the axial direction for each pair of cells. In the case of non-staggered fibers this requirement is fulfilled by constraining the surface E to remain parallel to the axis of rotation, which is formally equivalent to specifying symmetry-type boundary conditions in planar geometries. Following Tvergaard (1990), a pair of cells with offset fiber positions must be considered to define the boundary conditions for the case of staggered fiber arrangements, see Figure 3 (right). Each of these cells contains half a fiber plus the surrounding matrix and must deform in an identical way to the other, which implies that the arrangement must show point symmetry with respect to the pivot point marked as P in Figure 3. Together with the requirement that the total cross sectional area of the pair of cells must be independent of the axial position this symmetry leads to nonlinear relations for the radial microscopic displacements $u'$ and linear constraints for the axial microscopic displacements $v'$, which take the form

$$(r'_{\mathrm{U}} + u'_{\mathrm{U}})^2 + (r'_{\mathrm{L}} + u'_{\mathrm{L}})^2 = 2(r'_{\mathrm{P}} + u'_{\mathrm{P}})^2 \qquad \text{and} \qquad v'_{\mathrm{U}} + v'_{\mathrm{L}} = 2v'_{\mathrm{P}} \quad . \tag{2.5}$$

Here the subscripts U and L refer to pairs points on face E that are symmetric with respect to the pivot point P, their radial coordinates in the undeformed state being denoted as $r'_{\mathrm{U}}$ and $r'_{\mathrm{L}}$, respectively.

Nonlinear constraints of this type are not available in most standard FE codes or are cumbersome to use. For most applications this difficulty can be circumvented by linearizing the B.C.s for the radial displacements in eqn. (2.5) without major loss in accuracy. This results in expressions that are formally analogous to antisymmetry boundary conditions, see eqn.(3.3) in (Böhm, 2004), for the surface E,

$$u'_{\mathrm{U}} + u'_{\mathrm{L}} = 2u'_{\mathrm{P}} \qquad \text{and} \qquad v'_{\mathrm{U}} + v'_{\mathrm{L}} = 2v'_{\mathrm{P}} \quad . \tag{2.6}$$

Over the last decade, axisymmetric cell models have been a mainstay of numerical studies of aligned short fiber reinforced composites, especially when damage at the microscale is involved, see e.g. Povirk et al. (1991), Zhou et al. (2000), and Tvergaard (2001). Typically, descriptions using staggered fiber arrangements allow a wider range of microgeometries to be covered and give somewhat more realistic descriptions of actual composites.

Axisymmetric cells can also be used to model the thermomechanical responses of composites that contain aligned fibers of different sizes, different (axisymmetric) shapes, or different aspect ratios. This can be achieved by coupling two different sub-cells instead of two equal ones, compare Böhm et al. (1993), or by using sub-cells that contain multiple fibers, see Tvergaard (2004). Such extensions of the axisymmetric cell scheme allow a considerable range of phase arrangement effects to be studied at limited numerical cost.

## 3 Models for Composites Reinforced by Nonaligned Short Fibers

Composites reinforced by nonaligned short fibers can show a wide range of fiber orientation distribution functions, see e.g. Allen and Lee (1990). These ODFs can be evaluated experimentally or via process simulation. Axisymmetric orientation distributions give rise to statistically transversally isotropic overall behavior, a special case being planar random fiber orientations. More general ODFs lead to lower overall symmetries. An important special case are composites reinforced by randomly oriented fibers, which are statistically isotropic.

Analytical descriptions of composites reinforced by nonaligned short fibers typically make use of configurational averaging procedures, which may encompass aspect ratio averaging in addition to orientational averaging. Orientational averaging of tensor valued variables can be done by direct (numerical) integration, see e.g. Pettermann et al. (1997), or on the basis of expansions of the ODF in terms of generalized spherical harmonics (Viglin expansions), compare e.g. Advani and Tucker (1987). The latter approach can be formulated in a number of ways, e.g. via texture coefficients or texture matrices, compare Siegmund et al. (2004). For a discussion of a number of issues relevant to configurational averaging see Eduljee and McCullough (1993).

### 3.1 Orientation Averaging Models

A number of modeling approaches are based on the assumption that the contribution of a given fiber to the overall stiffness and strength of the composites can be approximated as depending solely on its direction cosine with respect to the applied load and on its length, interactions between neighboring fibers being neglected. The Fukuda–Kawata model (Fukuda and Kawata, 1974) for the stiffness and the Fukuda–Chou model (Fukuda and Chou, 1982) for the strength of short fiber reinforced composites are based on summing up stiffness contributions of individual fibers crossing an arbitrary normal section. These contributions are obtained by shear lag models and evaluated on the basis of fiber orientation and length distribution functions, compare also (Jayaraman and Kortschot, 1996). The results are typically given as modified rules of mixtures with fiber direction and fiber length corrections.

In laminate analogy approaches, see e.g. Halpin and Pagano (1969) and Fu and Lauke (1998), nonaligned composites are modeled as stacks of layers each of which is reinforced by unidirectional fibers of some given orientation. The overall stiffness and strength properties of the composite are then obtained on the basis of classical lamination theory. Methods of this type are well suited for modeling short fiber reinforced composites with aggregate-type microgeometries.

A related technique for obtaining the overall elastic properties of dispersed short fiber reinforced composites consists in orientational and fiber length averaging over elastic tensors pertaining to aligned systems of appropriate fiber orientation. Such elastic tensors can be obtained from mean field estimates, see e.g. Siegmund et al. (2004), or from unit cells that may contain multiple aligned fibers (Gusev et al., 2002a).

### 3.2 Mean Field Methods and Bounds

Mori–Tanaka methods have been adapted to describing the elastic behavior of rather general microstructures with nonaligned fibers (the orientations of which are described via ODFs) that may have different aspect ratios (described via LDFs) and may consist of different materials.

A typical starting point for such "extended Mori–Tanaka" algorithms are dilute stress concentration tensors for fibers having some given orientation with respect to the global coordinate system, $\bar{\mathbf{B}}_{\rm dil}^{(\rm f)\angle}$, which can be denoted as

$$\bar{\mathbf{B}}_{\rm dil}^{(\rm f)\angle} = \mathbf{T}_\sigma^\angle \left[\mathbf{I} + \mathbf{E}^{(\rm m)}(\mathbf{I} - \mathbf{S})(\mathbf{C}^{(\rm f)} - \mathbf{C}^{(\rm m)})\right]^{-1} \mathbf{T}_\sigma^{\angle -1} \quad . \tag{3.1}$$

Equation 3.1 is based on an expression for dilute stress concentration tensors due to Hill (1965) and follows the notation used in (Böhm, 2004). Elasticity tensors are denoted as $\mathbf{E}$, compliance tensors as $\mathbf{C}$, the Eshelby tensor as $\mathbf{S}$, the unit tensor as $\mathbf{I}$, and $\mathbf{T}_\sigma^\angle$ stands for the stress transformation tensor from the local (reinforcement based) to the global coordinate system. A phase averaged dilute fiber stress concentration tensor can then be obtained by orientational averaging as $\overline{\mathbf{B}}_{\rm dil}^{(\rm f)} = \left\langle \rho \, \bar{\mathbf{B}}_{\rm dil}^{(\rm f)\angle} \right\rangle$, where the orientation distribution function $\rho$ is assumed to be normalized to give $\langle \rho \rangle = 1$. The core statement of the Mori–Tanaka approach according to Benveniste (1987) then takes the form

$$\bar{\boldsymbol{\sigma}}^{(\rm f)} = \overline{\mathbf{B}}_{\rm M}^{(\rm f)} \boldsymbol{\sigma}^{\rm a} = \overline{\mathbf{B}}_{\rm dil}^{(\rm f)} \overline{\mathbf{B}}_{\rm M}^{(\rm m)} \boldsymbol{\sigma}^{\rm a} = \overline{\mathbf{B}}_{\rm dil}^{(\rm f)} \bar{\boldsymbol{\sigma}}^{(\rm m)} \quad , \tag{3.2}$$

where $\boldsymbol{\sigma}^{\rm a}$ is the macroscopic stress. $\bar{\boldsymbol{\sigma}}^{(\rm f)}$ and $\bar{\boldsymbol{\sigma}}^{(\rm m)}$ stand for the phase averaged stresses in fibers and matrix, respectively. The phase averaged Mori–Tanaka concentration tensors can then be expressed as

$$\overline{\mathbf{B}}_{\rm M}^{(\rm m)} = [\xi^{(\rm m)}\mathbf{I} + \xi^{(\rm f)}\overline{\mathbf{B}}_{\rm dil}^{(\rm f)}]^{-1} \qquad \text{and} \qquad \overline{\mathbf{B}}_{\rm M}^{(\rm f)} = \overline{\mathbf{B}}_{\rm dil}^{(\rm f)}[\xi^{(\rm m)}\mathbf{I} + \xi^{(\rm f)}\overline{\mathbf{B}}_{\rm dil}^{(\rm f)}]^{-1} \tag{3.3}$$

in analogy to standard Mori–Tanaka methods, compare Böhm (2004).

Extended Mori–Tanaka methods for modeling composites with nonaligned reinforcements were reported e.g. by Dunn and Ledbetter (1997), Pettermann et al. (1997), Mlekusch (1998), and Huang (2001). They differ mainly in the algorithms used for the orientational averaging. Such approaches, while typically very useful in practical applications, are of an ad-hoc nature. As pointed out by Benveniste et al. (1991) and Ferrari (1991) they can give rise to nonsymmetric effective "elastic tensors", especially when used with anisotropic phase properties. The reason for this behavior lies in the fact that Mori–Tanaka methods inherently describe ellipsoidal inclusions in ellipsoidal arrangements (Ponte Castañeda and Willis, 1995).

In addition to providing estimates for the overall thermoelastic behavior of composites containing nonaligned reinforcements, extended Mori–Tanaka theories can also be used for evaluating the microscopic stresses and strains in individual fibers in dependence on the fiber orientation. For this purpose the expression

$$\bar{\mathbf{B}}_{\rm M}^{(\rm f)\angle} = \bar{\mathbf{B}}_{\rm dil}^{(\rm f)\angle} \overline{\mathbf{B}}_{\rm M}^{(\rm m)} \tag{3.4}$$

is used, which is obtained by applying eqn. (3.3) at the fiber level. Here $\bar{\mathbf{B}}_{\rm M}^{(\rm f)\angle}$ is the Mori–Tanaka stress concentration tensor of a fiber with some given orientation; for details see Duschlbauer et al. (2003). From eqn. (3.4) it is evident that in extended Mori–Tanaka methods the orientation dependence of the fiber stresses is introduced solely via the orientation of a given fiber, whereas the interactions with other fibers, which are described via $\overline{\mathbf{B}}_{\rm M}^{(\rm m)}$, only serve to scale the stress levels. Estimates for the stresses in individual fibers obtained with eqn. (3.4) were found to be in good agreement with unit cell results for moderate volume fractions and elastic contrasts, compare Figure 5.

Extended Mori–Tanaka methods based on orientational averaging have also been extended into the elastoplastic range via secant plasticity schemes, compare e.g. Bhattacharyya and Weng (1994) or Dunn and Ledbetter (1997), and an incremental approach was proposed for modeling random short fiber reinforced composites that are subject to matrix plasticity and fiber debonding (Lee and Simunovic, 2000).

Estimates for the effective thermomechanical properties of nonaligned short fiber composites that are not subject to the inconsistencies of extended Mori–Tanaka methods can be generated with the Hashin–Shtrikman estimates proposed by Ponte Castañeda and Willis (1995). These are based on "ellipsoid-in-ellipsoid" phase arrangements and employ different Eshelby tensors for evaluating the ellipsoids describing the fiber positions and for those describing the shapes of the reinforcements. These estimates are limited, however, to reinforcement volume fractions for which the inner ellipsoids (reinforcement shapes) are subvolumes of the outer "safety" ellipsoids (reinforcement positions and orientations), compare Bornert (2001). Accordingly, the range of unrestricted applicability of the method is rather small for strongly prolate or oblate reinforcements that show a considerable degree of misalignment.

For the special case of randomly oriented fibers of a given aspect ratio, "symmetrized" dilute strain concentration tensors may be constructed, which are known as Wu tensors (Wu, 1966). They can be combined with Mori–Tanaka methods, see e.g. Benveniste (1987), or classical self-consistent schemes (Berryman, 1980) to describe composites with randomly oriented phases that have matrix–inclusion or interwoven microtopologies, respectively. Another mean field approach for composites reinforced by random fibers, the Kuster–Toksőz model (Kuster and Toksőz, 1974), is essentially a dilute approximation applicable to matrix–inclusion topologies.

Due to the overall isotropic behavior of composites reinforced by randomly oriented fibers, their overall elastic behavior must comply with the "standard" Hashin–Shtrikman bounds (Hashin and Shtrikman, 1963). A discussion of Hashin–Shtrikman-type bounds for the elastic responses applicable to more general fiber orientation distributions and to aggregated short fiber composites can be found in Eduljee and McCullough (1993).

## 3.3 Unit Cell Models for Nonaligned Short Fiber Composites

Whereas a considerable number of unit cell studies of materials reinforced by aligned discontinuous fibers have been reported, until recently the literature contained few reports on periodic models that deal with nonaligned fibers. Among them are three-dimensional analyses of composites reinforced by alternatingly tilted fibers (Sørensen, 1991) and plane-stress models describing planar random short fibers (Courage and Schreurs, 1992).

Recently models have been reported that use three-dimensional multi-fiber unit cells for modeling composites reinforced by short fibers that are oriented randomly (Böhm et al., 2002) or according to some specified ODF (Lusti et al., 2002). While representing a fairly obvious modeling strategy, such approaches tend to pose considerable challenges in generating appropriate fiber arrangements at nondilute volume fractions, in meshing them, and in solving for the corresponding microfields. Accordingly, the first studies of this type were restricted to low fiber volume fractions and to the linear elastic range, where the BEM answers well, see e.g. Banerjee and Henry (1992).

Appropriate arrangements of nonaligned short fibers can be generated by random insertion methods for moderate fiber volume fractions. Figure 4 shows such a multi-fiber unit cell that contains 15 randomly oriented and positioned cylindrical fibers of aspect ratio $a = 5$ at a fiber volume fraction of $\xi^{(f)}$=0.15. The cell supports periodicity boundary conditions and is discretized by quadratic tetrahedral finite elements. Rearrangement of fibers by Monte Carlo methods is typically required to overcome geometrical frustration at higher fiber volume fractions and/or higher fiber aspect ratios. The resulting phase arrangements tend to show aggregation-like effects in the form of nearly aligned clusters of fibers as the volume fraction is increased. Recently, software tools for generating unit cell geometries for various discontinuously reinforced composites have become available, such as the FE-based micromechanics code PALMYRA (MatSim Ges.m.b.H., Zurich, Switzerland), which can generate and mesh unit cells that contain nonaligned ellipsoids or "capped cylinders" as reinforcements and can evaluate their overall elastic, thermal expansion, and transport properties.

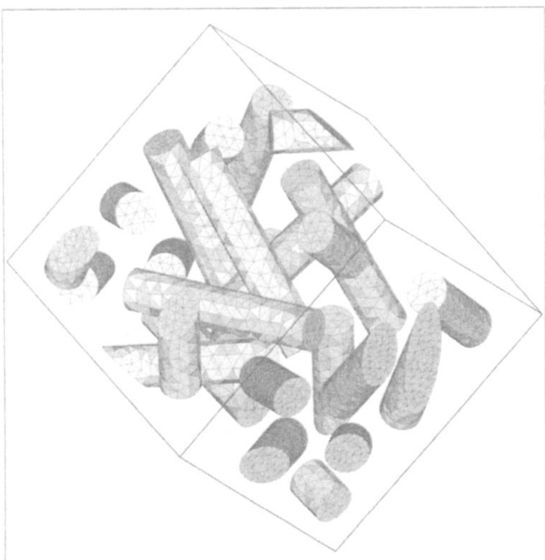

**Figure 4.** Unit cell for a short fiber reinforced MMC ($\xi^{(f)} = 0.15$ nominal) containing 15 cylindrical short fibers of aspect ratio $a = 5$ in a quasi-random arrangement suitable for using periodicity boundary conditions (Böhm et al., 2002).

In the following, some results obtained from multi-fiber unit cell models of random short fiber composites are discussed. Figure 5 shows a comparison of the maximum principal stresses in individual randomly oriented fibers in an SiC/Al MMC as evaluated by an extended Mori–Tanaka method (Duschlbauer et al., 2003) and by unit cells of the type shown in Figure 4. The load case is uniaxial tension in the elastic range and the stresses are plotted over the angle between loading direction and fiber axis. The unit cells used were generated in such a way that either ellipsoidal or cylindrical fibers of aspect ratio $a = 5$ can be employed with identical positions and orientations, thus allowing a direct evaluation of fiber shape effects. Excellent agreement between the analytical and

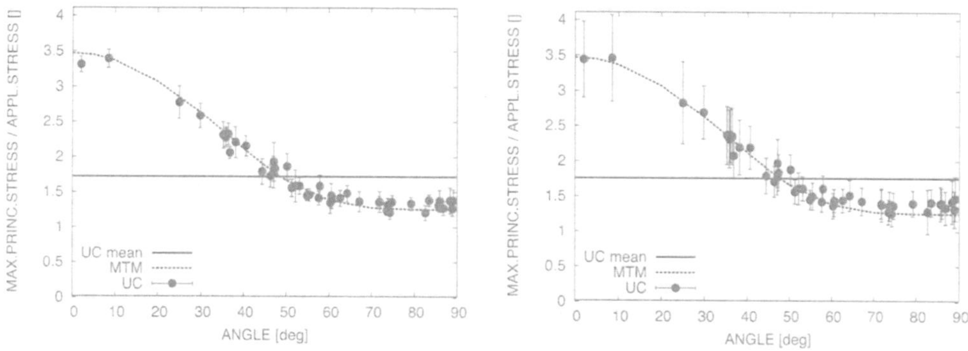


**Figure 5.** Angular dependence of the maximum principal stresses in randomly oriented fibers of aspect ratio $a = 5$ in an SiC/Al MMC subjected to uniaxial tension in the elastic range. Mori–Tanaka results are shown as dashed lines and unit cell predictions are displayed in terms of the phase average (solid line) as well as the mean values (solid circles) and standard deviations (error bars) within individual fibers.

numerical predictions is evident in terms of the average stresses in individual fibers (filled circles). Evidently, inter-fiber stress fluctuations are rather limited in the considered regime. However, the cylindrical fibers (right) give rise to noticeably higher intra-fiber stress fluctuations (represented by error bars) than the ellipsoidal fibers (left).

In Table 1 predictions are listed for the overall elastic properties of the same MMC consisting of randomly oriented SiC fibers in a matrix of Al2618. Analytical estimates are given in terms of the Hashin–Shtrikman (H/S) bounds (Hashin and Shtrikman, 1963), the

**Table 1.** Overall elastic properties of a SiC/Al MMC reinforced by randomly oriented fibers ($a = 5$, $\xi^{(f)}$=0.15) as predicted by the Hashin–Shtrikman (H/S) bounds, by the Mori–Tanaka theory, (MTM), the self-consistent scheme (CSCS) of Berryman (1980), by the Kuster–Toksőz model (KTM), and by multi-fiber unit cells of the type shown in Figure 4 (MFUC), which contain 15 spheroidal or cylindrical fibers.

| | $E$ [GPa] | $\nu$ [ ] |
|---|---|---|
| SiC fibers | 450.0 | 0.17 |
| Al2618-T4 matrix | 70.0 | 0.30 |
| H/S LB | 87.6 | 0.246 |
| H/S UB | 106.1 | 0.305 |
| MTM | 89.8 | 0.285 |
| CSCS | 91.2 | 0.284 |
| KTM | 90.3 | 0.285 |
| 3/D MFUC (sph) | 89.4 | 0.285 |
| 3/D MFUC (cyl) | 90.0 | 0.284 |

Mori–Tanaka method (MTM) of Benveniste (1987), the self-consistent scheme (CSCS) of Berryman (1980), and the Kuster–Toksőz (KTM) estimates (Kuster and Toksőz, 1974). Results from multi-fiber unit cells (MFUC) are listed both for spherical (sph) and cylindrical (cyl) fibers. Again satisfactory agreement between the different approaches is evident in spite of the rather low number of fibers per unit cell. Thermal conduction analyses of planar random short fiber reinforced composites with $\xi \approx 0.208$ and $a = 10$ gave comparably good agreement between Mori–Tanaka results and predictions from multi-fiber unit cells in terms of the temperature gradients in individual fibers (Duschlbauer, 2003).

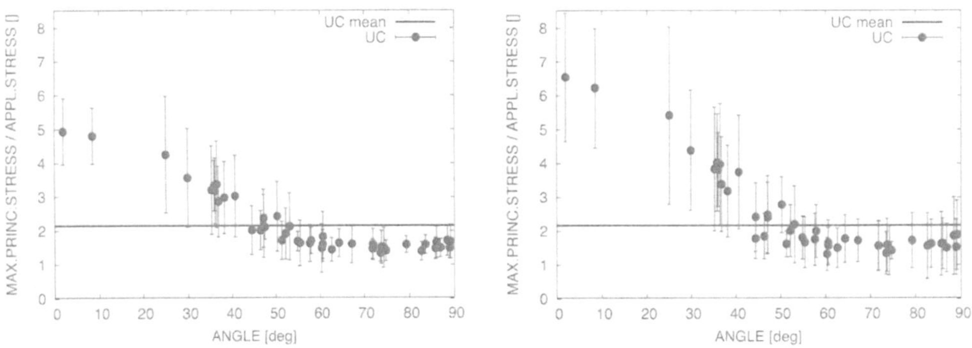


**Figure 6.** Angular dependence of the maximum principal stresses in randomly oriented fibers of aspect ratio $a = 5$ in a SiC/Al MMC subjected to uniaxial tension in the elastoplastic range. The unit cell predictions are displayed in terms of the phase average (solid line) as well as the mean values (solid circles) and standard deviations (error bars) within individual fibers.

Finally, Figure 6 shows the angular dependence of the maximum principal stresses predicted by a 15-fiber unit cell model for uniaxial tensile loading at approximately 2.45 times the initial yield stress of the aluminum alloy, in which state the matrix is fully yielded. Both the average stresses and their variations in the individual fibers can be seen to be markedly larger for the cylindrical reinforcements (right), i.e. both the intra-fiber and the inter-fiber stress fluctuations show a clear dependence on the fiber shape in this regime (Böhm et al., 2004). The extremely high stresses present in fibers oriented within approximately 30° of the loading direction are especially noteworthy; for cylindrical fibers they exceed the phase average of the maximum principal stress in the fibers by more than a factor of three.

## 4 Closing Remarks

Despite the challenges posed by the complex microgeometries of short fiber reinforced composites, a considerable number of continuum mechanical modeling approaches of different levels of complexity have been reported over the last 30 years for studying these materials.

For composites reinforced by aligned short fibers reliable analytical and numerical methods are available for predicting the homogenized elastic stiffnesses and coefficients of thermal expansion as well as the elastoplastic overall responses. Microscopic damage in such materials has been studied at considerable depth.

Analytical models for composites reinforced by nonaligned short fibers, most of which lack fully satisfactory theoretical foundations, can provide useful data on the elastic and thermoelastic behavior of these materials in most cases. Numerically based multi-fiber unit cell models are very promising for studying both the overall and microscopic responses of such composites, but they are rather demanding in terms of modeling and computer resources, especially in the inelastic range. Numerical and analytical modeling concepts developed for short fiber reinforced composites can typically be adapted for studying materials with platelet-like reinforcements. Analogous methods can be employed for studying the conduction behavior of short fiber reinforced materials.

## Bibliography


S.G. Advani and C.L. Tucker. The use of tensors to describe and predict fiber orientation in short fiber composites. *J. Rheol.*, 31:751–784, 1987.

D.H. Allen and J.W. Lee. The effective thermoelastic properties of whisker-reinforced composites as functions of material forming parameters. In G.J. Weng, M. Taya, and H. Abé, editors, *Micromechanics and Inhomogeneity*, pages 17–40, New York, NY, 1990. Springer–Verlag.

P.K. Banerjee and D.P. Henry. Elastic analysis of three-dimensional solids with fiber inclusions by BEM. *Int. J. Sol. Struct.*, 29:2423–2440, 1992.

Y. Benveniste. A new approach to the application of Mori–Tanaka's theory in composite materials. *Mech. Mater.*, 6:147–157, 1987.

Y. Benveniste, G.J. Dvorak, and T. Chen. On diagonal and elastic symmetry of the approximate effective stiffness tensor of heterogeneous media. *J. Mech. Phys. Sol.*, 39:927–946, 1991.

J.G. Berryman. Long-wavelength propagation in composite elastic media, II. Ellipsoidal inclusions. *J. Acoust. Soc. Amer.*, 68:1820–1831, 1980.

A. Bhattacharyya and G.J. Weng. Plasticity of isotropic composites with randomly oriented packeted inclusions. *Int. J. Plast.*, 10:553–578, 1994.

H.J. Böhm. A short introduction to continuum micromechanics. *This volume*, pages 1–40, 2004.

H.J. Böhm, A. Eckschlager, and W. Han. Multi-inclusion unit cell models for metal matrix composites with randomly oriented discontinuous reinforcements. *Comput. Mater. Sci.*, 25:42–53, 2002.

H.J. Böhm, W. Han, and A. Eckschlager. Multi-inclusion unit cell studies of reinforcement stresses and particle failure in discontinuously reinforced ductile matrix composites. *Comput. Model. Engng. Sci.*, 5:5–20, 2004.

H.J. Böhm, F.G. Rammerstorfer, and E. Weissenbek. Some simple models for micromechanical investigations of fiber arrangement effects in MMCs. *Comput. Mater. Sci.*, 1: 177–194, 1993.

M. Bornert. Homogénéisation des milieux aléatoires: bornes et estimations. In M. Bornert, T. Bretheau, and P. Gilormini, editors, *Homogénéisation en mécanique des materiaux 1. Matériaux aléatoires élastiques et milieux périodiques*, pages 133–221, Paris, 2001. Editions Hermès.

A.R. Clarke, G. Archenhold, and N.C. Davidson. A novel technique for determining the 3d spatial distribution of glass fibres in polymer composites. *Compos. Sci. Technol.*, 55:75–91, 1995.

B. Clyne and P.J. Withers. *An Introduction to Metal Matrix Composites.* Cambridge University Press, Cambridge, 1993.

W.M.G. Courage and P.J.G Schreurs. Effective material parameters for composites with randomly oriented short fibers. *Comput. Struct.*, 44:1179–1185, 1992.

H.L. Cox. The elasticity and strength of paper and other fibrous materials. *Brit. J. Appl. Phys.*, 3:72–79, 1952.

S.K. De and J.R. White, editors. *Short Fibre–Polymer Composites.* Woodhead Publishing, Cambridge, 1996.

M.L. Dunn and H. Ledbetter. Elastic–plastic behavior of textured short-fiber composites. *Acta mater.*, 45:3327–3340, 1997.

D. Duschlbauer. *Computational Simulation of the Thermal Conductivity of MMCs Under Consideration of the Inclusion–Matrix Interface.* PhD thesis, TU Wien, Vienna, Austria, 2003.

D. Duschlbauer, H.E. Pettermann, and H.J. Böhm. Mori–Tanaka based evaluation of inclusion stresses in composites with nonaligned reinforcements. *Scr. mater.*, 48:223–228, 2003.

R.F. Eduljee and R.L. McCullough. Elastic properties of composites. In R.W. Cahn, P. Haasen, and E.J. Kramer, editors, *Materials Science and Technology Vol.13: Structure and Properties of Composites*, pages 381–474, Weinheim, 1993. VCH.

M. Ferrari. Asymmetry and the high concentration limit of the Mori–Tanaka effective medium theory. *Mech. Mater.*, 11:251–256, 1991.

S.Y. Fu and B. Lauke. An analytical characterization of the anisotropy of the elastic modulus of misaligned short-fiber-reinforced polymers. *Compos. Sci. Technol.*, 58: 1961–1972, 1998.

H. Fukuda and T.W. Chou. A probabilistic theory of the strength of short-fibre composites with variable fibre length and orientation. *J. Mater. Sci.*, 17:1003–1007, 1982.

H. Fukuda and K. Kawata. On Young's modulus of short fiber composites. *Fibre Sci. Technol.*, 7:207–222, 1974.

H. Fukuda and Y. Takao. Thermoelastic properties of discontinuous fiber composites. In Kelly A. and Zwebern C., editors, *Comprehensive Composite Materials, Vol.1*, pages 377–401, Oxford, 2000. Pergamon Press.

A.A. Gusev, M. Heggli, H.R. Lusti, and P.J. Hine. Orientation averaging for stiffness and thermal expansion of short fiber composites. *Adv. Engng. Mater.*, 4:931–933, 2002a.

A.A. Gusev, H.R. Lusti, and P.J. Hine. Stiffness and thermal expansion of short fiber composites with fully aligned fibers. *Adv. Engng. Mater.*, 4:927–931, 2002b.

J.C. Halpin and N.J. Pagano. The laminate approximation for randomly oriented fibrous composites. *J. Compos. Mater.*, 3:720–724, 1969.

J.C. Halpin and S.W. Tsai. Effects of environmental factors on composite materials. Technical Report AFML–TR–67–423, Wright–Patterson AFB, Dayton, OH, 1969.

Z. Hashin and S. Shtrikman. A variational approach to the theory of the elastic behavior of multiphase materials. *J. Mech. Phys. Sol.*, 11:127–140, 1963.

R. Hill. A self-consistent mechanics of composite materials. *J. Mech. Phys. Sol.*, 13: 213–222, 1965.

G.K. Hu. A method of plasticity for general aligned spheroidal void or fiber-reinforced composites. *Int. J. Plast.*, 12:439–449, 1996.

J.H. Huang. Some closed-form solutions for effective moduli of composites containing randomly oriented short fibers. *Mater. Sci. Engng.*, A315:11–20, 2001.

M.S. Ingber and T.D. Papathanasiou. A parallel-supercomputing investigation of the stiffness of aligned short-fiber-reinforced composites using the boundary element method. *Int. J. Num. Meth. Engng.*, 40:3477–3491, 1997.

K. Jayaraman and M.T. Kortschot. Correction to the Fukuda–Kawata Young's modulus theory and the Fukuda–Chou strength theory for short fibre-reinforced composite materials. *J. Mater. Sci.*, 31:2059–2064, 1996.

J.W. Ju and L.Z. Sun. Effective elastoplastic behavior of metal matrix composites containing randomly located aligned spheroidal inhomogeneities. Part I: Micromechanics-based formulation. *Int. J. Sol. Struct.*, 38:183–201, 2001.

G. Korb, W. Buchgraber, and T. Schubert. Thermophysical porperties and microstructure of short carbon fibre reinforced Cu-matrix composites made by electroless copper coating or powder metallurgical route respectively. In *Twenty-Second IEE/CPMT International Electronics Manufacturing Technology Symposium*, pages 98–103, New York, 1998. IEEE.

G.T. Kuster and M.N. Toksőz. Velocity and attenuation of seismic waves in two-phase media: I. Theoretical formulation. *Geophysics*, 39:587–606, 1974.

H.K. Lee and S. Simunovic. Modeling of progressive damage in aligned and randomly oriented discontinuous fiber polymer matrix composites. *Composites B*, 31B:77–86, 2000.

A. Levy and J.M. Papazian. Elastoplastic finite element analysis of short-fiber-reinforced SiC/Al composites: Effects of thermal treatment. *Acta metall. mater.*, 39:2255–2266, 1991.

H.R. Lusti, P.J. Hine, and A.A. Gusev. Direct numerical predictions for the elastic and thermoelastic properties of short fibre composites. *Compos. Sci. Technol.*, 62: 1927–1934, 2002.

B. Mlekusch. Thermoelastic properties of short-fibre-reinforced thermoplastics. *Compos. Sci. Technol.*, 59:911–923, 1998.

M. Narkis, A. Vaxman, A. Siegmann, and S. Kenig. Short-fiber thermoplastic composites: Fiber fracture during melt processing. In S.M. Lee, editor, *International Encyclopedia of Composites, Vol.5*, pages 141–155, New York, NY, 1991. VCH Publishers.

S.R. Nutt and A. Needleman. Void nucleation at fiber ends in Al–SiC composites. *Scr. metall.*, 21:705–710, 1987.

H.E. Pettermann, H.J. Böhm, and F.G. Rammerstorfer. Some direction dependent properties of matrix–inclusion type composites with given reinforcement orientation distributions. *Composites B*, 28B:253–265, 1997.

H.E. Pettermann, A.F. Plankensteiner, H.J. Böhm, and F.G. Rammerstorfer. A thermo-elasto-plastic constitutive law for inhomogeneous materials based on an incremental Mori–Tanaka approach. *Comput. Struct.*, 71:197–214, 1999.

P. Ponte Castañeda and J.R. Willis. The effect of spatial distribution on the effective behavior of composite materials and cracked media. *J. Mech. Phys. Sol.*, 43:1919–1951, 1995.

G.L. Povirk, A. Needleman, and S.R. Nutt. An analysis of the effect of residual stresses on deformation and damage mechanisms in Al–SiC composites. *Mater. Sci. Engng.*, A132:31–38, 1991.

R. Pyrz. Microstructural description of composites, statistical methods. *This volume*, pages 173–234, 2004.

B.W. Rosen. Mechanics of composite strengthening. In B.W. Rosen, editor, *Fiber Composite Materials*, pages 37–75, Metals Park, OH, 1965. American Society of Metals.

T. Siegmund, R. Cipra, J. Liakus, B. Wang, M. LaForest, and A. Fatz. Processing–microstructure–property relationships in short fiber reinforced carbon–carbon composite system. *This volume*, pages 235–258, 2004.

N.J. Sørensen. A planar type analysis for the elastic-plastic behaviour of continuous fibre-reinforced metal-matrix composites under longitudinal shearing and combined loading. DCAMM Report No.**418**, Technical University of Denmark, Lyngby, Denmark, 1991.

C.L. Tucker and E. Liang. Stiffness predictions for unidirectional short-fiber composites: Review and evaluation. *Compos. Sci. Technol.*, 59:655–671, 1999.

V. Tvergaard. Analysis of tensile properties for a whisker-reinforced metal-matrix composite. *Acta metall. mater.*, 38:185–194, 1990.

V. Tvergaard. Three-dimensional analysis of ductile failure in metal reinforced by staggered fibers. *Modell. Simul. Mater. Sci. Engng.*, 9:143–155, 2001.

V. Tvergaard. Breakage and debonding of short brittle fibres among particulates in a metal matrix. *Mater. Sci. Engng.*, A369:192–200, 2004.

E. Weißenbek and F.G. Rammerstorfer. Influence of the fiber arrangement on the mechanical and thermo-mechanical behavior of short-fiber reinforced MMCs. *Acta metall. mater.*, 41:2833–2843, 1993.

G.J. Weng. Explicit evaluation of Willis' bounds with ellipsoidal inclusions. *Int. J. Engng. Sci.*, 30:83–92, 1992.

J.R. Willis. Bounds and self-consistent estimates for the overall moduli of anisotropic composites. *J. Mech. Phys. Sol.*, 25:185–202, 1977.

T.T Wu. The effect of inclusion shape on the elastic moduli of a two-phase material. *Int. J. Sol. Struct.*, 2:1–8, 1966.

C. Zhou, W. Yang, and D. Fang. Damage of short-fiber reinforced metal matrix composites considering cooling and thermal cycling. *J. Engng. Mater. Technol.*, 122:203–208, 2000.

# Thermomechanical Cohesive Zone Models for the Analysis of Composite Failure under Thermal Gradients and Transients

A. Hattiangadi and T. Siegmund

School of Mechanical Engineering, Purdue University, West Lafayette, Indiana, U.S.A.

**Abstract.** The coupled thermomechanical numerical analysis of damaged composite structures loaded by temperature transients and gradients is described. The numerical approach presented is based on the framework of cohesive zone models. Traction-separation laws are coupled to heat conduction across the cracks. An implementation of the model into a finite element framework is described. Three examples of application are discussed: (i) interface crack growth; (ii) crack bridging; (iii) application to photothermal imaging.

## 1 Introduction

Engineering components considered in high temperature applications have to be designed to withstand expected and hypothetical events such as thermal shock and/or high heat flux in which thermal gradients arise. Only if realistic failure criteria are implemented into fully coupled thermo-mechanical simulations the required detailed prediction of life cycle limiting events can be achieved.

It is well established that cracks and delaminations are discontinuities that impede heat flow and subsequent redistribution temperature. This results in local thermal stresses that can induce and promote crack growth. While this process is fundamentally obvious, the actual failure analysis – including crack initiation, growth and arrest - of engineering components under consideration of these aspects is complex. In the past all approaches to this problem have been concerned with detailed stress analysis based on input from a separate heat transfer analysis. Fully coupled thermo-mechanical analyses for non-uniform and non-steady heat flow conditions with arbitrarily growing cracks have not been performed. A methodology that simultaneously accounts for both material failure as well as the heat transfer problem with moving boundaries due to crack growth is currently not available.

Past investigations of the structural integrity of components subjected to high heat flux and transients were based on applications of the framework of linear-elastic fracture mechanics. Early investigations by Florence and Goodier (1959, 1963) and Sih (1962), were concerned with conditions of steady state and remote uniform heat flux, and assumed full isolated cracks in infinite homogeneous isotropic bodies. The energy release rate, $G$, for the mode II crack tip field present in such a case is given by:

$$G = \frac{(1+\nu)\pi E a^3}{16(1-\nu)}\left(\alpha \frac{\partial \theta}{\partial y}\right)^2 \qquad (1)$$

where $E$ is Young's modulus, $\nu$ Poisson's ratio, $\alpha$ the coefficient of thermal expansion and $(\partial\theta/\partial y)$ the temperature gradient normal to the crack. This result, and subsequent others derived by Barber (1979) and Chen and Huang (1992), demonstrated that the energy release rate associated with thermal effects could indeed be substantial. Subsequently, other investigators extended the linear elastic fracture mechanics analyses to cracks at bi-material interfaces, (Lee and Shul (1991), Budde and Gao (1992), Itou (1993), Yan and Ting (1993), Chao and Chang (1994), Lee and Park (1995), Lee and Erdogan (1998), Qian et al. (1997, 98), Miller and Chona (1998), as well as to arbitrary cracks geometries, Hasebe et al. (1986). Recently, Hutchinson and Evans (2002) investigated delamination cracks in thermal barrier coatings. More importantly in the context of the present analysis, crack heat flow was included into the analyses by Martin-Moran et al. (1983), Barber and Comninou (1983), Kuo (1990), Lee and Park (1995), Hutchinson and Lu (1995), Miller and Chona (1998) and Qia et al. (1998). The results from this group of investigations clearly demonstrate a very strong dependence of the energy release rate on the crack conductance. For example, Hutchinson and Lu (1995) analyzed plate of thickness, $H$, which contains a long crack, and is loaded by an applied temperature difference ($\theta_{top}$-$\theta_{bot}$) between the top and bottom of the plate. They predict the maximum value of the energy release rate to occur for a crack at a distance $0.211H$ from the hot surface. With the thermal conductivity of the plate material, $k_s$, the energy release rate for this situation is given by:

$$G = \frac{EH\left[\alpha\left(\theta_{top} - \theta_{bot}\right)\right]^2}{32\left(1 + Bi\right)^2} \tag{2}$$

where $Bi$ is the Biot number defined as $Bi=(H\, h_c)/k_s$, with $h_c$ the conductance across the crack. In contrast to the results from Equation (1), Hutchinson and Lu (1995) predict cracks to be under mixed mode loading, $45^\circ<\psi<90^\circ$, as long as the crack remains less than 0.5 $H$ from the hot surface. Equation (2) also clearly demonstrates a strong dependence of the structural response on the crack conductance, and thus the necessity to incorporate a correct description of $h_c$ into failure predictions of structures under high heat flux loading. As in Equation (2), crack conductance, however, has only been considered in simplified ways in the past. It was generally assumed to be constant, not only along the entire crack but also independent of crack opening. Since neither variable crack conductance values nor crack growth was considered, sequential solutions of the temperature and displacement problems were sufficient. If, however, integrated multi-physics life cycle predictions of critical components are the goal, these simplifying assumptions have to be overcome.

The methods currently available for the analysis of fracture under thermal loading are in a significant contrast to the state of knowledge in fracture analysis under pure mechanical loading. There, the development of physically mechanism based material separation models for use in continuum mechanics crack growth simulations has greatly enhanced the capabilities of failure predictions for a broad range of materials. One of the promising approaches developed within these efforts is the so-called cohesive zone model. Its success stems from the fact that the cohesive zone model approach allows one to incorporate a broad range of different physical processes of material separation within a single numerical method. Based on early ideas of Barenblatt (1962) and Dugdale (1960), cohesive zone models were used by several investigators for studies of fracture under pure mechanical loading, e.g., Needleman (1987, 1990), Tvergaard and Hutchinson (1992, 94), Xu and Needleman (1994), Camacho and Ortiz (1996), Siegmund and Needleman (1997), Chaboche et al. (1997), Foulk et al. (1998), Siegmund and Brocks (1999), de-Andres et al. (1999), Kolhe et al. (1999), Rahul-Kumar et al. (2000), and Pandolfi et al. (2000). Furthermore, in Gao and Klein (1998) and Klein and Gao (1998), a related approach was presented as the virtual internal bond model. Generally, two constitutive equations, a volumetric (hardening) constitutive equation and a cohesive (softening) constitutive equation describe the solid under consideration. The softening constitutive equations of the cohesive zone are phrased in terms of a traction-separation law. The main cohesive constitutive parameters in this law, cohesive strength and separation energy per unit area, are characteristic of the underlying material separation processes. A length-scale, i.e. separation energy per unit area divided by cohesive strength, is directly part of the constitutive equation. This cohesive length is the material separation distance after which the stress carrying capacity of the material falls to zero. As a consequence of the presence of a length scale in the constitutive equations, convergent results with respect to the finite element mesh can in principle be obtained. All past applications of cohesive zone models as outlined above were directed towards the analysis of mechanical loading only. It is the purpose of this article to demonstrate that the cohesive zone model approach is indeed promising in cases where fracture is coupled to and influenced by other physical processes. To enable coupled thermomechanical analyses which account for the creation of new free surfaces due to crack growth, or break down of crack bridging, and the subsequent changing heat transfer conditions, a description of conductance across the crack, need to be defined. Cohesive zone conductance laws capture the energy transport across cracks or delaminations, and provide additional coupling between thermal and stress analysis. Variations in cohesive zone conductance of the order of several magnitudes can occur depending on the stage of the processes associated with material failure. Material deterioration taking place in the cohesive zone leads to both large changes in stress carrying capability and conductance. As load transfer breaks down, the conductance in an open crack subsequently depends on the presence of a gas and/or radiative heat transfer. Crack face contact occurring at interfaces – under mixed mode loading and during unloading – is to be accounted for.

The present article focuses on the application of the approach to studies of fiber reinforced ceramic matrix composites and coating/substrate interfaces. Specifically, if crack growth in the composite itself is of concern, the influence of the fracture processes can be especially important due to extended crack bridging. Fiber reinforced ceramic matrix composites are promising candidates in high temperature applications for thermal protection systems for space vehicles, Freeman et al. (1997), Sumrall et al. (1999), subjected to heat flux of up to 100 W/cm$^2$, Papadopoulos et al. (1999); or actively cooled first wall liners in fusion reactors with expected peak heat fluxes of up to 15,000 W/cm$^2$, Tivey (1999). In addition to their thermal stability, fiber reinforced ceramic matrix composites also offer certain flexibility in microstructural design. With respect to the present investigation this is of interest since crack bridging – due to cross stitches, multidirectional microstructures or unbroken fibers in laminates – plays an important role in both load, and heat transfer.

The present contribution describes the general framework for coupled thermomechanical crack growth simulations within the cohesive zone model approach. The model is subsequently applied to three numerical studies. The first example is concerned with delamination growth at the interface between a carbon-carbon composite panel and a protective SiC coating. The second example studies bridged cracks in unidirectionally reinforced laminates. Finally, an application of the thermomechanical cohesive zone model to the analysis of the photothermal non-destructive evaluation method is discussed.

## 2 Formulation

For a solution of the coupled thermomechanical problem of solids containing cracks and interfaces both mechanical equilibrium and energy balance must be fulfilled. In the present approach, cracks and interfaces are considered as internal cohesive surfaces with (partial) load and heat transfer capabilities. Thus, in view of the finite element approximation, the mechanical equilibrium equation includes the contribution of the cohesive zone integrals over the internal surfaces, $S_{int}$. The principle of virtual work is written as:

$$\int_V \mathbf{s}:\delta\mathbf{F}\,\mathrm{d}V - \int_{S_{int}} \mathbf{T}_{\mathrm{CZ}}\cdot\delta\boldsymbol{\Delta}\,\mathrm{d}S = \int_{S_{ext}} \mathbf{T}_{\mathrm{e}}\cdot\delta\mathbf{u}\,\mathrm{d}S \tag{3}$$

with the nominal stress tensor, $\mathbf{s}=\mathbf{F}^{-1}\det(\mathbf{F})\sigma$, the Cauchy stress, $\sigma$, the deformation gradient, $\mathbf{F}$, and the displacement vector, $\mathbf{u}$. Traction vectors are related to $\sigma$ by $\mathbf{T}=\mathbf{n}\ \sigma$, with $\mathbf{n}$ being the surface normal. $\mathbf{T}_{\mathrm{e}}$ is the traction vector on the external surface of the body. The cohesive zone contribution on the internal surface, $S_{int}$, are given by the cohesive surface traction, $\mathrm{T}_{\mathrm{CZ}}$, and the displacement jump across the cohesive surface, $\boldsymbol{\Delta}=(\Delta_n\mathbf{n}+\Delta_t\mathbf{t})=\mathbf{u}_{+}-\mathbf{u}_{-}$, where $i$=+,– represent top and lower surfaces, with $\mathbf{u}_{+}$ and $\mathbf{u}_{-}$ the displacements of two initially bonded points on opposite crack surfaces.

The energy balance including the contribution of the cohesive zone on the internal surfaces, written in variational form and using Fourier's law, is expressed by:

$$\int_V \rho c_p(\partial\delta\theta)\delta\theta\,\mathrm{d}V - \int_{S_{int}} q_{CZ}\delta(\Delta\theta)\,\mathrm{d}S + \int_V \frac{\partial\delta\theta}{\partial \mathrm{x}}\mathbf{k}\frac{\partial\delta\theta}{\partial \mathrm{x}}\mathrm{d}V = \int_{S_{ext}} \delta\theta q_e\,\mathrm{d}S \tag{4}$$

with temperature, $\theta$, heat capacity, $c_p$, the conductivity matrix, $\mathbf{k}$, and the heat flux per external unit area into the body, $q_e$. The cohesive zone contributions are described by the integral over the internal surface, $S_{int}$, as the product of the cohesive zone heat flux, $q_{CZ}$, and the temperature jump across the cohesive surface, $\Delta\theta=\theta_{+}-\theta_{-}$.

The implementation of the mechanical part of the cohesive elements requires the evaluation of the derivative of the second term of equation (3). The virtual work of the cohesive forces, $\delta U_m$, can be expressed in terms of the normal and tangential components of tractions and displacement jumps as

$$\delta U_m = \int_{S_{int}} [T_n \quad T_t][\delta\Delta_n \quad \delta\Delta_t]^{\mathrm{T}}\,\mathrm{d}S \tag{5}$$

Substituting the virtual displacement jumps in terms of the cohesive element nodal shape functions, the first variation of $\delta U_m$ is

$$\mathrm{d}\delta U_m = \int_{S_{int}} [\mathrm{d}T_n \quad \mathrm{d}T_t]\cdot\delta\mathbf{A}_{\mathbf{m}}\mathrm{d}S \tag{6}$$

where $\delta \mathrm{A}_{\mathrm{m}} = \mathrm{N}_{\mathrm{m}}\cdot\delta\Delta$ and $\mathrm{N}_{\mathrm{m}}$ are cohesive element nodal shape function for mechanical terms. The incremental normal and tangential tractions, $\mathrm{d}T_n$ and $\mathrm{d}T_t$, are related to incremental displacement jumps by the cohesive zone Jacobian, $\mathbf{J}_{\mathbf{m}}$,

$$\begin{Bmatrix} \mathrm{d}T_n \\ \mathrm{d}T_t \end{Bmatrix} = \mathbf{J}_{\mathbf{m}} : \begin{Bmatrix} \mathrm{d}\Delta_n \\ \mathrm{d}\Delta_t \end{Bmatrix} \tag{7a}$$

$$\mathbf{J_m} = \begin{bmatrix} \dfrac{\partial T_n}{\partial \Delta_n} & \dfrac{\partial T_n}{\partial \Delta_t} \\ \dfrac{\partial T_t}{\partial \Delta_n} & \dfrac{\partial T_t}{\partial \Delta_t} \end{bmatrix} \tag{7b}$$

Expressing the increments of the tractions in equation (6) by use of $\mathbf{J_m}$, and writing the incremental displacement jump by use of the incremental cohesive element nodal displacement jump leads to the mechanical part of the tangent stiffness matrix, $\mathbf{K_m}$:

$$\mathbf{K_m} = \int_{S_{int}} \mathbf{N_m^T} : \mathbf{J_m} : \mathbf{N_m} \, \mathrm{d}S \tag{8}$$

In addition to the mechanical terms, the implementation of the thermomechanical cohesive element requires the evaluation of the first variation of the second term in the internal energy expressed in equation (4). The term, $\delta U_{th}$, is expressed via the virtual temperature jump as

$$\delta U_{th} = \int_{S_{int}} q_{CZ} \delta(\Delta\theta) \, \mathrm{d}S \tag{9}$$

The first variation of $\delta U_T$ has the form,

$$\mathrm{d}\delta U_{th} = \int_{S_{int}} \mathrm{d}q_{CZ} \delta(\Delta\theta) \, \mathrm{d}S \tag{10}$$

where $\mathrm{d}q_{CZ}$ is the incremental cohesive zone heat flux related to the increments in the jump in temperature and displacement across the cohesive zone:

$$\{\mathrm{d}q_{CZ}\} = \begin{bmatrix} \dfrac{\partial q_{CZ}}{\partial \Delta_n} & \dfrac{\partial q_{CZ}}{\partial \Delta_t} & \dfrac{\partial q_{CZ}}{\partial \theta_i} \end{bmatrix} \begin{Bmatrix} \mathrm{d}\Delta_n \\ \mathrm{d}\Delta_t \\ \mathrm{d}\theta_i \end{Bmatrix} \quad i=+,- \tag{11}$$

The cohesive zone flux, $q_{CZ}$, is the product of cohesive zone conductance, $h_{CZ}$, and the temperature jump,

$$q_{CZ} = h_{CZ} \Delta\theta . \tag{12}$$

Then, the right hand side of equation (11) can be further expressed as

$$\{\mathrm{d}q_{CZ}\} = \begin{bmatrix} \dfrac{\partial h_{CZ}}{\partial \Delta_n} \Delta\theta & \dfrac{\partial h_{CZ}}{\partial \Delta_t} \Delta\theta & h_{CZ} + \dfrac{\partial h_{CZ}}{\partial \theta_i} \end{bmatrix} \begin{Bmatrix} \mathrm{d}\Delta_n \\ \mathrm{d}\Delta_t \\ \mathrm{d}\theta_i \end{Bmatrix} . \tag{13}$$

Hence, the first variation of the virtual internal energy in terms of virtual temperature difference takes the form

$$\mathrm{d}\delta U_{th} = \int_{S_{int}} \begin{bmatrix} \dfrac{\partial h_{CZ}}{\partial \Delta_n} & \dfrac{\partial h_{CZ}}{\partial \Delta_t} \end{bmatrix} [\mathrm{d}\Delta_n \quad \mathrm{d}\Delta_t]^{\mathrm{T}} \, \delta(\Delta\theta) \mathrm{d}S + \int_{S_{int}} \left( h_{CZ} + \frac{\partial h_{CZ}}{\partial \theta_i} \right) \mathrm{d}\theta_i \, \delta(\Delta\theta) \mathrm{d}S \tag{14}$$

The first variation of $U_{th}$ can be further expressed in terms of the nodal shape functions as,

$$\mathrm{d}\delta U_{th} = \int_{S_{int}} \begin{bmatrix} \dfrac{\partial h_{CZ}}{\partial \Delta_n} & \dfrac{\partial h_{CZ}}{\partial \Delta_t} \end{bmatrix} [\mathrm{d}\Delta_n \quad \mathrm{d}\Delta_t]^{\mathrm{T}} \cdot \delta \mathbf{A_{tm}} \, \mathrm{d}S + \int_{S_{int}} \left( h_{CZ} + \frac{\partial h_{CZ}}{\partial \theta_i} \right) \mathrm{d}\theta_i \, \delta \mathbf{A_{th}} \mathrm{d}S \tag{15}$$

where $\delta \mathbf{A}_{\mathbf{tm}} = \mathbf{N}_{\mathbf{tm}} \cdot \delta\Delta\theta$, $\delta \mathbf{A}_{\mathbf{th}} = \mathbf{N}_{\mathbf{th}} \cdot \delta\Delta\theta$ and $\mathbf{N}_{\mathbf{tm}}$ and $\mathbf{N}_{\mathbf{th}}$ are cohesive element nodal shape functions for the coupling terms and the thermal part, respectively. The first term on the right hand side denotes the coupling between cohesive zone conductance and displacement jump, and the second term is due to the temperature dependence of the cohesive zone conductance. The coupling between thermal and mechanical equations is expressed in terms of the Jacobian, $\mathbf{J}_{\mathbf{tm}}$, and the thermal part is expressed in terms of $J_{th}$,

$$\{dq\}_{tm} = \mathbf{J}_{\mathbf{tm}} \cdot \begin{Bmatrix} d\Delta_n \\ d\Delta_t \end{Bmatrix} \tag{16a}$$

$$\mathbf{J}_{\mathbf{tm}} = \begin{bmatrix} \dfrac{\partial h_{CZ}}{\partial \Delta_n} & \dfrac{\partial h_{CZ}}{\partial \Delta_t} \end{bmatrix} \tag{16b}$$

$$\{dq\}_{th} = J_{th} \{d\theta_i\} \tag{17a}$$

$$J_{th} = \begin{bmatrix} h_{CZ} + \dfrac{\partial h_{CZ}}{\partial \theta_i} \end{bmatrix} \tag{17b}$$

Writing the incremental heat flux in equations (16) and (17) in terms of the Jacobians, $\mathrm{J}_{\mathrm{tm}}$, $J_{th}$, and the incremental temperature jumps, leads to the thermomechanical, $\mathrm{K}_{\mathrm{tm}}$, and thermal, $K_{th}$, contribution of the tangent stiffness matrix of the cohesive zone element in the form:

$$\mathbf{K}_{\mathbf{tm}} = \int_{S_{int}} \mathbf{J}_{\mathbf{tm}} : \mathrm{N}_{\mathrm{m}}^{\mathrm{T}} : \mathbf{N}_{\mathbf{tm}} \, dS \tag{18a}$$

$$K_{th} = \int_{S_{int}} J_{th} \mathrm{N}_{\mathrm{th}}{}^{\mathrm{T}} \mathrm{N}_{\mathrm{th}} dS \tag{18b}$$

Combining the mechanical, coupling and thermal parts from equation (8) and (18), the tangent stiffness matrix, $\mathbf{K}_{\mathbf{CZ}}$, for a single node neglecting the temperature dependence of the mechanical properties of the cohesive zone, can be expressed as,

$$\mathbf{K}_{\mathbf{CZ}} = \begin{bmatrix} \mathbf{K}_{\mathbf{m}\{2\mathrm{x}2\}} & 0 \\ \mathbf{K}_{\mathbf{tm}\{1\mathrm{x}2\}} & K_{th\{1\mathrm{x}1\}} \end{bmatrix}_{\{3\mathrm{x}3\}} \tag{19}$$

The numerical implementation of the cohesive zone model uses four node elements with linear interpolation functions for the displacement jumps, and nodal values for temperature jumps across the crack. The thermomechanical cohesive zone element has been implemented into the finite element program ABAQUS v6.3.1 through the UEL capability.

# 3 Cohesive Zone Laws

## 3.1 Interface failure

**Load transfer.** For the analysis of crack growth along interfaces – e.g. between a protective coating and a high temperature composite laminate – the present analysis makes use of a concept in which a prescribed traction separation law is used. While the authors have extensively made use of traction-separation laws based on exponential functions, here the investigation follows the work of Geubelle and Baylor (1998). This model uses a quasi-linear relationship between separation and traction and includes an internal residual property variable, $s$. The normal and tangential tractions, $T_n$ and $T_t$, are given in dependence of the displacement jumps across the crack by:

$$T_n = \frac{s}{1-s}\frac{\Delta_n}{\delta_n}\sigma_{\max} \quad \text{and} \quad T_t = \frac{s}{1-s}\frac{\Delta_t}{\delta_t}\tau_{\max} \quad \text{if} \quad s > 0 \tag{20}$$

The internal residual property variable is related to a measure of the normalized displacement jumps, and is assigned an initial value of $s_{ini}$.

$$s = \min\left[s_{ini}, \max(0, 1-\lambda)\right] \quad \text{with} \quad \lambda = \sqrt{\left(\frac{\Delta_n}{\delta_n}\right)^2 + \left(\frac{\Delta_t}{\delta_t}\right)^2} \tag{21}$$

Choosing $s_{ini}$ close to unity allows one to construct traction-separation response functions with high initial stiffness. This aspect of the traction-separation law is important in the present investigation since not only load transfer across the crack but also heat flux is of concern. In the coupling of material failure to heat transfer across the crack, it is important to realistically capture the crack opening profile. Furthermore, the use of the residual property variable ensures that the traction-separation response is irreversible and no healing of the interface occurs.

The material properties in the traction separation law are the cohesive strengths, $\sigma_{\max}, \tau_{\max}$, the cohesive lengths, $\delta_n, \delta_t$, as well as the cohesive energies, $\Gamma_n, \Gamma_t$, in normal and tangential direction, respectively. These three parameters are connected by:

$$\delta_n = \frac{2\Gamma_n}{\sigma_{\max} s_{ini}} \quad \text{and} \quad \delta_t = \frac{2\Gamma_t}{\tau_{\max} s_{ini}} \tag{22}$$

A contact surface is specified along the interface such that interpenetration is avoided. A linear relationship between contact pressure, $p_c$, and the overclosure is used:

$$p_c = \sigma_{\max} / (s_{ini}\delta_n)\Delta_n \quad . \tag{23}$$

**Heat transfer**. Heat flux across the cohesive zone, $q_{CZ}$, is described as a product of the temperature jump across the crack, $\Delta\theta$, and the cohesive zone conductance, $h_{CZ}$. Initially, in an undamaged state the interface possesses a conductance of $h_{i,ini}$. As damage is incurred by the interface and material separation progresses at the interface the conductance decreases

$$h_{CZ} = h_i = h_{i,ini} s \quad \text{if} \quad s > 0 . \tag{24}$$

For the initial crack, as well as for newly created traction free crack, the crack conductance for the open crack ($\Delta_n > 0$) is given in dependence of contributions due to radiation and conduction through the gas trapped in the crack:

$$h_{CZ} = h_c = h_r(\theta_+,\theta_-) + h_g(\Delta_n,\theta_+,\theta_-) = \\ \frac{S_b}{2/e_m - 1}\left[(\theta_+ - \theta_0) + (\theta_- - \theta_0)\right]\left[(\theta_+ - \theta_0)^2 + (\theta_- - \theta_0)^2\right] + . \\ \left[k_g + C_g\left(\frac{\theta_+ + \theta_-}{2}\right)\right]\frac{1}{\Delta_n} \quad \text{if} \quad s = 0 \tag{25}$$

Here, the Stephan Boltzman constant is $S_b$=5.67x10$^{-8}$ W/m$^2$K$^4$, the emissivity of the crack surfaces of, $e_m$, and the reference temperature, $\theta_0$. The conductivity of the gas trapped in the crack, depends on the temperature dependent thermal conductivity of the gas, $k_g + C_g(\theta)$, and the normal displacement jump across the crack.

For a closed crack, ($\Delta_n < 0$), a contact conductance relationship is prescribed following work by Mikic (1974):

$$\frac{1}{h_{contact}} = \frac{1}{1.55}\left(\frac{\omega}{k_{eq}\tan\vartheta}\right)\left[\frac{E_{eq}\tan\vartheta}{\sqrt{2}p_c}\right]^{0.94} \quad \text{if} \quad s = 0 \tag{26}$$

Thereby, the effective elastic modulus and the effective conductivity of the material combination of coating and substrate are given by:

$$1/E_{eq} = \left(1 - \nu_1^2\right)/E_1 + \left(1 - \nu_2^2\right)/E_2 \\ 1/k_{eq} = 1/k_1 + 1/k_2 \tag{27}$$

Surface roughness, $\omega$, and asperity slope values, $\vartheta$, are connected by $\tan\vartheta = \omega^{0.4}$.

### 3.2 Crack Bridging

**Load transfer**. In unidirectional reinforced laminates delamination cracks parallel to the reinforcement axis are commonly bridged by fibers. Three mechanisms are responsible for load transfer characteristics across bridged delamination cracks, *i.e.*:

- fiber pull-out,
- matrix spalling – the growth of cracks in the matrix due to stress concentrations caused by the presence of the bridging fibers,
- fiber fracture due to combined bending and axial loading, with a Weibull distribution describing the fiber failure statistics.

The model describing the load transfer across the fiber bridging zone is an extension of the mode I crack bridging model of Kaute et al. (1993). The present work generalizes this model to deal with arbitrary values of mode mixity, Hattiangadi and Siegmund (2004).

Figure 1 depicts the geometry and the main parameters used in the development of the cohesive zone model for fiber bridging. The normal and tangential cohesive zone tractions, $T_n(\Delta)$ and $T_t(\Delta)$, can be related to the force per fiber in the normal and tangential direction, $f_n(\Delta)$ and $f_t(\Delta)$, and the number of fibers per unit area of the crack, $n(\Delta)$ :

$$T_n(\Delta) = n(\Delta)f_n(\Delta) \text{ and } T_t(\Delta) = \frac{n(\Delta)}{2}f_t(\Delta) \tag{28}$$

The factor 1/2 in the equation for $T_t$ takes into account that during crack opening the bridging fibers appear in a criss-cross geometry with only half the number of fibers contributing to the tangential load transfer in either direction.

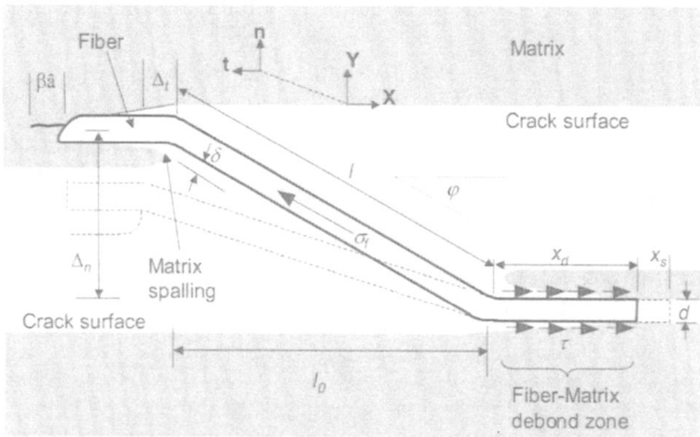


**Figure 1.** Load transfer mechanism across a bridged crack.

Under application of load, fibers debond from the matrix over a length $x_d$ and subsequently are pulled out. The present model assumes a negligible debonding energy at the fiber-matrix interface, a constant interface sliding stress, and average interface shear stresses. The resulting pull-out stress, $\sigma_p(\Delta)$, acting at the location where the fiber enters the matrix can then be determined from equilibrium to be

$$\sigma_p(\Delta) = \frac{4\tau\, x_d}{d} \tag{29}$$

where $d$ is the fiber diameter, and $\tau$ is the interface sliding stress. The fiber slippage length, $x_s(\Delta)$, is calculated from the average stress $\sigma_p(\Delta)/2$ acting over the debonding zone and the debonding length, $x_d$, as:

$$x_s(\Delta) = \frac{\sigma_p(\Delta)}{2E_f} x_d \tag{30}$$

where $E_f$ is the fiber modulus. As the crack opening increases, the strain $\varepsilon_f(\Delta)$ in the fiber is reduced due to the fiber slippage. The stress in the fiber due to pull-out can then be expressed by

$$\sigma_p(\Delta) = E_f\left(\frac{l(\Delta) - l_0}{l_0} - \frac{2x_s}{l_0}\right) \tag{31}$$

where $l(\Delta)$ is the current and $l_0$ the initial bridging length. If fiber slippage is extensive, the right hand side of equation (31) will dominate. The left hand side – corresponding to the stress expected if the fiber is tightly gripped (i.e. no debonding) – can be neglected in this case. Expressing the current bridging length as $l(\Delta) = [(l_0 + \Delta_t)^2 + \Delta_n^{\,2}]^{1/2}$, and substituting equation (30) into (31), $\sigma_p(\Delta)$ can be obtained as a function of material properties, $E_f$, $d$, $\tau$, $l_0$ and the values of the normal and tangential displacement jump:

$$\sigma_p(\Delta) = \sqrt{\frac{4\tau E_f}{d}\left[\sqrt{(l_0+\Delta_t)^2+\Delta_n^{\ 2}} - l_0\right]} \tag{32}$$

From the stress, $\sigma_p(\Delta)$, given by the fiber pull-out model, the components of the force carried per bridging fiber in normal and tangential direction, $f_n$ and $f_t$, are calculated as,

$$f_n(\Delta) = \sigma_p(\Delta)\frac{\pi d^2}{4}\tan\varphi \quad \text{and} \quad f_t(\Delta) = \sigma_p(\Delta)\frac{\pi d^2}{4} \tag{33}$$

where $\varphi$ is angle between the bridging fibers and the crack surface.

As the displacement jump across the crack increases, the forces carried by the bridging fibers increase until spalling of the matrix initiates. This condition imposes an upper limit on the forces that can be carried by the fibers. An approximate expression linking the maximum value of the normal component of the fiber force, max($f_n$), to the fracture toughness of the matrix, $K_{IC}^m$, (Kaute et al., 1993), can be obtained based on results of Tada et al., (1973) as:

$$\frac{\max(f_n)}{\beta\hat{a}} = K_{IC}^m\sqrt{\pi\hat{a}} \tag{34}$$

where $2\hat{a}$ is the length of cracks extending to either side of a fiber and $\beta\hat{a}$ is the mean crack length parallel to the fiber over which the forces $f_n$ acts.

During matrix spalling a crack can propagate catastrophically from the weak fiber matrix interface. Therefore, it may seem reasonable to link the critical crack-length, $\hat{a}_c$, beyond which matrix spalling occurs to the mean distance, $L$, between two neighbouring fibers as $\hat{a}_c = L - d/2$. In the case of parallel fibers, the distance $L$ and the fiber diameter $d$ are related through the fiber volume fraction $V_f$:

$$\hat{a}_c = d\left[\sqrt{\frac{\pi}{2\sqrt{3}}}\frac{1}{\sqrt{V_f}} - \frac{1}{2}\right] \tag{35}$$

Combining equations (33), (34) and (35) leads to a prediction of max($f_n$). Further crack opening occurs under constant values of $f_n$ and $f_t$ as well as of $\varphi$. The maximum value of $\varphi$ is obtained at the onset of spalling by inverting (32) and substituting max($f_n$) for $f_n$.

$$\varphi_{sp} = \max(\varphi) = \tan^{-1}\left[\frac{4\max(f_n)}{\sigma_p(\Delta)\pi d^2}\right] \tag{36}$$

The number of bridging fibers per unit area initially bridging the crack is $n_0$. As loading progresses this number subsequently diminishes due to the occurrence of fiber fracture. A Weibull statistics approach is used to predict the number of surviving fibers in the bridging zone. Fiber fracture is due to combined bending-axial loading. Using the result of Kaute et al. (1993) the bending stress can be expressed in terms of the angle, $\varphi$, and fiber modulus, $E_f$:

$$\sigma_b = \frac{E_f\tan^2\varphi}{2\delta/d} \tag{37}$$

where $\delta$ is the asymptotic distance of the fiber with reference to its axial direction, Figure 1. For the calculation of the bending stress, the ratio $\delta/d$ is assumed to be close to unity ($\delta/d$=0.9).

The combined stress due to tension in fibers during pull-out and bending at the root of the bridged fiber is $\bar{\sigma}(\Delta) = \sigma_p(\Delta) + \sigma_b$. Prior to matrix spalling the survival probability is given as

$$p_s = \exp\left[ -C_b \frac{l(\Delta)}{l_{ref}} \left( \frac{\bar{\sigma}(\Delta)}{\sigma_{ref}} \right)^m \right] \tag{38}$$

where $l_{ref}$ and $\sigma_{ref}$ are Weibull reference length and strength, respectively and $m$ is the Weibull modulus. $C_b$ is a dimensionless correction factor, ($C_b$<1), to allow for the comparison between bending and tensile stresses. As matrix spalling occurs, the survival probability defined in equation (38) is modified. As the bridging angle, $\varphi = \varphi_{sp}$, the spanning length $l(\Delta)$ in equation (38) is expressed in terms of the normal displacement jump, $\Delta_n$, as $l = \Delta_n \sin\varphi_{sp}$. Also, the bending stress, $\sigma_b$, reaches a maximum value when matrix spalling occurs. The number of bridging fibers per unit area is given as,

$$n(\Delta) = n_0 p_s(\Delta) \tag{39}$$

Finally, the traction-separation behavior in equation (28) is the outcome of combining equations (33), (34) and (39) which contain the material property information.

**Heat transfer**. In developing the thermal part of the fiber bridging cohesive zone law the following simplifying assumptions are made:

- One-dimensional heat transfer across the crack is considered and described by a thermal resistor model,
- Thermal resistance in the fiber debonding zone and constriction effects at locations where the fiber emerges from the matrix are neglected. The heat flux across the cohesive zone is the sum of the heat fluxes due to conduction through the bridging fibers, $q_f$, the gas trapped in the crack, $q_g$, and the radiation between the crack surfaces, $q_r$; Figure 2.

The cohesive zone conductance model combines the conductance of gas, fiber and radiation by a parallel arrangement of resistors, Donaldson et al. (1998), McDonald et al. (2000):

$$h_{CZ} = h_f + h_g + h_r \tag{40}$$

The conductance through fiber, $h_f$, depends on the current bridging length, $l$, which can be expressed in terms of the displacement jump, $\Delta$, the bridging angle, $\varphi$, and the volume fraction of bridging fibers in the crack, $V_f^C$. The model assumes a linear temperature dependence of the thermal conductivity of the fibers, $k_f + C_f(\theta)$, in dependence of the average temperature of the upper and lower crack surface. The contribution of the fibers to the cohesive zone conductance, $h_f$, is expressed as

$$h_f(\Delta, \theta_+, \theta_-) = V_f^C \left[ k_f + C_f \left( \frac{\theta_+ + \theta_-}{2} \right) \right] \frac{\sin\varphi}{\Delta_n} \tag{41}$$

The conductivity of the gas trapped in the crack, $h_g$, depends on the temperature dependent thermal conductivity of the gas, $k_g + C_g(\theta)$, the displacement jump across the crack and the volume fraction of gas in the crack, $(1 - V_f^C)$:

$$h_g(\Delta, \theta_+, \theta_-) = \left(1 - V_f^C\right) \left[ k_g + C_g \left( \frac{\theta_+ + \theta_-}{2} \right) \right] \frac{1}{\Delta_n} \tag{42}$$

The contribution to the cohesive zone conductance due to radiation, $h_r$, is dependent on the crack surface temperatures

$$h_r(\theta_+,\theta_-)=\frac{S_b}{2/e_m-1}\left[(\theta_+-\theta_0)+(\theta_--\theta_0)\right]\left[(\theta_+-\theta_0)^2+(\theta_--\theta_0)^2\right] \tag{43}$$

with the Stephan Boltzman constant, $S_b$=5.67x10$^{-8}$ W/m$^2$K$^4$, an assumed emissivity of the crack surfaces of, $e_m$=1.0, and the reference temperature, $\theta_0$.

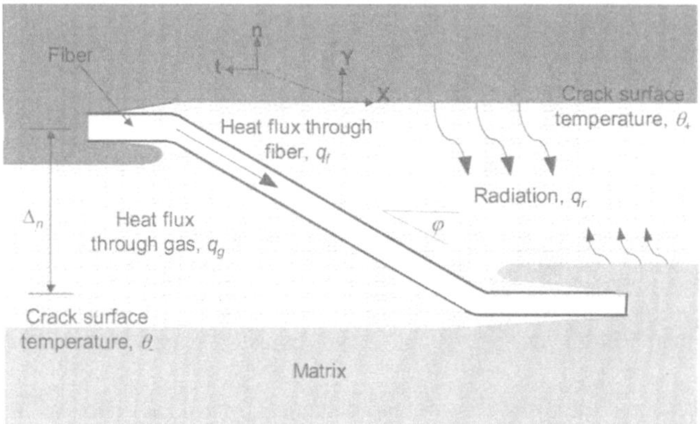


**Figure 2.** Heat transfer across a bridged crack

In describing the cohesive zone conductance, the current volume fraction of fibers in the cohesive zone, $V_f^C$, is to be defined. This quantity depends on the fiber diameter, the displacement jump across the cohesive zone, the bridging length, and the number of fibers in the cohesive zone. Combining these quantities, the volume fraction of fibers in the cohesive zone is expressed as

$$V_f^C=n_0 p_s\frac{d^2\pi}{4}\frac{l_0}{\Delta_n} \tag{44}$$

In deriving equation (44) the change of the bridging length during spalling was neglected. An upper limit for the value of $V_f^C$ is assumed as $\max\left(V_f^C\right)=\pi/4$. The thermal resistor model is irreversible by being coupled to the mechanical part of the cohesive zone model. The relative contributions of fibers and gas to $h_{CZ}$ change as material deterioration progresses.

# 4 Results

## 4.1 Interface Failure

**Material system.** The material system considered in the study of interface delamination under thermal loading consists of a carbon-carbon (C-C) composite laminate with a SiC coating. Such a material system is in use in the thermal protection system of the Space Shuttle leading edge, Columbia Accident Investigation Board, (2003).

The material properties for the orthotropic C-C composite laminate and the isotropic SiC coating are summarized in Table 1.

**Table 1.** Properties of SiC coating (density $\rho$=3100 kg/m$^3$)

| $\theta$ K | $E$ GPa | $\nu_T$ | $\alpha$ $10^{-6}$K$^{-1}$ | $k$ Wm$^{-1}$K$^{-1}$ | $c_p$ Jkg$^{-1}$K$^{-1}$ |
|---|---|---|---|---|---|
| 300 | 415 | 0.16 | 1.1 | 114 | 715 |
| 773 | 404 | 0.159 | 4.4 | 55.1 | 1086 |
| 1273 | 392 | 0.157 | 5.0 | 35.7 | 1240 |
| 1500 | 387 | 0.157 | 5.2 | 31.3 | 1282 |
| 1800 | 380 | 0.156 | 5.5 | 26.3 | 1336 |

**Table 2.** Properties of C-C composite (density ρ=1700 kg/m$^3$)

| $\theta$ K | $E_T$ Gpa | $E_P$ GPa | $G_T$ GPa | $G_P$ GPa | $\nu_T$ | $\nu_P$ | $k_T$ Wm$^{-1}$K$^{-1}$ | $K_P$ Wm$^{-1}$K$^{-1}$ | $\alpha_T$ $10^{-6}$K$^{-1}$ | $\alpha_P$ $10^{-6}$K$^{-1}$ | $c_p$ Jkg$^{-1}$K$^{-1}$ |
|---|---|---|---|---|---|---|---|---|---|---|---|
| 300 | 6 | 17 | 2.7 | 2.7 | 0.3 | 0.13 | 3.9 | 42 | 2.0 | 0.5 | 720 |
| 1300 | | | | | | | 5.0 | 43 | 2.5 | 0.5 | 2000 |
| 1800 | | | | | | | 5.8 | 45 | 2.9 | 0.6 | 2000 |

The values of the material parameters in the cohesive zone elements were assumed $\sigma_{\max} = \tau_{\max} = 30$ MPa, the cohesive lengths, $\delta_n = \delta_t = 5\mu$m, such that the cohesive energies, $\Gamma_n = \Gamma_t = 75$ Jm$^{-2}$, $s_{ini}$=0.85. The interface initially has a conductance of $h_{i,ini}$=1E$^4$W/m$^2$K. A surface roughness of 1 μm was assumed. The crack contains no conductive gas. Friction between the crack faces is neglected.

**Basic cohesive zone model response.** Figure 3 depicts the normal and tangential traction separation response for the cohesive zone law applied for the interface crack. The maximum traction values are obtained as the product of the initial value of the residual property variable and the cohesive strength. Under mixed mode loading the cohesive strength and the cohesive energy is considerably reduced from the values under either pure normal or tangential separation.

Figure 4 depicts the relationship between material separation and interface conductance. The conductance of the interface remains at $h_i = h_{i,ini} s_{ini}$ as long as the tractions are below their peak

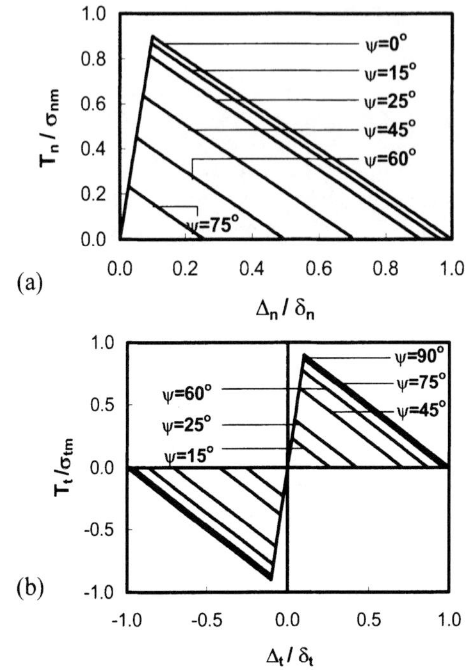


**Figure 3.** Traction separation law. (a) Normal separation; (b) Tangential separation.

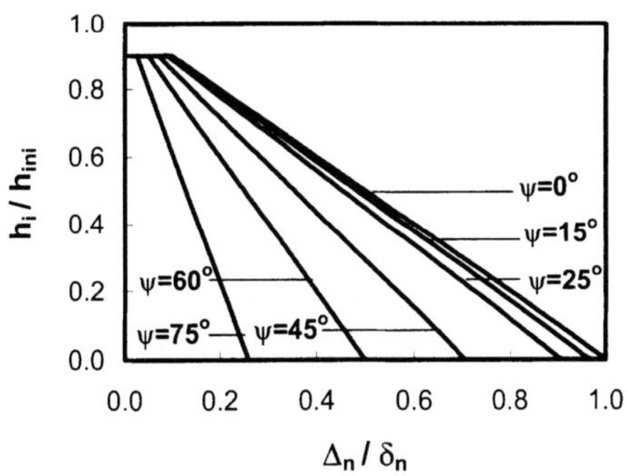


**Figure 4.** Interface conductance in dependence of material separation.

value, and subsequently the interface conductance is reduced linearly with the displacement jump across the crack. Mixed mode separation leads to a considerably faster reduction in interface conductance than normal separation only. The length scales introduced into the model by the cohesive zone are identical for the mechanical and thermal part.

**Crack growth under transient thermal loading**. In the specimen considered in the study of transient thermal loading is a carbon-carbon (C-C) composite laminate with a thickness of $H$=5 mm, and a SiC coating of thickness $H_c$=1 mm on both sides of the laminate. The laminate length is $2L$=50 mm. The laminate possesses an edge-type crack of length of $a/L$ = 1/3 at the center, located at the interface between the C-C laminate and the SiC coating close to the top-side of the laminate. Figure 5 depicts the specimen geometry. The conditions considered are those of a traction free laminate and no external displacement or traction boundary conditions are prescribed. The laminate is initially at a temperature of 300 K. The thermal boundary conditions consist of a prescribed heat flux at $y=H+H_c$. In the examples discussed below the applied heat flux is linearly increasing at a rate of 0.2 MW/m$^2$ per second. Radiative cooling of the specimen is accounted for at $y=H+H_c$. The emissivity for this surface is assumed as 0.7 with the surrounding temperature as 300 K.

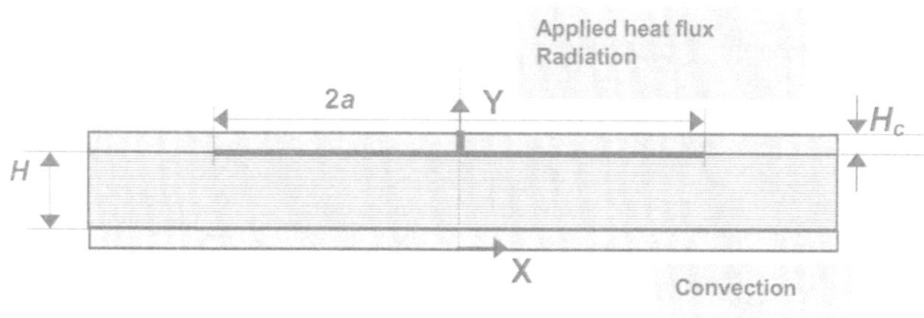


**Figure 5**: Geometry of the C-C composite laminate with SiC coating.

Convective cooling with a film coefficient of $1\times10^6$ W/m$^2$ to a sink of temperature 300 K is assumed at [$y$=0] such that the temperature at the bottom, [$y$=0], is kept nearly constant at the initial temperature $\theta_{bot}=\theta_0$. The sides [$x=\pm L$] are insulated. Only a half model of the specimen (symmetry plane at [$x$=0]) is considered with the displacement boundary conditions $u_y[(x=0),(y=0)]=0$. Transient temperature-displacement solutions are computed. Continuum elements used in the finite element model are four node plane strain temperature-displacement elements. Thermomechanical cohesive zone elements are placed along the interface, with $s$=0 for the initial crack.

Figure 6 shows contour plots of the heat flux through the specimen at times 4, s 5, s, 6 s, and 7 s. While initially a disturbance of the heat flux is present only at the initial crack, it can be clearly seen that the disturbance spreads out as the applied heat flux is increased. The specimen becomes insulating nearly over its entire extension. Finally, the applied heat flux is transferred over only a small area close to the edges which is then subjected to a considerable amount of heat flux.

Corresponding contour plots of temperatures are depicted in Figure 7. In addition to the temperature gradient through the undisturbed laminate, the central part of the laminate is experiencing an increase in temperature as a result of the insulating action of the initial crack.

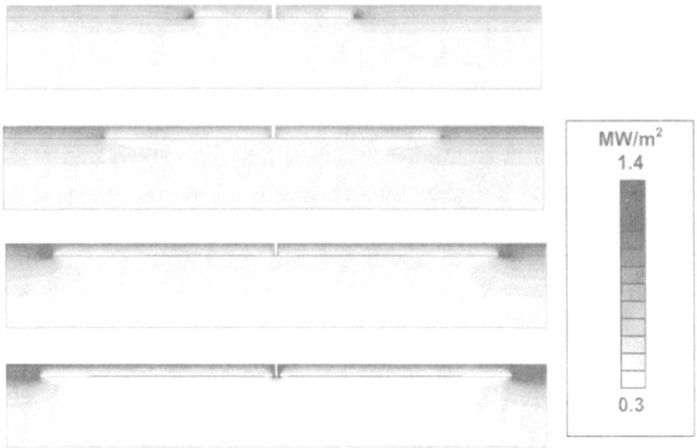


**Figure 6.** Heat flux in vertical direction through the SiC-C/C-SiC coated laminate under applied transient heat flux of 2 $MW/m^2s$. Times are 4 s, 5 s, 6 s, 7 s.

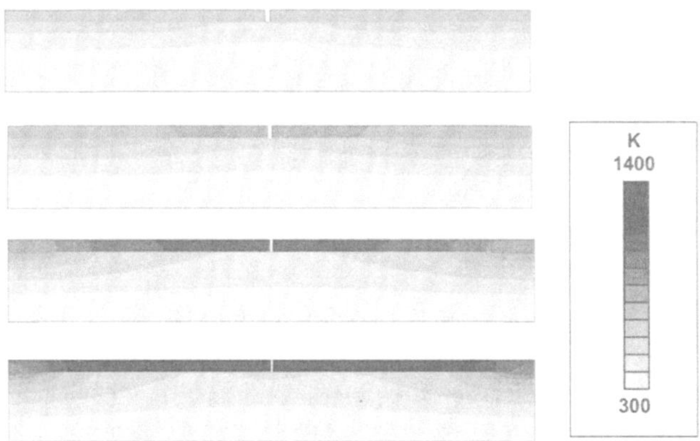


**Figure 7.** Temperatures in the SiC-C/C-SiC coated laminate under applied transient heat flux rate of 2 $MW/m^2s$. Times are 4 s, 5 s, 6 s, 7 s.

At longer times, as the crack progresses and the applied heat flux increases the temperature of the coating areas above the growing crack experience a considerable temperature increase. The area subjected to the highest temperatures nearly extends over the entire specimen in the end. An understanding of the crack propagation process can be obtained by investigating the stress state in the elements located adjacent to the interface. Stress components $\sigma_{12}$ and $\sigma_{22}$ as obtained in the C-C composite adjacent to the interface are shown in Figure 8. These results clearly demonstrate that the failure process is dominated by material separation in tangential direction.

This failure mode has two sources (i) the mismatch in thermal expansion between SiC and C-C as well as (ii) the gradient in temperature in dependence of the distance from the surface of the specimen at which the heat flux is applied.

The computed stress distribution is crack like, with a considerable peak in shear stresses close to the location of the current crack tip. The computed shear stresses remain below the shear cohesive strength of the interface due to the mixed mode character of the failure processes. The distribution of the normal traction changes considerable along the interface. Before crack initiation and during the initial stages of crack growth the crack is closed over a large portion such that contact conductance is important. However, the crack is open in the region close to the current crack tip. Behind the crack tip considerable negative normal tractions are present. Similar types of crack growth processes were predicted in a study of dynamic shear crack growth in Coker et al. (2003). At later stages in the crack growth process, the crack opens over its entire length. Contact conductance is no longer important and radiative heat transfer between the crack surfaces sets in.

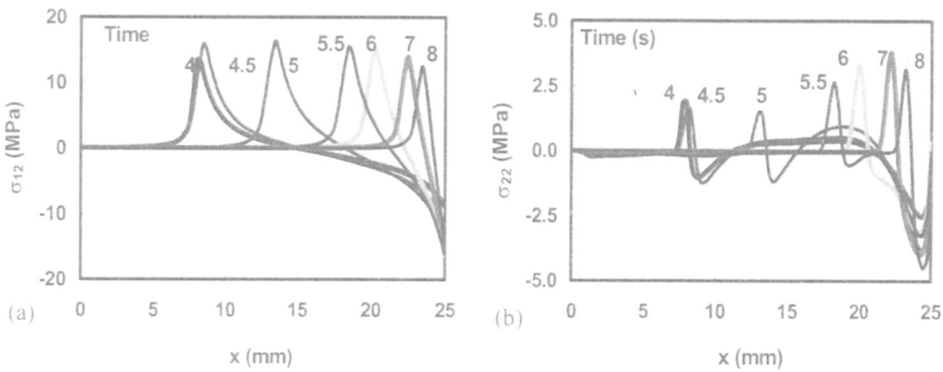


**Figure 8**: Stress distribution in the C/C composite adjacent to the interface. (a) Shear stress, (b) normal stress.

## 4.2 Crack Bridging

**Material system**. The thermomechanical bridging zone model is applied to a study of a ceramic matrix composite consisting of unidirectional aligned SiC fibers in an $Al_2O_3$ matrix.

The continuum properties of the composite are calculated from fiber and matrix properties. The temperature dependent thermomechanical material properties for typical SiC fibers and an $Al_2O_3$ matrix of 96% density were assembled from published data in Touloukian et al. (1967), Brennan et al. (1982), Simon and Bunsell (1984), Li and Bradt (1987), Feredia et al. (1995), Lu and Hutchinson (1995) and Munro (1997).

**Table 3.** Properties of SiC-$Al_2O_3$ CMC.

| $\theta$ | $E_T$ | $E_L$ | $G_T$ | $G_L$ | $v_T$ | $v_L$ | $k_T$ | $k_L$ | $\alpha_T$ | $\alpha_L$ |
|---|---|---|---|---|---|---|---|---|---|---|
| K | GPa | GPa | Gpa | GPa | | | $Wm^{-1}K^{-1}$ | $Wm^{-1}K^{-1}$ | $10^{-6}K^{-1}$ | $10^{-6}K^{-1}$ |
| 300 | 392.7 | 346.7 | 121.5 | 137.4 | 0.43 | 0.17 | 18.64 | 23.47 | 7.56 | 5.62 |
| 573 | 376.5 | 333.2 | 116.2 | 131.7 | 0.43 | 0.17 | 10.84 | 13.11 | 8.29 | 6.49 |
| 873 | 358.9 | 318.8 | 110 | 125.4 | 0.43 | 0.17 | 7.91 | 9.07 | 8.67 | 7.1 |
| 1500 | 324.3 | 289.1 | 98.8 | 112.9 | 0.43 | 0.18 | 5.68 | 5.92 | 8.89 | 7.48 |

Based on these properties, the effective properties of a composite with a volume fracture of $V_f$=0.3 were calculated by use of the Mori-Tanaka and Hatta-Taya theories, see (Nemat-Nasser and Hori, 1999, Nadeau and Ferrari, 1999, Hatta and Taya, 1985). The temperature dependent effective properties obtained by these procedures are summarized in Table 3. The thermomechanical CZM requires information on fiber, matrix, fiber-matrix interface properties, statistical information on fiber failure and reference parameters for size effects. The default parameters used in the present study are given in Table 4, compiled from data given in Kaute et al. (1993, 1995), Lu and Hutchinson (1995), Heredia et al. (1995) and Munro (1997).

**Table 4.** Bridging Model Parameters.

| | |
|---|---|
| Initial bridging span, $l_0$ [μm] | 200.0 |
| Fiber diameter, $d$ [μm] | 15.0 |
| Interface shear resistance, $\tau$ [MPa] | 15.0 |
| Matrix fracture toughness, $K_{IC}^m$ [MPa m$^{1/2}$] | 3.0 |
| Weibull modulus of fiber, $m$ | 3.1 |
| Weibull reference length, $l_{ref}$ [mm] | 25.0 |
| Weibull reference strength, $\sigma_{ref}$ [GPa] | 1.63 |
| Number of fibers/fracture surface area, $n_0$ [mm$^{-2}$] | 14.0 |
| Thermal conductivity of gas (air), $k_g$ [W/mK] | 0.032 - 0.1 (300 K - 1500 K) |

**Basic cohesive zone model response.** Figure 9(a) depicts the traction-separation response of the bridging model for pure normal separation, as well as for various values of mixed mode loading. The predicted normal (and similarly the tangential) tractions as a function of the displacement jumps are is a traction-separation curve that initially requires increasing load for material separation, then reaches a maximum traction value, and subsequently leads to a gradual loss in load carrying capacity. The maximum traction values reached can be interpreted as a cohesive strength.The displacement jump at the point where traction drops to near zero due to the survival probability of fibers approaching zero can be identified as mechanical cohesive length. Then, the present model shows the same characteristics as commonly used prescribed traction-separation laws. The mixed mode response is characterized in Figure 9(b) in terms of the predicted values of the cohesive zone energies. The resistance of the bridging zone to failure is considerable larger in shear loading than in normal loading. The predicted $\Gamma_t$ - $\Gamma_n$ envelope well matches with experimental failure data for composite laminates, as e.g. determined for laminated composites by Partridge and Singh (1995), Benzeggagh and Kenane (1996), Ducept et al. (1997) and Rikards et al. (1998).

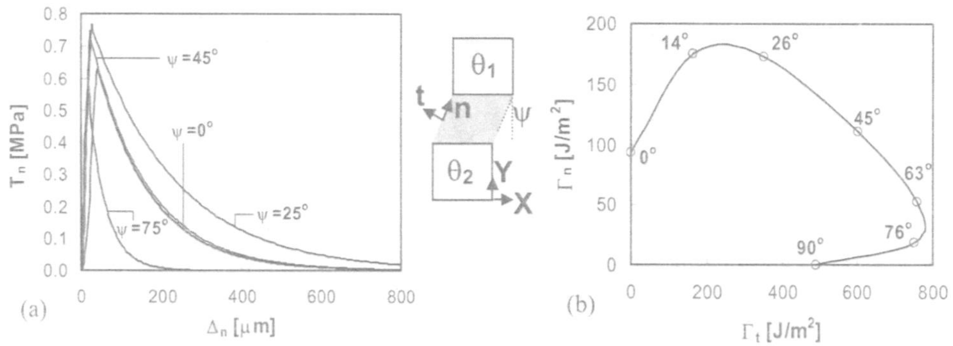


**Figure 9.** (a) Traction-separation response of the bridging model; (b) Cohesive energies for various mode mixities.

Figure 10 shows the dependence of the cohesive zone conductance on the normal displacement jump for the case of $\psi$=0°. In Figure 10(a) values of the cohesive zone conductance are given for heat transfer across the cohesive zone due to the presence of fibers only. Corresponding data for the volume fraction of fibers in the crack as well as for the survival probability are depicted. Initially for small values of $\Delta_n$, the bridging fibers do not contribute much to $h_{CZ}$ since their orientation is perpendicular to the heat flux. As the displacement jump increases, the fiber bridging angle starts to increase such that surviving fibers can contribute more to $h_{CZ}$. Nevertheless, due to the simultaneously occurring decay of $V_f^C$ the contribution of fibers actually drops with increasing opening. The initial value of $h_{CZ}$ can be calculated from equation (41) as $h_{CZ} = h_f = 0.0954$ MW/m$^2$K. This value of $h_{CZ}$ remains nearly constant until the displacement jump increases to $\Delta_n$=0.5 μm. Subsequently, $h_{CZ}$ drops, and for $\Delta_n$>70 μm is reduced to very small values. The value displacement jump at which the conductance drops to near zero can be identified as the thermal cohesive length. While the mechanical cohesive length is related to the survival probability of fibers, $p_s$, the thermal cohesive length is related to the fiber volume fraction in the crack. Due to its combined dependence on $1/\Delta_n$ and $p_s$ the quantity $V_f^C$ possesses much larger dependence on the material separation than the survival probability alone. For the present material data set, the mechanical cohesive zone length is approximately ten times the thermal cohesive zone length.

Figure 10(a) also compares the values of $h_{CZ} = h_f$ to those for a crack containing air at 1 atm. only, ($h_{CZ} = h_g$). In this case $h_{CZ}$ follows $1/\Delta_n$. As fibers are mostly aligned perpendicular to the heat flux, $h_g > h_f$. Model prediction compare favorable with the trends in the experimental study of McDonald et al. (2000). The cohesive zone conductance in dependence of $\Delta_n$ under mixed mode loading conditions is shown in Figure 10(b). For the case of a cohesive zone with fibers in a vacuum an increase in mode mixity leads to a faster decrease in $h_{CZ}$. The thermal cohesive zone length for $\psi \rightarrow 90°$ is considerably smaller than for $\psi$=0°, i.e. only 3.7 μm. However, for a cohesive zone containing air there is essentially no change in the cohesive zone conductance with mode mixity. At 1 atm. air provides the dominating factor in heat transfer, and this contribution is independent of $\psi$ and only depends on $\Delta_n$.

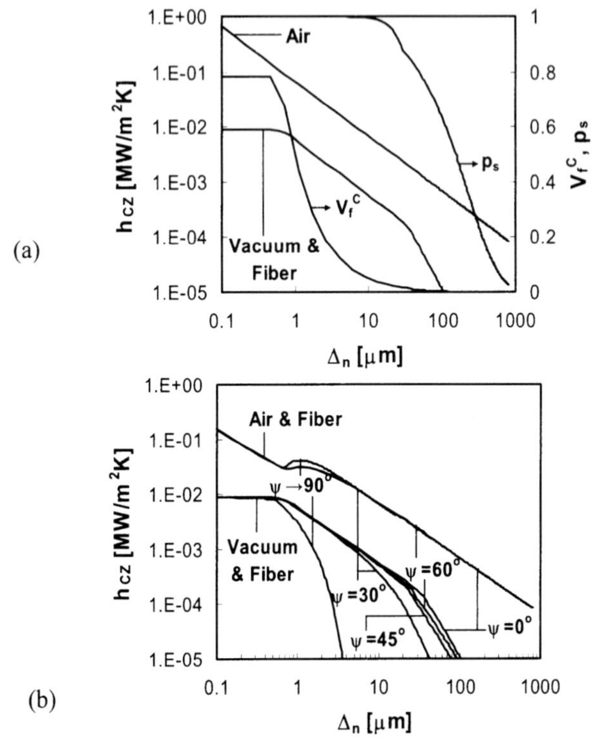


**Figure 10.** (a) Cohesive zone conductance-separation response of the bridging model with corresponding values of fiber survival probability, comparison to a crack containing air only; (b) Cohesive zone conductance for various mode mixities.

**Bridged delaminations.** Numerical studies are carried out for a laminate of thickness $H$=2 mm and a length of $L$=30 mm, Figure 11. A bridged crack of fixed length $2a$ is positioned at $H_I$=0.211$H$ from the upper hot laminate surface. The location of the crack provides the maximum value of energy release rate in specimen loaded by a temperature gradient and traction free boundary conditions (Hutchinson and Lu, 1995). Lengths for the bridged delamination crack are considered between $a/H$=4.0. Initially the laminate is at the reference temperature $\theta_0$=300 K. A temperature change linearly increasing in time is applied to the top surface of the specimen, [$y$=$H$], with the maximum temperature reaching max($\theta_{top}$)=1500 K. The temperature at the bottom, [$y$=0], is kept constant at the initial temperature $\theta_{bot}$= $\theta_0$. The sides [$x$=±$L$] are insulated. The displacement boundary conditions considered are those of a traction free laminate. The boundary conditions applied are $u_y$[($x$=0), ($y$=0)]=0. Coupled temperature-displacement solutions in steady state are computed for a half model of the laminate with the symmetry plane at [$x/a$=0].

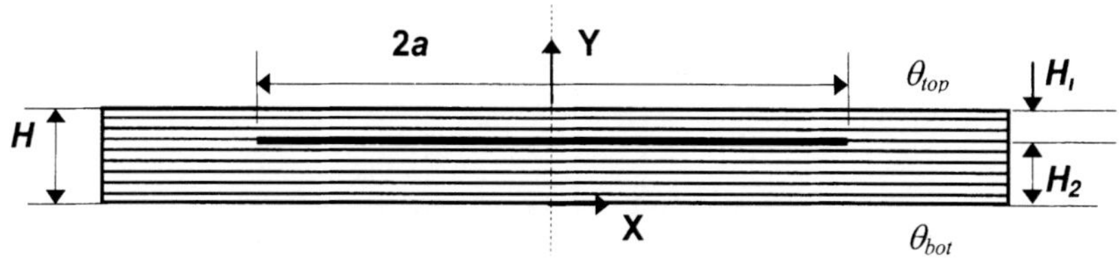


**Figure 11.** Geometry of laminate with bridged crack.

The length of the cohesive elements was chosen to be 50 μm, smaller than the cohesive zone lengths expected. Continuum elements used in the finite element model are four node plane strain temperature-displacement elements. A frictionless contact condition is employed at the crack surfaces. In contact, the thermal resistance of the cohesive zone is continuous to the thermal resistance during opening. The analysis of the laminates with buckling was facilitated by use of the STABILIZE option in ABAQUS. This introduces pseudo-damping proportional to the displacement velocity of the nodes, thus leading to a pseudo time integration of the buckling response. Default values provided in ABAQUS were reduced as much as possible to provide smallest impact on the analysis results.

As the applied temperature difference between the top and bottom of the laminate, $\theta_{top}$-$\theta_{bot}$, increases the thermal expansion differences due to the temperature gradient between the hot and the cold side of the laminate lead to bending of the specimen. The conditions pertaining to the delamination crack will impact the temperature distribution and thus the loading of the crack.

In Figure 12, the influence of different crack environments is documented for the crack length of $a/H$=4. The thermal response is depicted via the $q^*_{CZ}$ , the heat flux across the crack at $x/a$=0.0, and $1/(1+Bi)^2$ in Figures 12(a-b). The conditions $V_f^C$ =0.01 and $p_s$=0.1 are identified with symbols ∗ and ○, respectively. For the case with a crack with vacuum only, the crack is virtually insulating except at high temperatures where radiation becomes noticeable. For the laminate with bridging fibers in vacuum, again only a small increase in the conductance is observed relative to the pure vacuum case. For specimens with cracks containing air at 1 atm., the conductance is considerably increased and is initially close to that of the uncracked laminate. In the case of a crack containing air only, the initially large values of $q^*_{CZ}$ cannot be sustained up to large values of applied temperature difference, and $q^*_{CZ}$ breaks down already at an applied temperature difference of 131 K. As crack bridging is absent, the crack opening occurs without any restraints. For the crack containing both air and fibers, the most favorable picture emerges. The bridging fibers keep the crack closed such that air remains effective in transporting heat across the crack and thus delay the onset of the buckling delamination.

A heat flux across the crack is sustained to considerably higher values of applied temperature difference, i.e. $\theta_{top} - \theta_{bot} = 507$ K. Subsequently, in the presence of gas, $q^*_{CZ}$ remains above that for the crack with vacuum even at large values of applied temperature difference. Values of $1/(1+Bi)^2$ as computed at $x/a$=0.0 are given in Figure 12(b) for the four crack conditions. For the crack with vacuum $1/(1+Bi)^2$ is equal to one except for high values of applied temperature differences where radiation starts to contribute to crack conductance. For the other cases, however, $1/(1+Bi)^2$ approaches one at different values of applied temperature difference with the effect of crack bridging is again clearly visible.

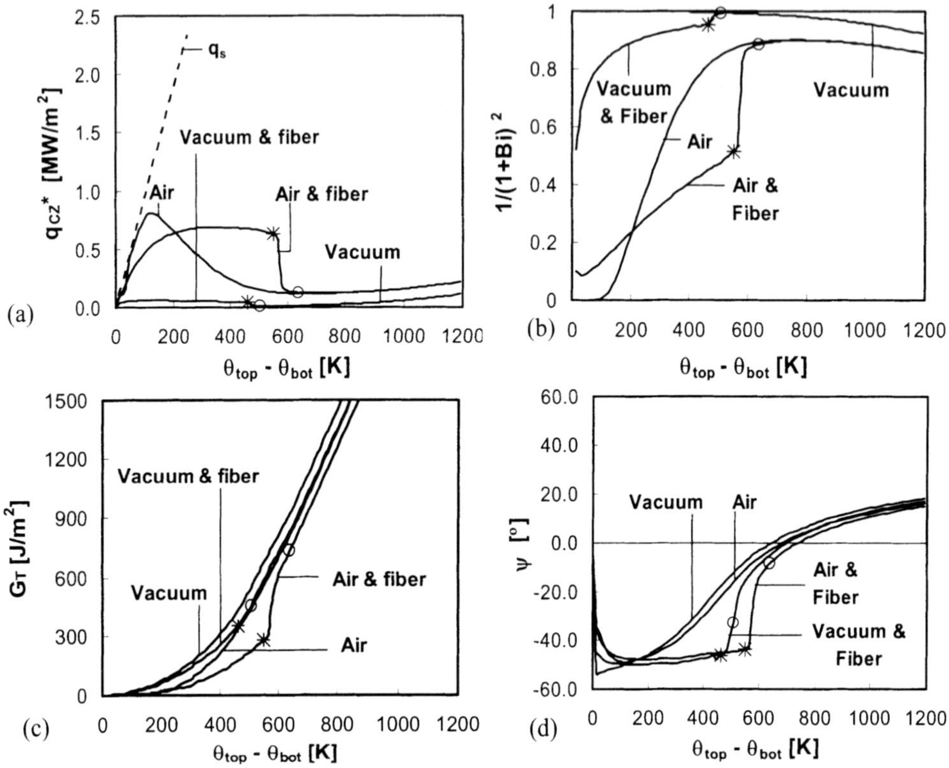


**Figure 12.** Predicted values of (a) $q^*_{CZ}$, (b) $1/(1+Bi)^2$, (c) $G_T$ and (d) $\psi$ in dependence of the applied temperature difference for various crack lengths *a/H*. All cracks with air, bridging fibers and radiation.

For bridged cracks the Biot number remains at large values up to considerably larger temperature differences than for the unbridged cracks indicating lower values of crack tip loading for these cases.

Computed values of energy release rate and mode mixity are depicted in Figures 12(c) and (d), respectively. For the crack with vacuum, due to the strongly insulating effect of the crack, the values of $G_T$ are non-zero already for small values of applied temperature difference, and increases steadily with an increase in $\theta_{top}$-$\theta_{bot}$. As fibers are included in the crack, the computed value of $G_T$ are slightly smaller than for the perfectly insulating crack due to the shielding effects of crack bridging as well as due to the somewhat reduced thermal stresses caused by larger values of $q^*_{CZ}$.

The action of the bridging fibers in delaying the onset of the formation of the buckling delamination is seen clearly in the mode mixity values. While breakdown in thermal bridging occurred at 464 K as characterized by $V_f^C$=0.01, thus leading to an insulating crack and an increased load magnitude at the crack tip.

For the crack containing bridging fibers the transition from the bending to the buckling mode of deformation occurs at larger values of $\theta_{top}$-$\theta_{bot}$. The breakdown in the mechanical bridging action sets in at 551 K and shielding remains effective in reducing the values of $G_T$ up to this value of applied temperature difference.

For a crack filled with air, but without bridging fibers, $G_T$ initially remains low as only a low level of thermal stresses are being induced by the conducting crack. However, as the opening of the crack progresses, the crack quickly becomes insulating. The resulting thermal stresses lead to a rapid rise in $G_T$. Finally, as bridging fibers are added to the crack $G_T$ remains at low levels for values of the applied temperature difference up to 507 K as a consequence of the reduced crack opening enabled by the presence of the crack bridging fibers. As $V_f^C$ =0.01 is reached at $x/a$=0, a pronounced change in the dependence of $G_T$ on $\theta_{top}$-$\theta_{bot}$ occurs. Related to the break down in $q^*_{CZ}$, the value of $dG_T/d(\theta_{top}$-$\theta_{bot})$ increases at this temperature difference due to the reduction in crack heat flux due to higher level of the resulting thermal stresses. Still, the mechanical action of crack bridging remains effective and keeps $G_T$ to the lowest level for all four cases considered in Figure 12(c). Bridging fibers thus provide longer cracks to sustain higher temperature differences prior to buckling. The computed values of the mode mixity are given in Figure 12(d) in dependence on the applied temperature difference. For applied temperature differences below the breakdown of crack heat transfer the mode mixity is constant for the cracks with bridging fibers, $\psi = -47°$.

As the crack conductance starts to break down, the mode mixity starts to increase, then becomes positive at the onset of bridging breakdown, and finally reaches $\psi = 20°$. Maintaining high values of mode mixities is favourable since the high cohesive strength at high values of $\psi$ will delay the breakdown of bridging. For the cracks without crack bridging the initial and final values of the mode mixity are the same. Nevertheless, the change from negative to positive $\psi$ is steady over the temperature considered.

**Figure 13:** Contour plots of temperatures.

Figure 13 depicts contour plots of the temperature distribution on the deformed specimen configuration for $a/H$=4.0. The increase in crack length results in a change of the failure mechanism from bending to buckling delamination. Again, results are obtained for the cohesive zone to represent a bridged delamination cracks containing air at 1 atm. with radiation accounted for. While initially, the delamination crack is nearly closed, and the crack presents only a small disturbance to the temperature field, such that the overall deformation mode is again by bending. At a critical temperature difference, the crack opens rapidly and the formation of a buckling delamination is seen clearly, Figure 13(c). As a consequence the temperature jump across the crack increases and finally the crack becomes fully insulating such that the temperature jump across the crack is identical to the applied temperature difference, Figure 13(d).

The thermomechanical cohesive zone model is based on the micro-mechanism of the crack bridging processes with the mechanical and thermal response of the cohesive zone as the outcome of the boundary value problem. Consequently, the present cohesive model approach can be used to directly assess the influence of the microlevel crack bridging material parameters on the macroscopic specimen response. In terms of material design, this allows to identify material parameters important for a specific loading configuration. A parametric study is carried out in order to demonstrate the effect of different types of cohesive zone parameters on the response of the cracked laminate under thermomechanical loading. The bridging parameters investigated here are divided into two categories for which one example is given. Mechanical cohesive zone properties are represented by the fiber reference strength, $\sigma_{ref}$, and the interface sliding stress, $\tau$. These values of these parameters can change due to the presence of manufacturing conditions or environmental degradation. Parameters affecting both the thermal as well as the mechanical behavior of the cohesive zone are representative of changes in the composite microstructure. As the fiber misalignment in the nominally unidirectional reinforced laminate changes, the number of bridging fibers per crack surface area left behind by the matrix crack, $n_0$, as well as the length of the bridging fiber segments, $l_0$, change. Results in the parametric study stem from computations for laminates with $a/H$=4.0. The cohesive zone represents conductance through the available fibers and air, respectively, as well as radiative heat transfer.

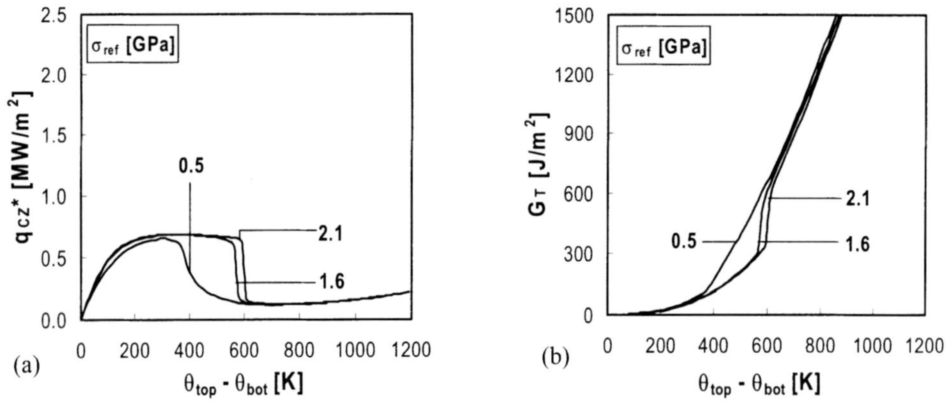


**Figure 14.** Thermomechanical crack bridging behavior in dependence of the fiber strength. (a) Crack heat flux, (b) Energy release rate.

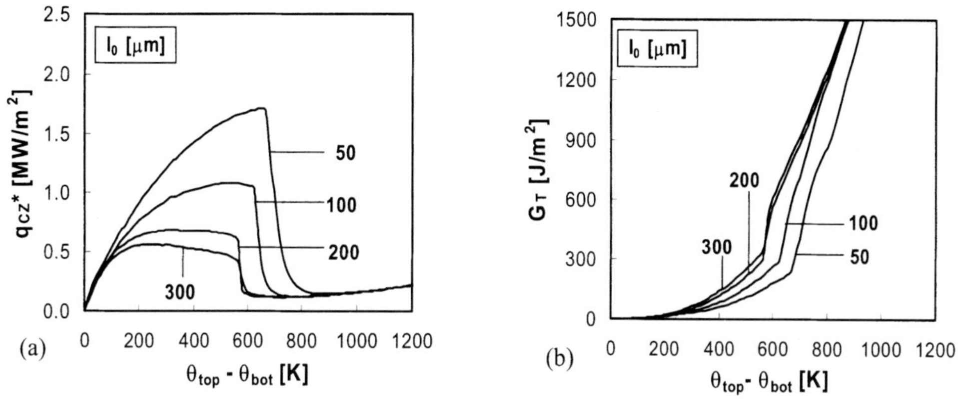


**Figure 15.** Thermomechanical crack bridging behavior in dependence of the spanning length. (a) Crack heat flux, (b) Energy release rate.

Changes in the fiber reference strength, $\sigma_{ref}$, lead to changes in the cohesive strength but leave the initial stiffness of the traction-separation response unchanged. The heat flux at the crack center for the bridged crack and the values for $G_T$ are depicted in Figure 14 for reference strengths of $\sigma_{ref}$=0.5, 1.6 and 2.1 GPa, respectively. As fiber strength is changed no changes in the initial heat flux values and the maximum heat flux across the delamination crack is predicted. The only impact on $q^*_{CZ}$ is that the breakdown in heat flux depends on the fiber strength. While the breakdown temperature difference increases by approximately 200 K as $\sigma_{ref}$ increases from 0.5 GPa to 1.6 GPa, a further increase to 2.1 has less additional effect. Higher values of $\sigma_{ref}$ lead to a more pronounced formation of the buckling delamination of the specimen.

Predictions for four different values of the initial spanning length, $l_0$, are given in Figure 15. A decrease in the initial spanning length increases the initial stiffness of the traction-separation response, increases the cohesive strength and the fiber bridging angle $\varphi$, and changes the thermal response by decreasing the volume fraction of fibers in the crack, $V_f^C$. As $l_0$ is decreased from 300 to 50 μm, the crack conductance increases considerably despite the reduction in $V_f^C$. This behavior can be contributed to increased bridging angles for short spanning lengths, as well as the reduced amount of crack opening due to the increased stiffness of the traction-separation response. The increase in cohesive strength with smaller $l_0$ values explains the delay in the onset of buckling delamination with reduced $l_0$. The resulting larger heat flux values lead to lower thermal stresses, and lower values of $G_T$ during the bending dominated region and a delay in the onset of the buckling delamination.

### 4.3 Application to Non-Destructive Evaluation

In consideration concerned with crack growth in fiber reinforced composite materials crack bridging plays a major role. Fibers extending across a crack or delamination provide partial load transfer across the crack, thus reducing the crack opening displacement as well as the driving force towards further crack extension. In general, bridged cracks are internal defects, and thus difficult to investigate by visual inspection methods. It has been recognized that changes in the transport properties of the composite under investigation can be used to detect such defects. Electrical resistance changes, and photothermal techniques – as discussed in the present investigation – have been suggested to this purpose. For such methods to be successful for the investigation of bridged cracks, changes in the crack conductance are to be related to the breakdown of crack bridging during mechanical loading.

For a non-destructive method to provide quantitative information about the actual the state of the bridging zone, models describing the thermal conductance of the cracked specimen are needed. For composites with bridged cracks correlations linking the crack conductance to the crack opening displacement and the number of fibers per area are required. As demonstrated already in this paper, the interdependence of these quantities is non-linear, and can also influenced by the actual geometry of the specimen under consideration. To improve the understanding of quantitative photothermal techniques for the characterization of bridged cracks and delaminations, numerical simulations of the nondestructive evaluation process are performed. At the center of these computations is the thermomechanical crack bridging which is applied in simulations of the photothermal imaging of a composite laminate with a bridged crack.

The specimen considered is a notched four-point bend specimen in which the presence of an initial bridged delamination is assumed. The specimen has a thickness of 2 mm, and the initial crack was located 0.422 mm from the top surface of the laminate. The length of the specimen was 30 mm and the crack length $2a$=16 mm. Specimen geometry and loading are described in Figure 16. The reinforcements are again aligned parallel to the delamination.
A total of twenty thermal cycles were computed, with a steady state reached after approximately 10s. The temperature response, $\theta^*$, recorded at the location remote from the crack, $x/a$=1.5, was used as the reference signal. The quantity $(\theta' - \theta_0)/(\theta^* - \theta_0)$, with $\theta_0$ the initial temperature of the specimen, was used to characterize the thermal response of the specimen.

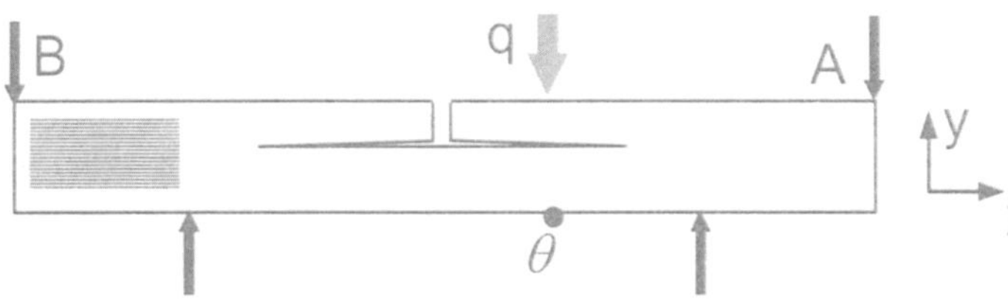


**Figure 16.** Specimen geometry used in the simulation of the photothermal imaging of bridged delamination cracks.

Figure 17 summarizes the results of the computations. The crack opening reaches a maximum at the center of the specimen and decreases as the crack tip is approached. Simultaneously, the fiber survival probability increases from the center of the specimen to the crack tip. For the amount of loading considered, 25% of the initial fibers are retained at the center of the specimen. Only a small change in $p_s$ occurs up to $x/a$=0.25. Then, in the interval $0.25<x/a<0.9$ the fiber survival probability increases rapidly to a value close to one. Important differences exist between the thermal response variable and the fiber survival probability. The normalized temperature amplitude remains nearly constant up to $x/a$=0.5, then increases rapidly until $x/a$. A value close to one is only reached at $x/a$<1.25, i.e. outside of the actual crack.

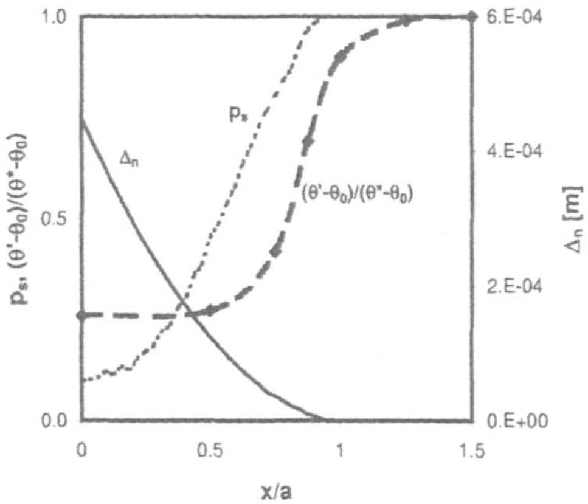


**Figure 17.** Predicted spatial dependence of crack opening, $\Delta_n$, fiber survival probability, $p_s$, and corresponding normalized amplitude of the temperature response $(\theta' - \theta_0)/(\theta^* - \theta_0)$. Applied displacement $u_{y,A}$ =–5×10$^{-4}$ m.

From the simulation it is evident that the predicted – and potentially a measured - temperature response cannot be directly related to the number of crack bridging fibers. The simulations predict some qualitative agreement between temperature response and fiber survival probability. To use the photothermal imaging techniques in a quanitative way, effects due to the details in the coupling between the mechanical and thermal part of the problem need to be accounted for. As loading increases, not only does the crack opening increases, but also the number of surviving fibers decreases. This quantity, however, depends also on the shear separation in the crack. The predictions - and potentially the measurements – of the photothermal response thus become dependent on the mixed mode loading conditions.

# Acknowledgements

The financial support of Sandia National Laboratories through the NSF-Sandia LCE program is gratefully acknowledged.

# References

A. de-Andres, J.L. Perez, and M. Ortiz. Elastoplastic finite element analysis of three-dimensional fatigue growth in aluminum shafts subjected to axial loading. *International Journal of Solids& Structures* 36:2231–2258, 1999.

J.R. Barber. Steady-state thermal stresses caused by an imperfectly conducting penny shaped crack in an elastic solid. *Journal of Thermal Stresses* 3:77–83, 1979.

J.R. Barber and M. Comninou. The penny-shaped interface crack with heat flow: 2. Imperfect contact. *Journal of Applied Mechanics* 50:770–776, 1983.

G.I. Barenblatt. The mathematical theory of equilibrium cracks in brittle fracture, *Advances in Applied Mechanics* 7:55–129, 1962.

L. Budde and S.M. Gao. Dynamic thermal stress in metal-adhesive bonds. *Journal of Adhesion Science and Technology* 6:1189–1204, 1992.

G.T. Camacho and M. Ortiz. Computational modeling of impact damage in brittle materials. *International Journal of Solids & Structures* 33:2899–2938, 1996.

C.K. Chao and R.C. Chang. Analytical solutions and numerical examples for thermoelastic interface crack problems in dissimilar anisotropic media. *Journal of Thermal Stresses* 17:285–299, 1994.

J.L. Chaboche, R. Girad, and A. Schaff. Numerical analysis of composite systems by using interphase/interface models. *Computational Mechanics* 20:3–11, 1997.

W.H. Chen and C.C. Huang. Three-dimensional thermal analysis of an infinite solid containing an elliptical surface crack. *International Journal of Fracture* 54:225–234, 1992.

D. Coker, A.J. Rosakis, and A. Needleman. Dynamic crack growth along a polymer composite–Homalite interface, *Journal of the Mechanics and Physics of Solids* 51:425–460, 2003.

Columbia Accident Investigation Board, 2003, www.caib.org.

D.S. Dugdale. Yielding in steel sheets containing slits, *Journal of the Mechanics and Physics of Solids* 8:100–104, 1960.

D.C. Freeman, T.A. Talay, and R.E. Austin. Reusable launch vehicle technology program, *Acta Astronautica* 41:777–790, 1997.

A.L. Florence and J.N. Goodier. Thermal stresses at spherical cavities and circular holes in uniform heat flow. *Journal of Applied Mechanics* 26:293–294, 1959.

A.L. Florence and J.N. Goodier. The linear thermoelastic problem of uniform heat flow disturbed by a penny-shape insulated crack. *International Journal of Engineering Sciences* 1:553–540, 1963.

H. Gao and P. Klein. Numerical simulation of crack growth in an isotropic solid with randomized internal cohesive bonds. *Journal of Mechanics and Physics of Solids* 46:187–218, 1998.

N. Hasebe, K. Tamai, and T. Nakamura. Analysis of kinked crack under uniform heat-flow. *Journal of Engineering Mechanics* 112:31–42, 1986.

A. Hattiangadi and T. Siegmund. *Journal of Mechanics and Physics of Solids,* in press, 2004.

P.H. Geubelle and J.S. Baylor. Impact-induced delamination of composites: a 2D simulation, *Composites Part B: Engineering* 29:589–602, 1998.

J.W. Hutchinson and A.G. Evans. On the delamination of thermal barrier coatings in a thermal gradient. *Surface and Coatings Technology* 149:179–184, 2002.

J.W. Hutchinson and T.J. Lu. Laminate delamination due to thermal gradients. *Journal of Engineering Materials and Technology* 117:386–390, 1995.

S. Itou. Thermal stresses around a crack in an adhesive layer between two dissimilar elastic half-planes. *Journal of Thermal Stresses* 16:373–400, 1993.

D.A.W. Kaute, H.R. Shercliff, and M.F. Ashby. Delamination, fiber bridging and toughness of ceramic matrix composites. *Acta Metallurgica et Materialia* 41:1959–1970, 1993.

P. Klein and H. Gao. Crack nucleation and growth as strain localization in a virtual-bond continuum. *Engineering Fracture Mechanics* 61:21–48, 1998.

A.Y. Kuo. Effects of crack surface heat conductance on stress intensity factors. *Journal of Applied Mechanics* 57:354–358, 1990.

K.Y. Lee and C.W. Shul. Determination of thermal stress intensity factors for an interface crack under vertical heat-flow. *Engineering Fracture Mechanics* 40:1067–1074, 1991.

K.Y. Lee and S.J. Park. Thermal-stress intensity factors for partially insulated interface crack under uniform heat-flow. *Engineering Fracture Mechanics* 50:475–482, 1995.

Y.D. Lee and F. Erdogan. Interface cracking of FGM coatings under steady-state heat flow. *Engineering Fracture Mechanics* 59:361–380, 1998.

T.J. Lu and J.W. Hutchinson. Effect of matrix cracking on the overall thermal conductivity of fiber-reinforced composites. *Philosophical Transactions of the Royal Society of London A* 351:595–610, 1995.

C.J. Martin-Moran, J.R. Barber, M. Comninou. The penny-shaped interface crack with heat flow: 1. Perfect contact. *Journal of Applied Mechanics* 50:29–36, 1993.

K.R. McDonald, J.R. Dryden, A. Majumdar, and F.W. Zok. Thermal conductance of delamination cracks in a fiber-reinforced ceramic composite. *Journal of the American Ceramic Society* 83:553–562, 2000.

B.B. Mikic. Thermal contact conductance: Theoretical considerations, *International Journal of Heat and Mass Transfer* 17:205–214, 1974.

T.C. Miller and R. Chona. Finite element analysis of a thermally loaded interface crack in a ceramic coating. *Engineering Fracture Mechanics* 59:203–214, 1998.

A. Needleman. A continuum model for void nucleation by inclusion debonding. *Journal of Applied Mechanics* 54:525–531, 1987.

A. Needleman. An analysis of decohesion along an imperfect interface. *International Journal of Fracture* 42:21–40, 1990a.

A. Needleman. An analysis of tensile decohesion along an interface. *Journal of Mechanics and Physics of Solids* 38:289–324, 1990b.

A. Pandolfi, P.R. Guduru, M. Ortiz, and A.J. Rosakis. Three dimensional cohesive-element analysis and experiments of dynamic fracture in C300 steel. *International Journal of Solids & Structures* 37:3733–3760, 2000.

P. Papadopoulos, E. Venkatapathy, D. Prabhu, M.P. Loomis, and D. Olynick. Current grid-generation strategies and future requirements in hypersonic vehicle design, analysis and testing. *Applied Mathematical Modeling* 23:705–735, 1999.

G. Qian, T. Nakamura, C.C. Berndt, and S.H. Leigh. Tensile toughness test and high temperature fracture analysis of thermal barrier coatings. *Acta Materialia* 45:1767–1784, 1997.

G. Qian, T. Nakamura, and C.C. Berndt. Effects of thermal gradient and residual stresses on thermal barrier coating fracture. *Mechanics of Materials* 27:91–110, 1998.

P. Rahul-Kumar, A. Jagota, S.J. Bennison, and S. Saigal. Interfacial failures in a compressive shear strength test of glass/polymer laminates, *International Journal of Solids and Structures* 37:7281–7305, 2000.

G.C. Sih. On the singular character of thermal stresses near a crack tip. *Journal of Applied Mechanics* 29:587–589, 1962.

T. Siegmund and A. Needleman. A numerical study of dynamic crack growth in elastic viscoplastic solids, *International Journal of Solids & Structures* 34:769–787, 1997.

T. Siegmund and W. Brocks. Prediction of the work of separation and implications to the modeling. *International Journal of Fracture* 99:97–116, 1999a.

T. Siegmund and W. Brocks. The role of cohesive strength and separation energy for modeling of ductile fracture. *American Standards for Testing of Materials STP 1360*, Editors: K. Jerina, P.C. Paris, pp. 139–151, 1999b.

J. Sumrall, C. Lane, and R. Cusic. Venturestar$^{TM}$: reaping the benefits of the X-33 program. *Acta Astronautica* 44:727–736, 1999.

V. Tvergaard and J.W. Hutchinson. The relation between crack growth resistance and fracture process parameters in elastic-plastic solids. *Journal of Mechanics and Physics of Solids* 40:1377–1397, 1993.

V. Tvergaard and J.W. Hutchinson. Effect of T-stress on mode I crack growth resistance in a ductile solid. *International Journal of Solids & Structures* 31:823–833, 1994.

R. Tivey, T. Ando, A. Antipenkov, V. Barabash, S. Chiocchio, G. Federici, C. Ibbott, R. Jakeman, G. Janeschitz, R. Raffray, M. Akiba, I. Mazul, H. Pacher, M. Ulrickson, and G. Vieider. ITER divertor, design issues and research and development. *Fusion Engineering Design* 46:207–220, 1999.

G. Yan and T.C.T. Ting. The r(-1/2)ln( r) singularity at interface cracks in monoclinic and isotropic bimaterials due to heat-flow. *Journal of Applied Mechanics* 60:432–437, 1993.

X.P. Xu and A. Needleman. Numerical simulations of fast crack growth in brittle solids. *Journal of Mechanics and Physics of Solids* 42:1397–1434, 1994.

# Deformation and Damage in Particle-Reinforced Composites: Experiments and Models

Javier LLorca

Department of Materials Science, Polytechnic University of Madrid.
E. T. S. de Ingenieros de Caminos, 28040 - Madrid, Spain

**Abstract.**

The optimization of the mechanical properties of composites requires a detailed simulation of the complex deformation and failure mechanisms which arise in these heterogeneous microstructures upon loading. It has been shown that damage in the form of particle fracture, interface decohesion or ductile matrix failure controls many important mechanical properties, and this has led to the development of micromechancial models which include damage. This article reviews the current progress in this area within the framework of the two leading simulation approaches: the homogeneization techniques and the finite element analysis of three-dimensional multiparticle unit cells. Although the simulation tools presented are general, they are focused on analysing the behavior of metal- or polymer-matrix composites reinforced with stiff and brittle ceramic particles.

## 1 Introduction

Composites are traditionally divided, attending to the reinforcement geometry, into fiber and particle-reinforced materials. Long fibers constitute the reinforcing phase in the former, while equiaxed particles, platelets, and whiskers are dispersed within the matrix in the latter. They present marked differences in processing and performances, and are normally studied separately. From the viewpoint of mechanical behavior, fiber-reinforced composites (FRC) take full advantage of the outstanding fiber properties (stiffness, strength) and exhibit the best mechanical properties in the fiber directions. Their behavior is, however, very anisotropic and FRC are normally used in the form of lamina (flat or curved) due to the problems associated with the manufacturing of three-dimensional fiber preforms. FRC require sophisticated processing routes as compared to the unreinforced counterparts, and their use is restricted to high added value applications where the improvement in performances justifies the cost. On the contrary, bulk shapes can be produced with particle-reinforced composites (PRC), which can be manufactured by standard processing techniques, reducing significantly the cost. Their properties are not as good as those obtained in FRC but they exhibit a more isotropic behavior and are widely used in the building, automotive, and petro-chemical industries, where moderate improvements in performances at affordable costs are looked for.

PRC based on metallic matrices normally use ceramic particles ($Al_2O_3$, SiC) or whiskers (SiC), which increase the material stiffness, wear and creep resistance, and to a minor extent, the strength. In the case of polymer-based composites, epoxy resins as well as thermoplastics are often loaded with stiff particles (silica, glass, $Al_2O_3$) to modify some physical properties (viscosity, color, density, etc.) or to reduce the shrinkage associated with curing. The strain incompatibility between the ductile matrix and the rigid particles generates important stress concentrations around the reinforcements in both metal and polymer-matrix composites, which lead to the nucleation of damage in the form of matrix, reinforcement or interface fracture, and to the early failure of the composite, limiting the actual composite strength to levels well below the ideal ones.

During the last decades, the micromechanical modelling of PRC was focused on the analysis of the linear and non-linear deformation modes which arise upon loading the heterogeneous microstructures. This approach provided a quantitative assessment of the influence of microstructural factors on the linear and non-linear behavior and supplied constitutive equations which may be used in commercial finite element codes to compute the response of structural elements subjected to complex loading conditions. However, realistic predictions of the ultimate strength of PRC should include the micromechanisms of damage. These notes review the recent developments in this area and are focused on polymer- and metal-matrix composites reinforced with stiff and brittle ceramic particles. After the introduction, the first section is devoted to the experimental evidence on the deformation and damage processes in these PRC. The basic micromechanical models to simulate reinforcement fracture, interface decohesion and matrix failure are presented afterwards. The two following sections are devoted to the simulation techniques, the first one within the framework of homogeneization models and the second one through the finite element analysis of three-dimensional multiparticle cells. A brief comment on the advantages and limitations of each methodology is presented in the final section.

Throughout the paper, bold lowercase greek letters stand for second order tensors and bold capital letters for fourth order tensors. In addition, the different products are expressed as $(\mathbf{A}:\boldsymbol{\sigma})_{ij} = \mathrm{A}_{ijkl}\sigma_{kl}$, $\mathbf{A} :: \mathbf{B} = \mathrm{A}_{ijkl}\mathrm{B}_{ijkl}$, and $(\boldsymbol{\sigma}\otimes\boldsymbol{\sigma})_{ijkl} = \sigma_{ij}\sigma_{kl}$. Finally, $\bar{\boldsymbol{\sigma}}$ represents the volumetric average of $\boldsymbol{\sigma}$.

## 2 Experimental Results

Damage initiation in PRC during deformation has been studied by various techniques, such as *in situ* optical or scanning electron microscopy of the specimen surfaces (Moloney et al., 1987; LLoyd, 1991; Brechet et al., 1991; Singh and Lewandowski, 1993; Maire et al., 1995), *post mortem* analyses of longitudinal sections (Cantwell and Roulin-Moloney, 1989; Friedrich and Karger-Kocsis, 1989; Hunt et al., 1991; LLorca et al., 1993; Whitehouse and Clyne, 1993; Zhao et al., 1994; Poza and LLorca, 1999), and X-ray tomography (Mummery et al., 1993; Buffiere et al., 1999; Babout et al., 2001).

The aforementioned investigations showed that damage was always nucleated at the ceramic reinforcements. The main mechanisms of void nucleation were particle fracture and decohesion at the matrix/reinforcement interface. Particle fracture was the dominant damage process in metal-matrix PRC at ambient temperature (LLoyd, 1991; Singh and Lewandowski, 1993; LLorca et al., 1993; Poza and LLorca, 1999). These materials usually contain 10 to 20 vol. % of ceramic particles ($Al_2O_3$, SiC) of an average size in the range 5 to 20 $\mu$m embedded in a medium-to-high strength metallic matrix. Reinforcement fracture during tensile deformation occurred as a result of the load transfer from the matrix to the ceramic particles, which were broken by one crack perpendicular to the loading axis although sometimes two or more parallel cracks were observed in elongated particles oriented in the loading direction (Figure 1). Quantitative microscopy analyses showed that large particles were more prone to fail than small ones (Figure 2). Particles whose average size (understood as the square root of the area) was below 5 $\mu$m were rarely fractured, while reinforcements larger than 17 $\mu$m were always broken. In addition, the fracture probability was higher for elongated reinforcements (low $D_{min}/D_{max}$ in Figure 2) than for equiaxed ones. It was also found that particle fracture occurred earlier during deformation in clusters of particles, where the local volume fraction of reinforcements was very high (see, for instance, LLorca et al. (1993) and Poza and LLorca (1999)).

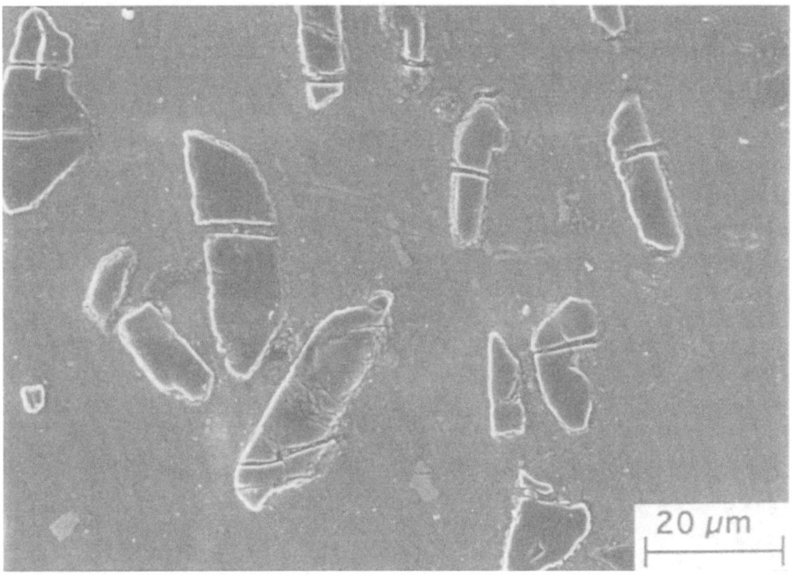


**Figure 1.** Damage nucleation by particle fracture during tensile deformation in a 2014 Al alloy reinforced with 15 vol. % of $Al_2O_3$ particles. The loading axis was vertical.

Damage initiation by decohesion at the matrix-reinforcement interface, as opposed to reinforcement fracture, was reported in several metal-matrix PRC. For instance, whisker-reinforced composites with the reinforcements aligned in the loading direction tended to nucleate voids at the whisker ends (Nutt and Duva, 1986). Similarly, particle-reinforced materials often nucleated voids at the ends of elongated reinforcements oriented parallel to the stress axis when the composite was made up of a very low strength alloy, such us commercially pure Al (Whitehouse and Clyne, 1993). The various decohesion patterns experimentally found are depicted in Figure 3, and it is worth noting that the ends of elongated reinforcements oriented parallel to the stress axis were favored sites for decohesion. In addition, the flattish edges of the jagged particles aligned normal to the

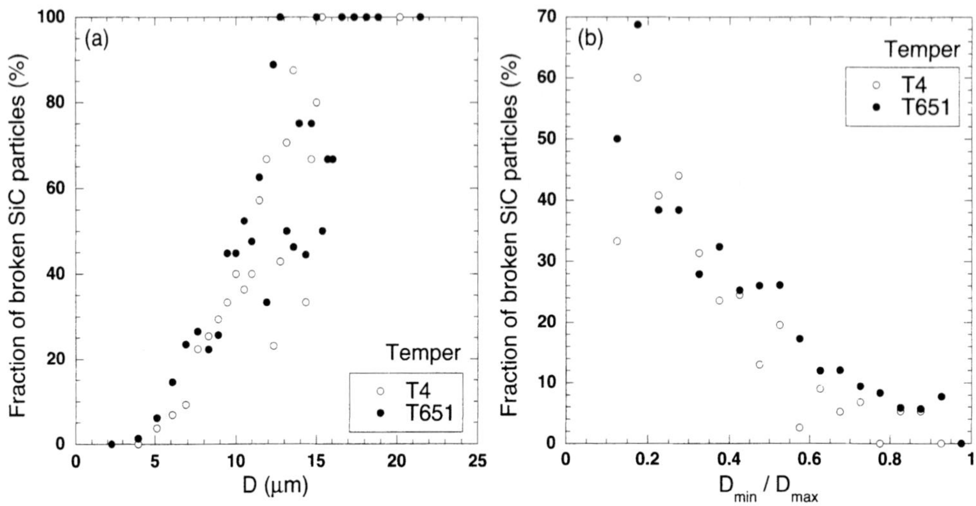


**Figure 2.** Fraction of SiC particles broken during tensile deformation in a 2618 Al alloy reinforced with 15 vol. % of SiC particles. (a) Influence of the average particle size, $D$. (b) Influence of the the particle aspect ratio, $D_{min}/D_{max}$. The composite heat treatment is shown in the plots.

stress axis led to decohesion earlier than in the case of rounded reinforcements, which also experienced decohesion at the polar regions. The experimental results also indicated that void nucleation started at lower strains at the end of short fibers and that the number of decohered particles at any given strain level was higher for the angular particles than for the rounded ones.

The prevalence of reinforcement fracture or interface decohesion as the dominant damage mechanism depends on the magnitudes of the reinforcement and interface strength. For instance, Downes and King (1991) analyzed the effect of reinforcement size on the void nucleation mechanism in an 8090 Al alloy reinforced with 20 vol. % SiC particles whose average sizes were 3, 6, 13 and 23 $\mu$m. Interface decohesion was dominant in the composites with 3 and 6 $\mu$m average particle sizes, whereas reinforcement fracture was prevalent in the composite with an average particle size of 23 $\mu$m. Evidently, the particle strength decreased with size while the interface strength was insensitive to this factor. Void nucleation was also seen to change from particle fracture to failure near to the interface with severe overaging, which weakened the interfacial bonding (Manoharan and Lewandowski, 1991) or during high temperature testing (Poza and LLorca, 1996). In this latter case, the load transferred to the reinforcement by the metallic matrix decreases with the matrix strength at high temperature, and the reinforcement stresses are not high enough to induce fracture. Although the interface stresses are also reduced, the build-up of plastic strains at the interface led finally to decohesion.

The micromechanisms of void nucleation in epoxy resins reinforced with rigid parti-

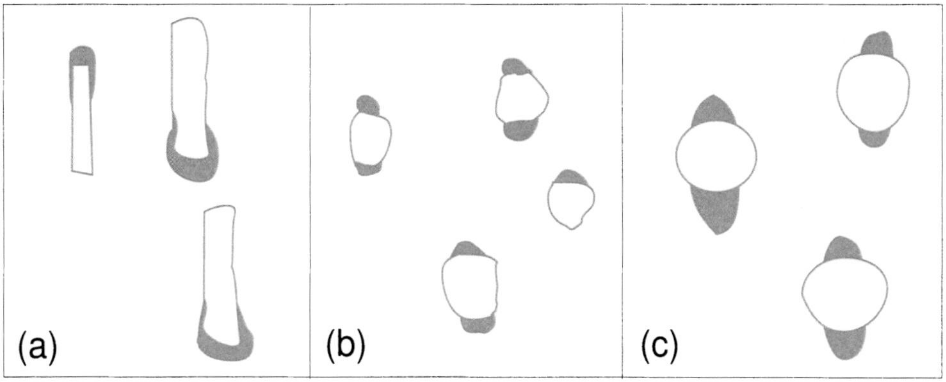


**Figure 3.** Matrix/reinforcement decohesion patterns under uniaxial tensile deformation for (a) short fibers. (b) angular particles and (c) rounded reinforcements. Loading axis is vertical.

cles were studied by Moloney et al. (1987), Cantwell and Roulin-Moloney (1989), and by Friedrich and Karger-Kocsis (1989) in short-fiber reinforced thermoplastics. They found a similar behavior: damage occurred by either reinforcement fracture or interface decohesion. As the matrix ductility in these materials is very limited, the dominant mechanism was determined by the relative magnitude of the reinforcement and interface strength. The use of coupling agents (such as silane) led to materials which failed by particle fracture, and conversely, composites treated with silicone-based mould-release agents nucleated voids at the interface during deformation. Likewise, weak fillers, such as aluminum trihydroxide, dolomite or hollow glass spheres promoted void nucleation by reinforcement fracture.

It should be noted finally that the void nucleation process, whether by reinforcement fracture or interface decohesion, began at the onset of yielding and continued during the plastic deformation of the composite. In the case of reinforcement fracture, the void nucleation rates were measured in several Al-matrix composites using various techniques (see, for instance, (Hunt et al., 1991; Whitehouse and Clyne, 1993; Zhao et al., 1994; Liu and Lewandowski, 1993; LLorca and Poza, 1993; Maire et al., 1995; Poza and LLorca, 1996) and it was found that the particle fracture rate increased in an approximately linear fashion with the applied plastic strain (Figure 4). The experimental data of void nucleation by interface decohesion were more scarce (Zhao et al., 1994; Maire et al., 1995; Babout et al., 2001) but also indicated that the fraction of decohered reinforcements increased linearly with the applied strain in all cases.

PRC experience substantial damage in the form of reinforcement fracture or interface decohesion prior to failure. Void nucleation is initially distributed homogeneously throughout the gage length upon tensile loading but localization develops sooner or later at a given section of the composite. Evidence of this behaviour was provided by *post mortem* analysis of fractured samples which found that the fraction of broken reinforcements was always greatest near the fracture surface (Hunt et al., 1991; LLorca et al.,

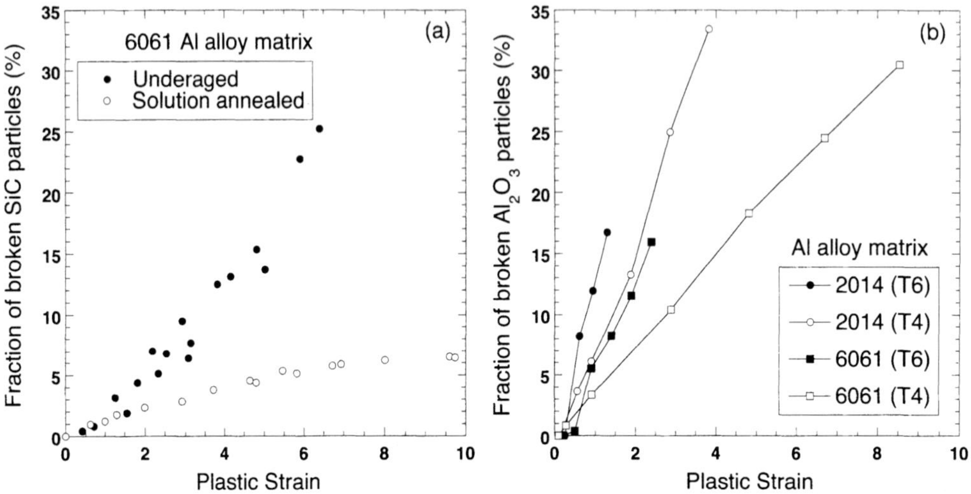


**Figure 4.** Fraction of broken reinforcements as a function of the applied plastic strain during tensile deformation. (a) 6061 alloy reinforced with 15 vol. % SiC particles (Liu and Lewandowski, 1993). (b) 6061 and 2014 Al alloys reinforced with 15 vol. % $Al_2O_3$ particles. The average reinforcement size was 12 $\mu$m (LLorca and Poza, 1993).

1993; Singh and Lewandowski, 1993; Maire et al., 1995), as shown in Figure 5. In addition, results obtained by means of *in situ* quantitative microscopy (LLoyd, 1991; Zhao et al., 1994) demonstrated that damage localization may begin very early during deformation and is not associated with the onset of plastic instability, which concentrates the deformation in a given section of the composite. On the contrary, plastic instability followed by the specimen fracture seemed to come about as a result of the reduction in strain hardening rate induced by the concentration of damage in a particular section.

The final failure of PRC takes place by the sudden coalescence through the matrix of the voids nucleated at the reinforcements. Although this result is general, the physical mechanism of matrix rupture depends very much on the microstructural features and nature (metal or polymer) of the matrix as well as on the test temperature and loading conditions. The analysis of the fracture surfaces and of longitudinal sections near the fracture surface showed the most characteristic fracture modes found in metal-matrix composites (Klimowicz and Vecchio, 1990; LLorca and Poza, 1993; Hadianfard et al., 1994). On the one hand, if the metallic matrix was free from inclusions, dispersoids, precipitates or any other void nucleating particles, void coalescence took place by the localization of the plastic deformation in the intervoid matrix until the matrix is necked to a knife between the voids. The fracture surfaces are characterized by only one population of large dimples, and either a broken or decohered particle can be found at the center of each dimple. The matrix material in the void shows evidence of extensive plastic deformation by shear and is often drawn out to form a sharp shear lip. On the

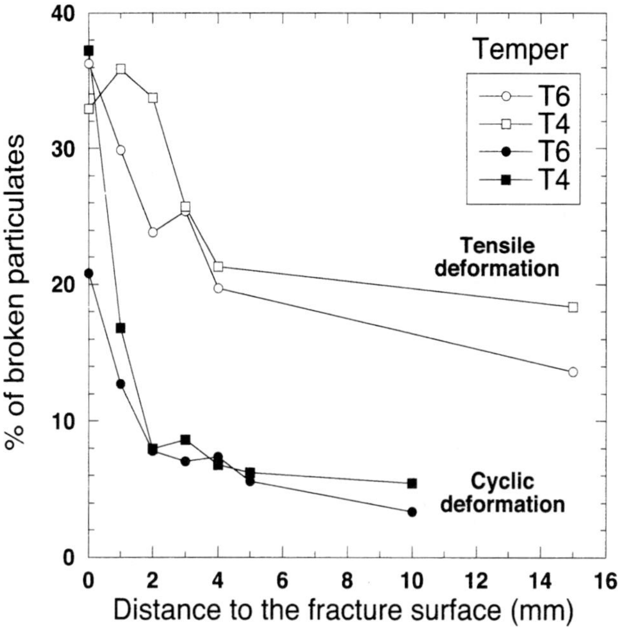


**Figure 5.** Fraction of broken ceramic particles as a function of the distance to the fracture surface in a 2014 Al alloy reinforced with 15 vol. % $Al_2O_3$ particles tested in monotonic tension and cyclic deformation. The open symbols correspond to the monotonic tensile tests whereas the solid symbols stand for specimens tested under fully-reversed cyclic deformation (Poza and LLorca, 1995).

other hand, any second-phase particles dispersed in the matrix nucleated microvoids as plastic deformation became localized in the matrix between the fractured or decohered reinforcements. Void coalescence occurred by the progressive growth and impingement of these microvoids, and the fracture surface showed two or more dimple populations of different diameter. The largest dimples formed around the ceramic reinforcements were linked by smaller ones formed around micron-sized intermetallic inclusions. These were in turn surrounded by submicron voids nucleated by dispersoids and precipitates. Evidently, this mechanism of void coalescence was reported in alloys aged to the peak strength or overaged, which contain large volume fractions of void nucleating particles in the matrix.

Finally, grain boundary cracking may also contribute to matrix failure. For instance, this mechanism was reported by Singh and Lewandowski (1993) in an overaged Al-Zn-Mg-Cu alloy reinforced with SiC particles and by Poza and LLorca (1999) in a peak-aged Al-Li alloy also reinforced with SiC particles. It is evident that grain boundary fracture during the last stages of deformation may be favoured if grain boundaries are embrittled during high temperature ageing or it may simply occur because the stress concentrations induced by dislocation pile-ups surpass the grain boundary strength.

## 3 Micromechanics of Fracture Processes

The experimental results have demonstrated that failure in PRC is due to three different mechanisms, which may be activated sequentially or in parallel: particle fracture, interface decohesion and matrix failure. The modelling the deformation and fracture processes in PRC is based upon realistic micromechanical models of these processes, which are presented below.

### 3.1 Particle fracture

The experimental observations indicated that the brittle particles failed in tension as a result from the load transfer from the matrix during deformation. The tensile strength of brittle materials is dictated by the largest flaw in the material, and it can be analyzed through a stress criterion based on the weakest link statistics. Each reinforcement is divided into $N$ elements whose volume is $V_i$. Let $n_\sigma$ be the number of defects per unit volume which lead to the unstable propagation of a crack under a maximum principal tensile stress $\sigma^I$. The failure probability of element $i$ is given by $F_i = n_\sigma V_i$. The reinforcement will fail when one element fails and $F_i$ would be small if the volume of each element is very small. Then, the survival probability, $\tilde{F}$, can be expressed as

$$\tilde{F} = \prod_{i=1}^{N}\Big[1 - n_\sigma V_i\Big] \approx \exp\Big\{-\sum_{i=1}^{N}\Big[n_\sigma V_i\Big]\Big\} \tag{1}$$

Weibull (1951) proposed an expression for $n_\sigma$ as,

$$n_\sigma = \frac{1}{V_0}\left[\frac{\sigma_i^I}{\sigma_0}\right]^m \tag{2}$$

where $\sigma_i^I$ is the maximum principal tensile stress acting on the element $i$, and $\sigma_0$, $V_0$ and $m$ are, respectively, the characteristic strength, the characteristic volume, and the Weibull modulus of the material. The particle fracture probability, $F$, is thus given by

$$F = 1 - \exp\left\{-\frac{V}{V_0}\left[\frac{\sigma_p^I}{\sigma_0}\right]^m\right\} \tag{3}$$

when the maximum tensile stress on the particle is constant and equal to $\sigma_p^I$.

The Weibull model has been widely used to characterize the tensile strength of ceramics and it was introduced by Wallin et al. (1987) to study the fracture strength of brittle particles embedded in ductile matrices (CuO in Cu, cementite in C steel, and Si in Al-Si alloy). Detailed quantitative microscopy studies verified the applicability of the Weibull statistics to model damage nucleation by particle fracture in $Al_2O_3$ (Mochida et al., 1991) and SiC-reinforced (LLorca et al., 1993; LLorca, 1995; Lewis and Withers, 1995) Al-matrix composites. It was found that the Weibull modulus of the particle was low (in the range $3 < m < 6$) and the dispersion in the reinforcement strength was partially attributed to the fact that the ceramic particles used as reinforcements are polycrystals manufactured by attrition which present a wide distribution of the defects. In addition, particles with very different aspect ratio, shape and orientation are dispersed in the matrix and all these factors influence the stresses borne by the reinforcements and the corresponding failure probability. Estimations of the reinforcement characteristic strength, $\sigma_0$, were difficult for these reasons but several independent studies on SiC particles embedded in Al matrices led to values in the range 1.0 to 1.8 GPa (LLorca, 1995; Lewis and Withers, 1995; Poza and LLorca, 1999).

### 3.2 Interface decohesion

The process of failure by interface decohesion involves the nucleation of a crack (usually from a preexisting defect) and its propagation along the reinforcement/matrix interface. These phenomena are equivalent to those found during reinforcement fracture, but while reinforcement fracture is a brittle process which leads to the complete fracture of the particle, cracks nucleated at the reinforcement/matrix interface can be arrested after a certain amount of crack propagation, leading to partial decohesion and increasing the complexity of the problem.

The nucleation of voids by interface decohesion at second phase particles was extensively studied in metallic alloys (Van Stone et al., 1985; Thomason, 1990). A necessary condition for interface decohesion (as well as for particle fracture) is that the elastic strain energy released during decohesion should be enough to create the new free surfaces at the interface. This energy criterion is commonly fulfilled but it is a sufficient condition only for particles of a few tenth of nm or smaller (Goods and Brown, 1979). Either a critical normal stress or a critical strain at the interface (or a combination of both) was usually taken as valid criteria for microvoid nucleation at the interface for larger particles. Both criteria were extensively used to analyze the interfacial decohesion in different systems (precipitate strengthened Al alloys, oxide dispersion strengthened copper, spheroidized and pearlitic steels, etc.) and a detailed discussion of their suitability to model interface decohesion in metallic alloys can be found in the cited references. For the purposes of this paper, it is enough to indicate that the predictions of void nucleation by interface decohesion based on these hypotheses were in reasonable agreement with the experimental observations.

More rigorous approximations to the study of interface decohesion in PRC were proposed by Needleman (1987). Crack nucleation and propagation at the interface was analyzed through a cohesive crack model, which is able to transmit normal and tangential stresses between the crack surfaces. The normal ($\sigma_n$) and shear ($\sigma_t$) interfacial stresses are derived from an elastic potential $\phi_{in}$, which depends only on the normal ($\delta_n$) and tangential ($\delta_t$) displacement difference across the interface. In two dimensions $\phi_{in}$ is expressed as,

$$\phi_{in} = \frac{27\sigma_{in}\delta_c}{4}\left[\frac{1}{2}\left(\frac{\delta_n}{\delta_c}\right)^2\left[1-\frac{4}{3}\frac{\delta_n}{\delta_c}+\frac{1}{2}\left(\frac{\delta_n}{\delta_c}\right)^2\right]+\frac{\gamma}{2}\left(\frac{\delta_t}{\delta_c}\right)^2\left[1-2\frac{\delta_n}{\delta_c}+\left(\frac{\delta_n}{\delta_c}\right)^2\right]\right] \quad (4)$$

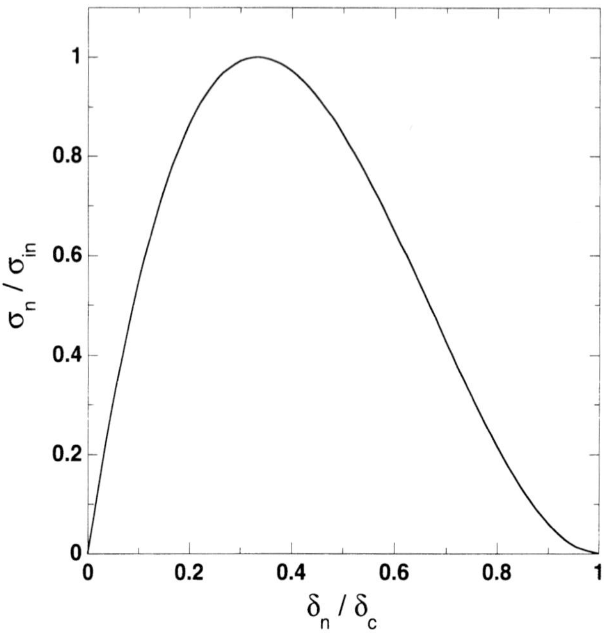


**Figure 6.** Normal stress transmitted through an interface crack, $\sigma_n$ (normalized by the interface strength, $\sigma_{in}$) as a function of the normal displacement across the interface, $\delta_n$ (normalized by the critical crack opening displacement, $\delta_c$) when the tangential displacements are zero.

for $\delta_n \leq \delta_c$. The stresses transmitted through the cohesive crack can be computed as

$$\sigma_n = -\frac{\partial \phi_{in}}{\partial \delta_n} \qquad \sigma_t = -\frac{\partial \phi_{in}}{\partial \delta_t} \tag{5}$$

for $\delta_n \leq \delta_c$ and $\sigma_n = \sigma_t = 0$ for $\delta_n \geq \delta_c$. The dependence of the normal stress, $\sigma_n$, with the normal displacement across the interface, $\delta_n$, is shown in Figure 6 for the particular case of $\delta_t = 0$. The stresses transmitted through the crack increase up to a maximum given by $\sigma_{in}$, the maximum normal stress supported by the interface when the tangential displacements are negligible, and then decrease to become zero at $\delta_n = \delta_c$. The interface toughness or fracture energy is given by $\Gamma_{in} = \phi_{in}(\delta)$, the area under the $\sigma_n$ - $\delta_c$ curve. The critical crack opening displacement, $\delta_c$, at which the cohesive stresses vanish is given by,

$$\delta_c = \frac{16}{9} \frac{\Gamma_{in}}{\sigma_{in}} \tag{6}$$

The last parameter in the model is $\gamma$, which is of secondary importance for tensile-dominated decohesion and is usually taken as equal to 1.

This constitutive model was used to analyze interface decohesion within the context of axisymmetric unit cell calculations by Nutt and Needleman (1987). They studied the influence of the interfacial strength, $\sigma_{in}$, on void formation in a 6061 Al matrix composite reinforced with 20 vol. % of SiC whiskers with an aspect ratio equal to 4. In agreement with the decohesion patterns experimentally observed, they found that the crack propagated through the whole whisker end in materials with a low strength interface. As the interface strength increased, decohesion was confined to a small region near the whisker corner where the stresses and strain were maxima, but the relaxation induced by the void nucleation prevented the propagation of the crack along the interface. Moreover, the numerical analysis indicated that interface decohesion was more sensitive to the interfacial strength than to the toughness (Tvergaard, 1990), and thus that a strength criterion could be a good approximation to this problem in many cases.

### 3.3 Ductile matrix failure

Final rupture in PRC always occurs through the matrix, the only continuous phase. The experimental evidence presented above showed that matrix failure takes place through very different processes, such as shear fracture (in polymer-based composites), grain boundary cracking, or the nucleation, growth and coalescence of voids (in metal-based composites). This latter mechanism is by far the most important and several authors have applied it to PRC models originally developed to study the ductile tearing of metallic alloys containing void-nucleating particles.

A simple approximation is based on the Brown and Embury (1973) model, which assumed that adjacent voids grow by plastic extension to the point when their length is equal to their spacing. The ligaments are no longer subjected to lateral plastic constraint at this point (i.e. $45^0$ shear lines can be drawn) and the voids spontaneously coalesce by localized necking. If the strain increment during coalescence is negligible, the macroscopic strain required for void coalescence, $\epsilon^c$, is given by

$$\epsilon^c = \sqrt{\frac{\pi}{6f}} - \sqrt{\frac{2}{3}} - 1 \tag{7}$$

where the application to PRC assumes that $f$ stands for the fraction of reinforcement particles which have nucleated voids by either interface decohesion or particle fracture. An alternative model was developed by Whitehouse and Clyne (1993), who proposed that voids grow to a size dictated by the adjacent reinforcing particle and that void coalescence occurs when a certain fraction of the available cavitation sites have been activated.

More recently, Gammage (2002) proposed a new approximation based on the Thomason (1990) plastic limit-load failure model. When particle spacing is below a critical distance, the work hardening rate of the matrix locally is enhanced by the high hydrostatic stresses induced by the plastic constraint. The plastic constraint is lost, however, if the particles fracture, and the stress acting on the matrix between two cracked particles can exceed the critical value necessary to initiate localized plastic flow in the interparticle matrix. Assuming that the fracture particles behave as penny-shaped cracks of equivalent diameter and that no stress partitioning takes place between the two cracked halves of the particles and the matrix, the critical condition for interparticle matrix failure can be obtained by setting the far-field work hardening rate, $d\sigma/d\epsilon$, equal to the stress acting between two penny-shaped cracks of diameter $d$ (the broken particle diameter) separated a distance $\lambda$ (the interparticle distance). Mathematically,

$$\frac{d\sigma}{d\epsilon} = \sigma\left[1 + \alpha\sqrt{\frac{d}{\lambda}}\right] \tag{8}$$

where $\sigma$ stands for the far-field stress in the matrix, and $\alpha \approx 2$ takes into account the stress concentration between the cracks.

A unified description of the process of void nucleation, growth and coalescence in the matrix can be found in the Gurson model, which can be readily introduced into computational analyses. Assuming that $f$ is the volume fraction of voids in the matrix, the matrix behaviour can be described by the flow potential $\Phi$ given by Gurson (1977)

$$\Phi = \frac{\sigma_{eq}^2}{\bar{\sigma}^2} + 2q_1 f^* \cosh\left(\frac{3q_2\sigma_h}{2\bar{\sigma}}\right) - 1 - {q_1}^2 f^{*2} = 0 \tag{9}$$

where $\bar{\sigma}$ is the matrix flow stress, $\sigma_h$ the hydrostatic stress component, and $\sigma_{eq}$ the Von Mises equivalent stress. It should be noted that $\Phi$ becomes the Von Mises flow potential when $f^* = 0$. The parameters $q_1$ and $q_2$ were introduced to bring predictions of the model into closer agreement with full numerical analyses of a periodic array of voids Tvergaard (1981, 1982). In addition, the function $f^*(f)$ was introduced by Tvergaard and Needleman (1984) to account for the effects of rapid void coalescence at failure. It has the form

$$f^* = f \qquad \text{if} \quad f \le f_c \tag{10a}$$

$$f^* = \frac{f_u^* - f_c}{f_f - f_c}(f - f_c) \quad \text{if} \quad f \ge f_c \tag{10b}$$

where the constant $f_u^*$ is the value of $f^*$ at zero stress, i.e. $f_u^* = 1/q_1$. As $f \to f_f$, $f^* \to f_u^*$ and the material loses all stress carrying capacity.

Changes in the void volume fraction during deformation result from void nucleation as well as from the growth of existing voids.

$$df = df_{growth} + df_{nucleation} \tag{11}$$

and the growth of existing voids with deformation is determined from,

$$df_{growth} = (1 - f)\, d\bar{\epsilon}_v \tag{12}$$

where $d\bar{\epsilon}_v$ is the volumetric component of the plastic strain, which is different from zero for the Gurson potential. The void nucleation rate can be expressed as a function of the effective plastic strain, $\bar{\epsilon}$, as

$$df_{nucleation} = \frac{f_N}{S_N\sqrt{2\pi}} \exp\left[-\frac{1}{2}\left(\frac{\bar{\epsilon} - \epsilon_N}{S_N}\right)^2\right] d\bar{\epsilon} \tag{13}$$

where $f_N$ is the fraction of void nucleating particles in the matrix, $\epsilon_N$ stands for the average nucleation strain and $S_N$ for the corresponding standard deviation of the normal distribution.

# 4 Homogeneization Models

Homogeneization techniques were the workhorse to simulate the mechanical behavior of composites and multiphase materials before the arrival of digital computers. They are still widely used as a result of the simplicity of their basic assumptions and of their accuracy to describe the overall or "effective" composite behavior. The first attempts to describe the deformation of PRC including damage were carried out within the framework of very simple, unidirectional homogeneization models, which were aimed to describe the composite behaviour under uniaxial tension. More sophisticated approaches were later developed based on mean-field approximations (basically the Mori-Tanaka and the self-consistent schemes), and more recently, on ensemble-volume averaging procedures. They are presented in this section after a short introduction on the use of mean-field approximations in the context of non-linear deformations.

## 4.1 Brief introduction to mean-field approximations

Mean-field approximations assume that the stress and strain fields in each phase $i$ can be represented by the volume-averaged values, $\bar{\boldsymbol{\sigma}}_i$ and $\bar{\boldsymbol{\epsilon}}_i$, which can be computed by integration over the representative volume element $V$ as

$$\bar{\boldsymbol{\sigma}}_i = \frac{1}{V_i} \int_{V_i} \boldsymbol{\sigma}_i(\vec{x}) \, dV \tag{14a}$$

$$\bar{\boldsymbol{\epsilon}}_i = \frac{1}{V_i} \int_{V_i} \boldsymbol{\epsilon}_i(\vec{x}) \, dV \tag{14b}$$

where $\Sigma V_i = V$ and $\vec{x}$ expresses the position of a material point within phase $i$. In turn, the average stress and strain in the composite, $\bar{\boldsymbol{\sigma}}$ and $\bar{\boldsymbol{\epsilon}}$, are obtained by integration of the corresponding stresses and strains in each phase within $V$. This operation is called homogeneization and it is expressed as

$$\bar{\boldsymbol{\sigma}} = \sum c_i \, \bar{\boldsymbol{\sigma}}_i \qquad \text{and} \qquad \bar{\boldsymbol{\epsilon}} = \sum c_i \, \bar{\boldsymbol{\epsilon}}_i \tag{15}$$

where $c_i$ stands for the volume fraction of phase $i$.

The effective stress and strain are related to the average stress and strain in each phase through the respective stress and strain concentration tensors, $\mathbf{A}_i$ and $\mathbf{B}_i$, as

$$\bar{\boldsymbol{\sigma}}_i = \mathbf{B}_i : \bar{\boldsymbol{\sigma}} \qquad \text{and} \qquad \bar{\boldsymbol{\epsilon}}_i = \mathbf{A}_i \, \bar{\boldsymbol{\epsilon}} \tag{16}$$

which evidently satisfy the following equations

$$\sum c_i \, \mathbf{A}_i = \mathbf{I} \qquad \text{and} \qquad \sum c_i : \mathbf{B}_i = \mathbf{I} \tag{17}$$

where $\mathbf{I}$ is the unit tensor of fourth order. The elastic stiffness tensor of the PRC, $\mathbf{L}$, formed by a matrix containing different types of reinforcing particles can be obtained easily from (15), (16) and (17) as

$$\mathbf{L} = \mathbf{L}_m + \sum c_j \left(\mathbf{L}_j - \mathbf{L}_m\right) : \mathbf{A}_j \tag{18}$$

where $\mathbf{L}_m$ stands for the matrix elastic stiffness tensor, and $c_j$, $\mathbf{L}_j$ and $\mathbf{A}_j$ are, respectively, the volume fraction, elastic stiffness tensor, and strain concentration factor of each type of reinforcing particle. Evidently, the overall composite stiffness depends on the matrix and reinforcement properties and volume fractions as well as on the reinforcement strain concentration tensors which, in turn, depend on the volume fraction, shape, and constitutive equation of each phase. Their simplest form is given by $\mathbf{A}_j = \mathbf{I}$, the well-known isostrain model, which is a very poor approximation for PRC.

More realistics values of the particle strain concentration factors can be obtained following various approximations. The simplest one — within the framework of linear elasticity — is based on the pioneer work of Eshelby (1957), who analyzed the stress distribution in an elastic and isotropic ellipsoidal inclusion embedded in an elastic, isotropic and infinite matrix which is subjected to a remote strain $\boldsymbol{\epsilon}$. Eshelby showed that the strain field within the inclusion, $\boldsymbol{\epsilon}_j$ was constant and it was given by (Eshelby, 1957; Hill, 1965)

$$\boldsymbol{\epsilon}_j = \mathbf{A}_j^{dil} \bar{\boldsymbol{\epsilon}} \qquad \text{with} \qquad \mathbf{A}_j^{dil} = \left[\mathbf{I} + (\mathbf{S}_j : \mathbf{L}_m^{-1}) : (\mathbf{L}_j - \mathbf{L}_m)\right]^{-1} \tag{19}$$

where $\mathbf{L}_m$ and $\mathbf{L}_j$ stand for matrix and inclusion stiffness tensors, respectively, and $\mathbf{S}_j$ is the Eshelby's tensor for the inclusion, whose components depend on the inclusion shape as well as on the matrix elastic constants. There are expressions for Eshelby's tensor when the matrix is transversally isotropic or under more general conditions (Laws and McLaughlin, 1979; Mura, 1987). The superindex *dil* in the strain concentration tensor of the inclusion indicates that this expression is only exact when $c_p \to 0$, according to the hypotheses in the work of Eshelby.

There are many variations of Eshelby's method to take into account the distortion in the inclusion stress field induced by the presence of neighbor particles. They can be used to compute the elastic properties of the composite material when the reinforcement volume fraction is finite. One of the most popular ones is the Mori-Tanaka method (Mori and Tanaka, 1973), which was reformulated by Benveniste (1987) in the context of the strain concentration tensors. Following Benveniste (1987), the particle strain concentration tensor, $\mathbf{A}_j^{mt}$, can be obtained by interpolation between the strain concentration tensors for dilute conditions ($\mathbf{A}_j^{dil}$) and when $c_j \to 1$ ($\mathbf{A}_p \approx \mathbf{I}$), and it is given by

$$\mathbf{A}_j^{mt} = \mathbf{A}_j^{dil} : \left[(1 - c_j)\mathbf{I} + c_j \mathbf{A}_j^{dil}\right]^{-1} \tag{20}$$

and the effective elastic properties are obtained directly using this expression for the reinforcement strain concentration factor in equation (18).

Another mean-field approximation to the elastic properties of multiphase materials is the self-consistent method, which is particularly appropriate when the various phases are distributed forming an interpenetrating network. This model was developed by Kroner (1958) to compute the effective elastic properties of polycrystalline solids. All the phases in the composite are assumed to be embedded in an effective medium, whose properties are precisely those of the composite which are sought. The corresponding strain concentration tensor for each phase ($\mathbf{A}_j^{sc}$) is obtained from Eshelby's dilute solution (equation (19)) substituting the matrix elastic constants by those of the effective medium, $\mathbf{L}$. Mathematically

$$\mathbf{A}_j^{sc} = \left[\mathbf{I} + (\mathbf{S}_j : \mathbf{L}^{-1}) : (\mathbf{L}_j - \mathbf{L})\right]^{-1} \tag{21}$$

Introducing equation (21) in (18), it leads to

$$\mathbf{L} = \mathbf{L}_m + \sum c_j(\mathbf{L}_j - \mathbf{L}_m) : \left[\mathbf{I} + (\mathbf{S}_j : \mathbf{L}^{-1}) : (\mathbf{L}_j - \mathbf{L})\right]^{-1} \tag{22}$$

where equation (22) stands for a non-linear set of equations for the components of $\mathbf{L}$, which can be solved numerically to obtain the elastic constants of the composite medium. It should be noted that Eshelby's tensor, $\mathbf{S}_j$, depends on the particle shape as well as on the elastic constants of the effective medium given by $\mathbf{L}$.

The methodologies described above can be extended to study the composite behavior if one or more phases exhibit a non-linear behavior, and a detailed review of the different methods can be found in (Suquet, 1997) and (Castaneda and Suquet, 1998). The non-linear analysis can be addressed from two different viewpoints, which are briefly described below. Secant formulations deal with the plastic deformation within the context of non-linear elasticity, and the relationship between average stress, $\bar{\boldsymbol{\sigma}}$, and the average strain, $\bar{\boldsymbol{\epsilon}}$, in the composite is given by a secant stiffness tensor, $\mathbf{L}^s$, which —in the typical case of a non-linear matrix reinforced with a volume fraction $c_p$ of elastic particles— is expressed as (see equation 18)

$$\bar{\boldsymbol{\sigma}} = \mathbf{L}^s(\bar{\boldsymbol{\epsilon}}) : \bar{\boldsymbol{\epsilon}} \qquad \text{where} \qquad \mathbf{L}^s = \mathbf{L}_m^s + c_p(\mathbf{L}_p - \mathbf{L}_m^s) : \mathbf{A}_p^s \tag{23}$$

where $\mathbf{A}_p^s$ is the "secant" strain concentration tensor of the particles, which is computed using any of the elastic approximations presented above with $\mathbf{L}_m^s$ as the matrix elastic stiffness tensor. $\mathbf{L}_m^s$ depends on the reference equivalent strain in the matrix, $\hat{\epsilon}_{eq}$, as (Berveiller and Zaoui, 1979; Tandon and Weng, 1988),

$$\mathbf{L}_m^s = 3k_m\mathbf{J} + 2\mu_m^s(\hat{\epsilon}_{eq})\mathbf{K} \tag{24}$$

where $J_{ijkh} = \delta_{ij}\delta_{kh}/3$ and $K_{ijkh} = 0.5(\delta_{ik}\delta_{jh} + \delta_{ih}\delta_{jk}) - J_{ijkh}$ stand for the orthogonal volumetric and deviatoric projections, respectively, and $k_m$ and $\mu_m^s$ are, respectively, the bulk and secant shear modulus of the matrix. Evidently, $\mu_m^s$ varies during deformation and is a function of the reference equivalent matrix strain, $\hat{\epsilon}_{eq}$, which can computed from the volume-averaged stress tensor in the matrix as

$$\hat{\epsilon}_{eq}^2 = \frac{2}{3}(\mathbf{K} : \bar{\boldsymbol{\epsilon}}_m) : (\mathbf{K} : \bar{\boldsymbol{\epsilon}}_m) \tag{25}$$

The determination of the effective composite repose under an effective strain given by $\bar{\boldsymbol{\epsilon}}$ begins by choosing a trial value of $\mu_m^s$. Once the elastic constants of both phases are known, it is possible to compute the matrix and particles strain concentration tensors with any of the linear methods described above as well as the overall composite secant stiffness and the average strain tensor in each phase. The effective strain in the matrix is thus determined through equation (25) and $\mu_m^s$ is updated , the whole procedure being repeated until convergence is achieved.

The extension of the mean-field methods to the elasto-plastic regime using the incremental theory of plasticity is straightforward in theory although the details of the numerical implementation are much more complex and cannot be detailed here. Basically, the analysis starts from a known situation, where the stresses and strains in each phase and in the effective medium have been computed in the previous step. The stress and strain increments in the next step are obtained from the average stress and strain rates in the composite, $\dot{\bar{\boldsymbol{\sigma}}}$ and $\dot{\bar{\boldsymbol{\epsilon}}}$. They are related through the tangent (or instantaneous) stiffness tensor, $\mathbf{L}^t$, and this is expressed in the case of an elasto-plastic matrix reinforced with a volume fraction $c_p$ of elastic particles as

$$\dot{\bar{\boldsymbol{\sigma}}} = \mathbf{L}^t(\bar{\boldsymbol{\epsilon}}) : \dot{\bar{\boldsymbol{\epsilon}}} \qquad \text{where} \qquad \mathbf{L}^t = \mathbf{L}_m^t + c_p(\mathbf{L}_p - \mathbf{L}_m^t) : \mathbf{A}_p^t \tag{26}$$

where $\mathbf{A}_p^t$ is the "tangent" strain concentration tensor of the particles, which is computed using any of elastic approximations with $\mathbf{L}_m^t$ as the matrix elastic stiffness tensor. $\mathbf{L}_m^t$ depends on the average accumulated plastic strain in the matrix, $\bar{\epsilon}_p$ and it is anisotropic because of the matrix plastic flow. It has been shown, however, that the composite response predicted by the incremental methods is too stiff when the general, anisotropic form of the $\mathbf{L}_m^t$ is used (Castaneda and Suquet, 1998; Doghri and Ouaar, 2003; González and LLorca, 2000). Nonetheless, the tangent matrix stiffness tensor adopts an isotropic form given by

$$\mathbf{L}_m^t = 3k_m\mathbf{J} + 2\lambda_m^t(\epsilon_p)\mathbf{K} \tag{27}$$

where $\lambda_m^t$ stands for the tangent shear modulus of the matrix in the case of proportional loading, which provides much better results (González and LLorca, 2000; Doghri and Ouaar, 2003). The first incremental version of mean-field approximations was presented by Hill (1965), and this seminal work led to the development of a large number of similar approaches (Hutchinson, 1970; Petermann et al., 2002; Doghri and Ouaar, 2003).

Secant methods are -strictly speaking- limited to radial (or approximately radial) loading paths of each phase in the stress space, while models based on the incremental theory of plasticity can take into account accurately the effect of non-proportional loading and thus, are more adequate to simulate damage, which always leads to partial unloading of one or various phases during deformation. However, secant approaches usually provide better approximations of the composite behavior than tangent methods in the plastic regime and are significantly easier to implement. A detailed discussion of the advantages and limitations of each methodology is beyond the scope of this paper and can be found in the cited references (Castaneda and Suquet, 1998; González and LLorca, 2000; Doghri and Ouaar, 2003), but both secant and incremental formulations have been used to model elasto-plastic deformation and damage in PRC.

### 4.2 One-dimensional models

The first attempts to study the deformation of PRC including damage by particle fracture were carried out within the framework of one-dimensional isostrain models which simulated with accuracy the uniaxial deformation of these materials despite their simplicity (Kiser et al., 1996; González and LLorca, 1996; Poza and LLorca, 1999). The composite was represented in terms of a three dimensional array of hexagonal cells with one reinforcement at the centre (Figure 7). The reinforcement was either intact or fractured, the latter being broken by a penny-shaped crack perpendicular to the loading axis. Assuming that the displacements at the cell boundaries (and thus the strains) were equal for cells containing intact or broken reinforcements, and that both types of cells were homogeneously distributed within the composite, the average composite stress, $\sigma$, along the loading direction can be calculated as

$$\sigma = \sigma_i (1 - c_d) + \sigma_d c_d \tag{28}$$

where $c_d$ was the fraction of cells with fractured particles and $\sigma_i$ and $\sigma_d$ stood for the average stress borne by intact and damaged cells. $\sigma_i$ and $\sigma_d$ were computed as a function of the strain in the loading direction through the finite element analysis of axisymmetric cylindrical cells, which can be regarded as an approximation to the hexagonal cells. Moreover, the average stress in the loading direction borne by the reinforcements was obtained from the finite element simulations of intact cells, and it was used to determine the volume fraction of broken particles, $c_d$, at any time during deformation through equation (3) assuming that the particle strength follows the Weibull statistics.

An example of the results obtained with this approach can be found in Figure 8 (González and LLorca, 1996). The stresses carred by the intact and damaged cells (computed from the finite element analysis of axisymmetric cylindrical cells) are marked with solid and open circles, and they stand for the behavior of the intact and fully-damaged composite, respectively. The model predictions taking into account the progressive fracture of the particles are much closer to the experimental tensile stress-strain curve. The evolution of the fraction of broken particles as a function of the applied strain, as obtained from the model, is also ploted in the Figure. The model was developed further by LLorca and González (1998) by mixing cells containing different reinforcement volume fraction and particles with different shape (cylinders or spheres) while the main hypotheses (isostrain deformation of all the cells and uniaxial deformation) were maintained.

Although this model provided the first estimation of the damage evolution during deformation under uniaxial deformation, a more refined approximations was developed by Maire et al. (1997) to relax the isostrain condition, which tended to provide a too stiff response, and to include the effect of the stress redistribution originated by the reinforcement fracture. The former problem was solved by idealizing the composite microstructure as an interpenetrating network of two kind of spheres (Figure 9)), which stood for the behavior of the intact or damaged composite. The spheres representing the intact material were made up of a spherical particle embedded in a matrix shell. The particles in the damaged spheres were broken by a crack perpendicular to the loading axis, and damage in the composite was given by $c_d$, the volume fraction of damaged spheres. The elasto-plastic response along the loading axis of the intact and damaged

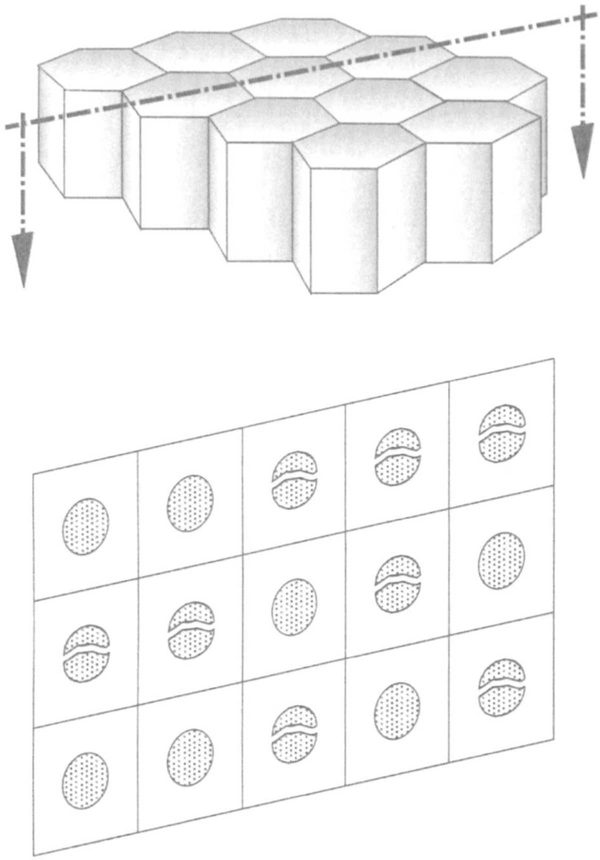

**Figure 7.** Composite idealization as a three dimensional array of hexagonal cylinders, each containing an intact or broken spherical reinforcement at the centre.

spheres was determined from the finite element analysis of axisymmetric unit cells, as in the previous model, but the strain partition between the intact and damaged regions was obtained using an incremental (tangent) and one-dimensional formulation of the self-consistent model. The effective response of the composite material including damage (path AC in Figure 10) was decomposed into two steps, AB and BC. The volume fraction of damaged phase, $c_d$, remained constant in the initial step AB. The second step, BC, includes the stress relaxation associated with the transformation of a certain amount of intact to damaged spheres while the total strain stays constant. The tangent modulus of the composite during the first step (slope of the line AB) was computed through a one-dimensional simplification of equation (22), which is valid if the Poisson's ratio in both phases is equal to 1/2. Under such conditions, the effective tangent modulus only depends on the tangent moduli of each phase and of the volume fraction of damaged material, $c_d$. The unloading along the path BC was determined from an expression

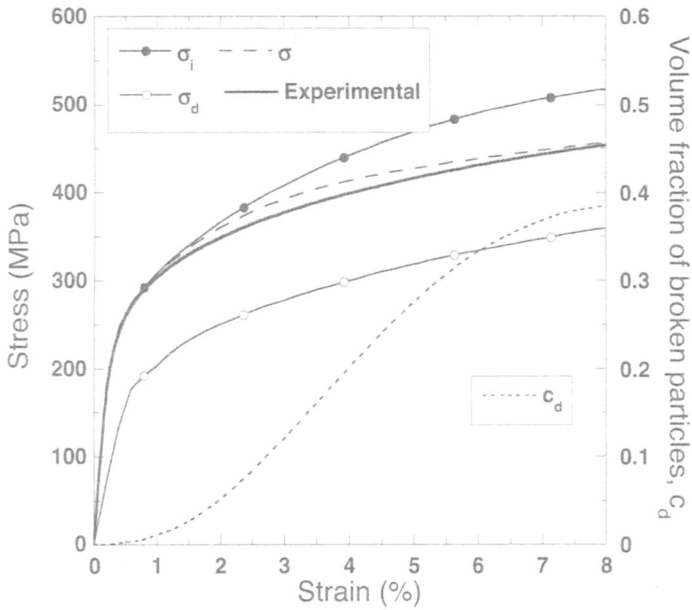


**Figure 8.** Tensile stress-strain curve of a 2618 Alloy reinforced with 15 vol. % SiC particles. Experimental results and predictions using the uniaxial isostrain model (González and LLorca, 1996).

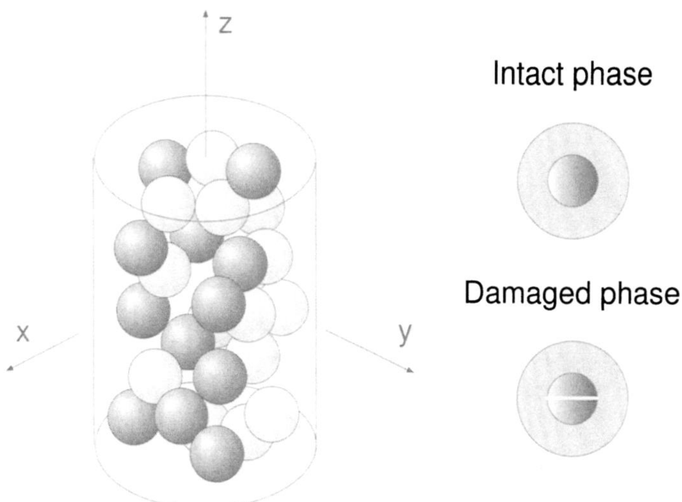


**Figure 9.** Idealized representation of the composite microstructure (Maire et al., 1997).

derived from the self-consistent method, which provides the stress variation as a result of a phase transformation of an infinitesimal volume fraction $dc_d$ from the intact to the damaged state. The increment in the volume fraction of damaged material, $dc_d$, was determined by a Weibull function of known parameters, as in previous models.

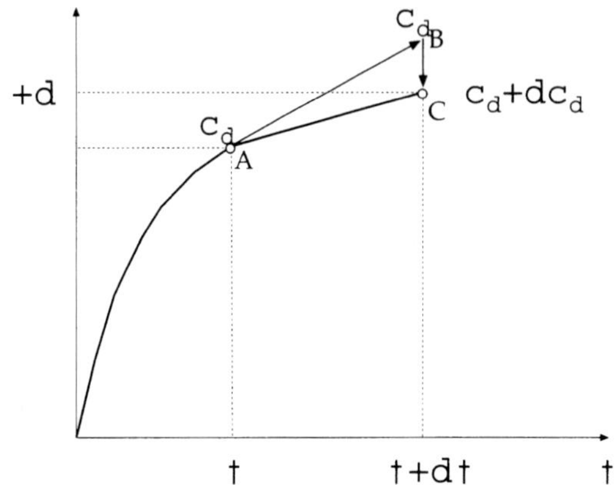


**Figure 10.** Scheme of the incremental analysis (Maire et al., 1997).

This model provided a better approximation to the composite stress-strain curve than those obtained with the isostrain approach but tended to overestimate the ductility because it did not take into account the fracture of the matrix in the last stages of deformation. This phenomenon was included by Gammage (2002), who assumed that the matrix failure occurred if the stress acting on the matrix between two cracked particles exceeded the critical value necessary to initiate localized plastic flow in the interparticle matrix. For any given applied strain, the stress, $\sigma_d$, and the hardening rate in the damaged regions, $d\sigma_d/d\epsilon$, are known, and the critical interparticle distance leading to matrix failure, $\lambda_c$ is given by,

$$\lambda_c = \frac{4d}{\left[\frac{d\sigma_d/d\epsilon}{\sigma_d} - 1\right]^2} \tag{29}$$

according to equation (8), where $d$ is the particle diameter. If the overall particle distribution follows the Poisson distribution, it is possible to determine the distribution function for spacing between broken particles as

$$f(\lambda) = \frac{8c_d\lambda}{d^2} \exp\left[\frac{-4c_d\lambda^2}{d^2}\right] \tag{30}$$

which provides the fraction of cells containing broken particles whose distance is below $\lambda_c$ and where matrix failure has taken place. This is introduced in the self-consistent model by reducing the tangent modulus of the damaged phase by a linear factor $F_d$ given by

$$F_d = 1 - \left[\int_0^{\lambda_c} f(\lambda)d\lambda\right] / \left[\int_0^{\infty} f(\lambda)d\lambda\right] \tag{31}$$

The model predictions of the tensile stress-strain curve using the original self-consistent model and the modification of (Gammage, 2002) are plotted in Figure 11 for a 2618 Al alloy reinforced with 18 vol. % SiC particles, together with the experimental results. Both models predict very well the experimental stress-strain curve as well as the composite hardening rate at at low strains. However, the original self-consistent model tends to overpredict the composite hardening rate in the last stages of deformation because damage by matrix failure is not taken into account. On the contrary, the modified model captured very well the sudden drop in the composite hardening rate due to the matrix fracture just before failure.

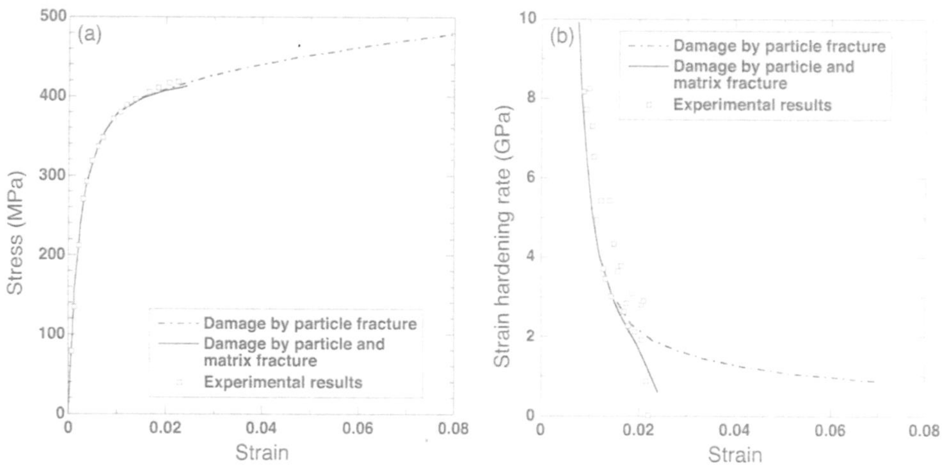


**Figure 11.** (a) Tensile stress *vs.* strain curve and (b) strain hardening rate *vs.* strain curves of a 2618 Alloy reinforced with 18 vol. % SiC particles. Experimental results and predictions using the uniaxial self-consistent models.

### 4.3 Mean-field secant models

Despite their success in predicting the tensile behaviour, the models presented in the previous section were limited to uniaxial deformation. The first multiaxial models were developed within the context of the Mori-Tanaka mean-field approximation using a secant approach to introduce the non-linear deformation.

The first step in this direciton was given by Mochida et al. (1996), who analyzed the effect of damaged particles on the degradation of the elastic modulus in $Al/Al_2O_3$ composites. They took into account only reinforcement damage, and assumed three different damage micromechanisms: reinforcement shattering, reinforcement fracture through a crack perpendicular to the loading axis, and decohesion at the particle/matrix interface. The effective elastic constants were obtained using the Mori-Tanaka approximation for a three-phase composite, leading to

$$\mathbf{L} = \mathbf{L}_m + c_i(\mathbf{L}_i - \mathbf{L}_m) : \mathbf{A}_i^{mt} + c_d(\mathbf{L}_d - \mathbf{L}_m) : \mathbf{A}_d^{mt} \tag{32}$$

where $c_i$ and $c_d$ stand, respectively, for the volume fraction of intact and damaged reinforcements, and the elastic constants of each phase were represented by the corresponding stiffness tensors $\mathbf{L}_m$, $\mathbf{L}_i$, and $\mathbf{L}_d$. The three terms in equation (32) represent the contributions of the matrix, the intact and the damaged particles to the effective elastic constants. Shattered damaged reinforcements were assumed to behave as voids (Figure 12), with the same volume and shape as the intact reinforcements, and null elastic constants ($\mathbf{L}_d = \mathbf{0}$). Particles broken by a crack perpendicular to the loading axis were substituted by penny-shaped cracks (Figure 12). Finally, interphase decohesion is taken into account by assuming that decohered particles cannot transmit any stress in the loading direction while the strains in the transverse directions are null (Figure 12). Evidently, the model can accommodate the coexistence of the three types of damage by including the contribution of each one to the effective elastic constants in equation (32), and it gives an accurate estimation of the effect of damage in the composite stiffness. It does not allow, however, a prediction of the evolution and accumulation of damage during deformation.

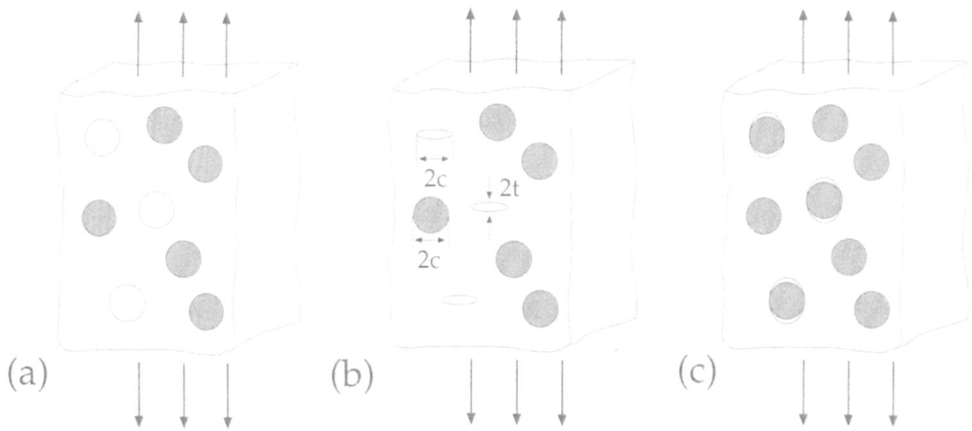


**Figure 12.** Patterns of reinforcement damage according to Mochida et al. (1996). (a) Shattered particles behave as voids. (b) Broken particles are replaced by penny-shaped cracks perpendicular to the loading axis. (c) Interface decohesion.

The initial analyses of progressive damage in composite materials using the secant method were due to Bourgeois (1994) and Zhao and Weng (1996). Bourgeois (1994) analyzed the tensile deformation of an Al alloy reinforced with SiC particles including the effect of particle fracture. The metallic matrix behaved as an elasto-plastic solid following the deformation theory of plasticity. The intact reinforcements were taken as linear elastic and isotropic ellipsoidal inclusions, while the damage reinforcements were substituted by penny-shaped cracks, following the previous work of Mochida et al. (1996) (Figure 13). The scheme of the analysis starts at point A, where the average stress and strain in the composite are known (Figure 13) as well as the volume fraction of broken particles, $c_d$. The average stress tensor in the intact particles is then computed using the corresponding stress concentration tensor, and the maximum principal stress in the intact particles is obtained. This value is used to determine the new volume fraction of broken particles, $c_d^{i+1}$, assuming that the particle strength is dictated by a Weibull law, equation (3), whose parameters are known. The new secant constants are computed using equation (27), and the next point (B in Figure 13) in the stress-strain curve —corresponding to an applied stress $\sigma_{i+1}$— is computed using the new secant elastic constants of the composite and the Mori-Tanaka approximation. This procedure is repeated to obtain the complete stress-strain curve. (Zhao and Weng, 1996) used the same approximation to study damage by partial interface decohesion. Their analysis assumed that interface fracture was strength controlled, and partial decohesion occurred when the tensile stress in the particle along the loading axis surpassed a critical value, which they identified as the interfacial strength. It was assumed that the interfacial strength followed the Weibull statistics to account for the fact that interface fracture occurs progressively during deformation. Decohered particles were replaced by anisotropic inclusions which cannot transmit normal and shear stress in the debonded direction but whose normal and shear stiffnesses were not zero in the perpendicular ones.

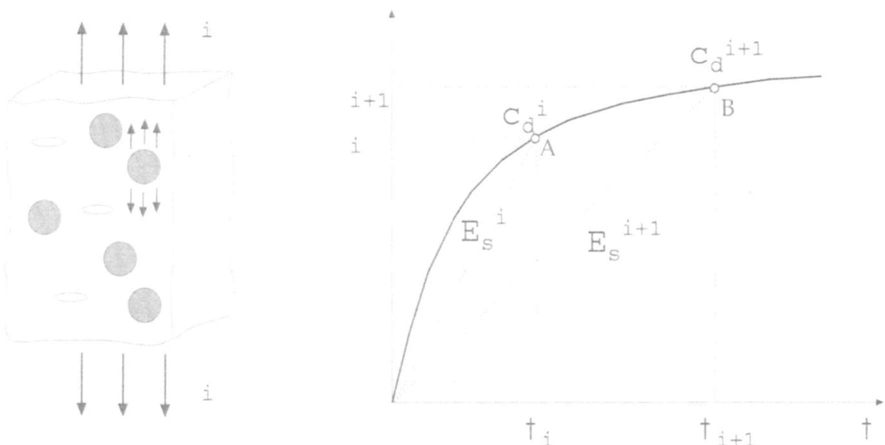


**Figure 13.** (a) Idealized representation of the composite microstructure. (b) Scheme of the analysis (Bourgeois, 1994).

These models provided the first estimation of the damage evolution during deformation within a three-dimensional framework but presented two main limitations. Firstly, the representation of broken particles as penny-shaped cracks or of partly-decohered particles as anisotropic inclusions with zero stiffness along the loading axis were crude simplifications. Secondly, they did not take into account the stress redistribution which occurs as a result of reinforcement fracture or interface decohesion. This latter problem was overcome by Estevez et al. (1999), who extended the one-dimensional model of Maire et al. (1997) to three dimensions. The composite was represented as an interpenetrating network of three types of spheres made up matrix, particles and particles broken by a penny-shaped crack perpendicularly to the loading axis. The matrix behaved as an elasto-plastic solid following the deformation theory of plasticity and the intact reinforcements as linear elastic and isotropic solids. The normal (shear) stiffness of the broken reinforcements along (perpendicular to) the loading axis were set equal to zero. The effective behavior of this three phase composite was computed using the secant version of the self-consistent approximation. The evolution of damage during deformation and the corresponding stress relaxation were computed following the same scheme as that used by Maire et al. (1997). Comparison between the results obtained with the one-dimensional and the three-dimensional models for uniaxial deformation showed negligible differences, although the composite representation was slightly different.

### 4.4 Mean-field tangent models

Tohgo and Chou (1991) were the first to analyze the elasto-plastic deformation of a PRC including damage through a tangent method. Their analysis was based on the Mori-Tanaka approximation and the composite was represented as a homogeneous distribution of spherical inclusions and voids, which stood for the intact and damaged particles dispersed within the elasto-plastic matrix. No difference was made between damage by particle fracture or interface decohesion: it was controlled by the maximum principal stress in the intact particles following the Weibull statistics, and damaged reinforcements were substituted by voids. The increment in the fraction of broken reinforcements in each step of the analysis was controlled by the corresponding increase in the maximum principal stress on the particles. However, as it was not known beforehand, the stress increment computed in the previous step was used to simplify the computations, and the stress carried by new damaged particles was distributed between the matrix and the intact particles.

More recently, González and LLorca (2000) studied the deformation with damage of a composite material under arbitrary loading conditions within the framework of Hill's self-consistent model and the incremental formulation of the theory of plasticity. The composite was represented by an interpenetrating network of two spherical phases, which stood for the intact and damaged regions in the composite as in Figure 9. Both phases were assumed to behave as isotropic, elasto-plastic solids following the Von Mises yield criterion and the incremental theory of plasticty and damage was understood as the progressive transformation of the intact phase in the damaged one. The incremental process to compute the mechanical response of the composite is analogous to that shown in Figure 10. From the mathematical point of view, the effective strain hardening, $d\bar{\boldsymbol{\sigma}}$, due an infinitesimal increment of the effective strain, $d\bar{\boldsymbol{\epsilon}}$, can be computed as

$$\frac{d\bar{\boldsymbol{\sigma}}}{d\bar{\boldsymbol{\epsilon}}} = \left[\frac{\partial\bar{\boldsymbol{\sigma}}}{\partial\bar{\boldsymbol{\epsilon}}}\right]_{c_d} + \left[\frac{\partial\bar{\boldsymbol{\sigma}}}{\partial c_d}\right]_{\boldsymbol{\epsilon}} \left[\frac{\partial c_d}{\partial\boldsymbol{\epsilon}}\right] \tag{33}$$

The first term in (33) includes the contribution from each phase to the strain hardening in the absence of any phase change. The second term stands for the stress redistribution induced by an infinitesimal variation of the intact phase —which is transformed into the damaged phase— while the boundary conditions remain constant.

This expression for the effective strain hardening assumes that the processes take place consecutively during each increment of deformation. The material is deformed initially without any change in the volume fraction of the phases and its behavior can be determined from the classic self-consistent model developed by Hill (1965). The changes in the stress and strain in the intact phase during this process lead to a small increment $dc_d$ in the volume fraction of the damaged phase, and the second term in (33) accounts for the corresponding stress redistribution. It can be calculated under two hypotheses. Firstly, the transformation from intact to damage phase of a volume fraction $dc_d$ occurs instantaneously as compared to the variation of the boundary conditions, which can be assumed to remain constant. Secondly, the phase change leads to an elastic stress redistribution in both phases and in the effective material. As a result, the stress relaxation can be determined by deriving the constitutive equation of the composite provided by Hill's model in the elastic regime.

This model was applied to determine the tensile stress-strain curve in various Al-matrix composites reinforced with SiC particles, where the dominant damage mechanism was the progressive fracture of the reinforcements controlled by a Weibull model (González and LLorca, 2000). The constitutive equation of each phase and the average principal stress acting on the reinforcements was obtained from the finite element analysis of axisymmetric cylindrical cells containing intact or broken particles, as in other investigations (Kiser et al., 1996; Maire et al., 1997; LLorca and González, 1998). The predicted stress-strain curves were in good agreement with the experimental ones as well as the predictions for the volume fraction of broken particles, which were compared with the *post mortem* values obtained on tested samples.

Moreover, the effect of superposed hydrostatic pressure on the stress-strain curve of PRC was assessed with this self-consistent model, which was used to simulate the deformation of a composite reinforced with 20 vol. % ceramic particles subjected to a uniform hydrostatic pressure followed by uniaxial tension (González and LLorca, 2002). The evolution of the composite flow stress, $\bar{\sigma}$, with the plastic strain, $\bar{\epsilon}_p$, is plotted in Figure 14 for superposed hydrostatic pressures between 0 and 600 MPa, together with the corresponding strain hardening rates, $d\bar{\sigma}/d\bar{\epsilon}_p$, showing that the composite hardening capacity increased with the hydrostatic pressure. This result cannot be attributed to the behavior of the intact and damaged regions because the Von Mises yield criterion is insensitive to the hydrostatic stress component, and the differences in hardening rate were due to the progress of damage during deformation. This is shown in Figure 14(a) where the fraction of broken reinforcements, $c_d$, is plotted as a function of the equivalent plastic strain for the materials depicted in Figure 14(b). The superposed hydrostatic pressure reduced markedly the tensile stresses in the particles, and damage by particle fracture practically disappeared in the composites tested under 600 MPa of hydrostatic pressure. These observations are in excellent agreement with the experimental results of Liu et al. (1989) and Lewandowski and Lowhaphandu (1998), who also found an increase in flow stress and a reduction in damage by particle fracture as the hydrostatic pressure increased.

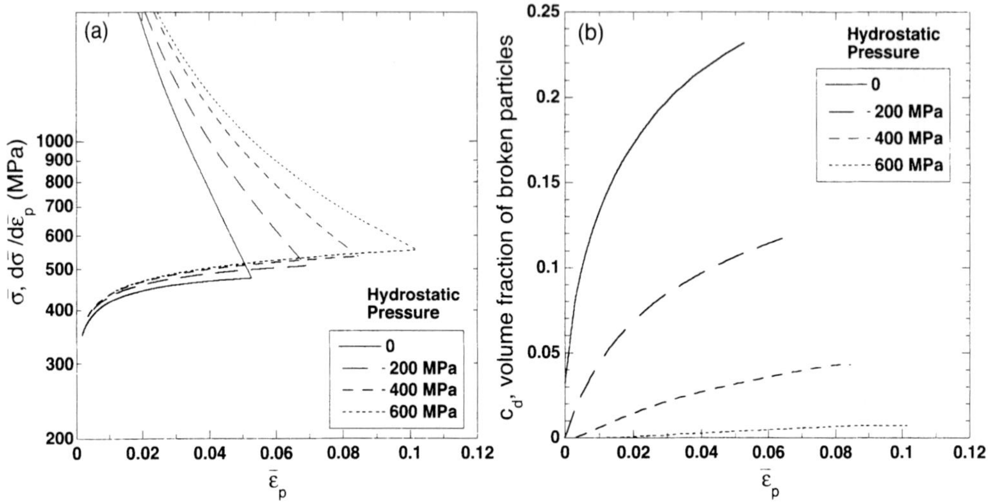


**Figure 14.** Effect of superposed hydrostatic pressure on (a) composite flow stress and strain hardening rate. (b) Fraction of broken reinforcements (González and LLorca, 2002).

### 4.5 Other tangent models

Ju and Sun (2001) developed a new homogenization approach to study the deformation of PRC composites made up of an elasto-plastic matrix reinforced with spheroidal particles. A novel formulation was derived for the local stress distribution in the matrix in the presence of spheroidal inclusions. The ensemble-averaging procedure was employed to derive the tangent stiffness of the composite taking into the account the the particle-matrix influences as well as the associative plastic flow rule. This new approach was used to study the effect of progressive interface decohesion (Sun et al., 2003) or particle fracture (Sun et al., 2003) on the tensile deformation of PRC. Particles damaged by interface decohesion were modeled as fictitious orthotropic inclusions with zero tensile stiffness perpendicularly to the loading axis, while broken particles are represented by penny-shaped cracks. The damage criterion in both cases was given by the maximum average tensile stress on the particle, and the critical stress is assumed to follow the Weibull statistics.

## 5 Multiparticle Unit Cell Models

Together with the homogeneization techniques, periodic unit cell models have been widely used in the last decade to simulate the elasto-plastic deformation of PRC. They assume that the particles are placed in a regualr lattice(cubic, BCC, FCC, etc.), and this microstructure —due to the symmetries— can be represented by a simple unit cell containing one or two particles with periodic boundary conditions, the composite behaviour being simulated through the finite element analysis of the periodic unit cell (Christman et al., 1989; Bao et al., 1991). This methodology provides the detailed local stress and strain fields within the unit cell, an important advantage to simulate damage, which is normally triggered by the maximum local values, while the homogeneization models only give average magnitudes. Periodic unit cell models of this kind were used to study the effect of damage by matrix void growth (LLorca et al., 1991; Tvergaard, 1998) and interface decohesion (Tvergaard, 1990). While these models were very useful to assess the effect of particle shape and volume fraction on damage initiation in PRC, their ability to predict the effective properties afterwards was very limited because damage was spread throughout the composite by the periodic boundary conditions. This is opposed to the experimental observations, which show that damage is nucleated and grows preferentially in given regions of the microstructure.

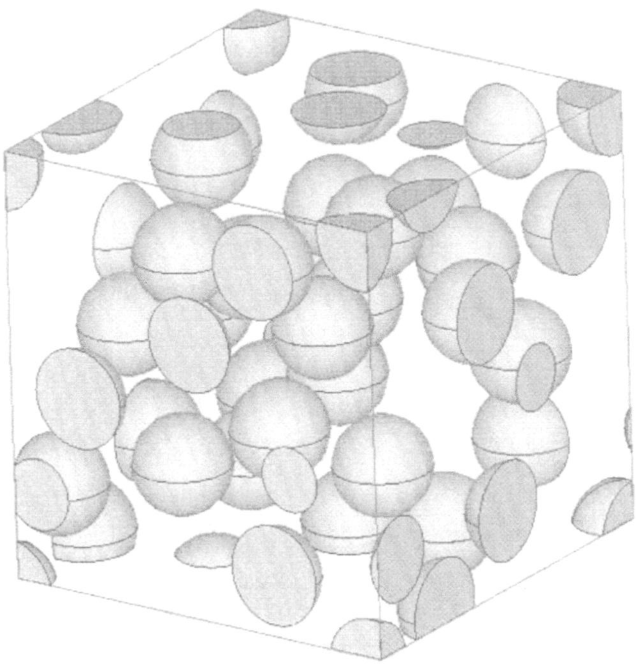

**Figure 15.** Multiparticle cubic unit cell made up of a random and homogeneous dispersion of thirty spherical particles. Notice that the microstrucutre is periodic and spheres intersecting the cube faces were split into an appropriate number of parts and copied on the opposite sides of the cube

This drawback of the single-particle unit cells model can be partially overcome using multiparticle unit cells, as shown in Figure 15. Several studies (Gusev, 1997; Michel et al., 1999; Böhm and Han, 2001; Segurado and LLorca, 2002; Segurado et al., 2002) have shown that multiparticle cells containing several dozens of particles with periodic boundary conditions are large enough to capture with great accuracy the effective properties of PRC subjected to elasto-plastic deformation, and were able to quantify the influence of reinforcement clustering on the overall behavior as well as on the local stresses in the matrix and particles (Segurado et al., 2003). The applicability of this new methodology to study damage in PRC is dictated by the compromise between two factors. On the one hand, the accuracy of the solution increases with the number of particles in the unit cell. The minimum size of the representative volume element is surprisingly small for elastic and elasto-plastic deformation (Drugan and Willis, 1996; Drugan, 2000; Segurado et al., 2002) but this is not the case in the presence of damage due to the localization of the deformation. On the other hand, computer time and memory increase very rapidly with the unit cell size, limiting the maximum number of particles that can be included in the model. Taking into account the present power of the digital computers, the typical volume element (containing up to 50 particles) is not large enough to capture accurately the effective response of the PRC up to complete failure when damage is included. However, multiparticle unit cell simulations can provide very interesting insights on the effect of particle spatial distribution on the onset and the initial evolution of damage in PRC, which cannot be obtained with other techniques.

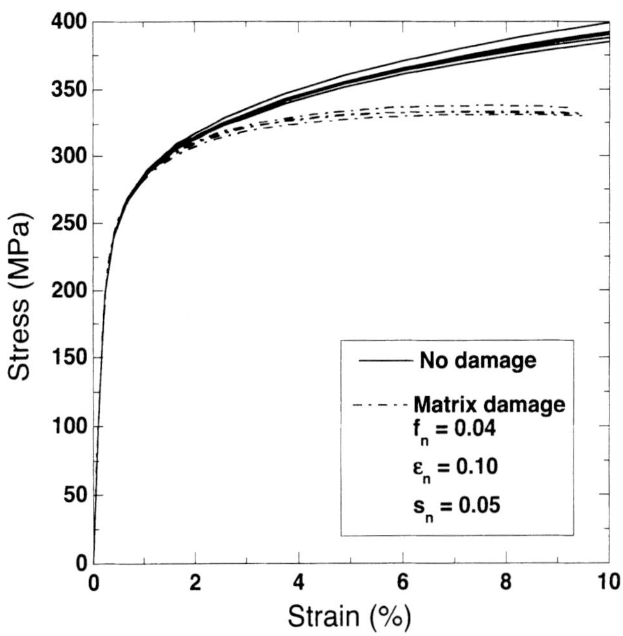


**Figure 16.** Numerical simulations of the tensile stress-strain curves of a composite reinforced with 20 vol. % ceramic spheres including the effect of damage in the matrix *via* the modified Gurson model.

LLorca and Segurado (2004) simulated the tensile deformation of an Al matrix composite reinforced with elastic ceramic particles by means of the three-dimensional finite element analysis of periodic multiparticle cells containing a random dispersion of thirty identical spheres (Figure 15). Damage in the form of void nucleation, growth and coalescence in the matrix was included with the aid of the modified Gurson model, as presented in section 3.3, equations (9-13). Reasonable values of the fraction of void nucleating particles in the matrix, $f_N$, the average nucleation strain, $\epsilon_n$, and the standard deviation of the nucleation strain, $S_N$ were used, and are shown in Figure 16 together with the stress-strain curves corresponding to six different unit cells. It is worth noting that the behavior of the different cells shows a minimum scatter, indicating that the size of the volume element considered was large enough to simulate the effective composite behavior even in the presence of damage. The Figure also shows the marked influence of matrix damage, which reduces significantly the flow stress and led to a maximum in the stress-strain curve. The contour plot of the void volume fraction on the surface of one cell at the end of the analysis (tensile strain 9.5%) is shown in Figure 17 and indicates that damage was localized in certain zones within the cell, corresponding to matrix regions between spheres closely packed along the deformation axis. This sphere arrangement induced a marked strain concentration in the matrix, leading to the early nucleation of voids at very low far-field strains. These voids grew very quickly, driven by high tensile hydrostatic stresses generated between particles aligned in the loading direction, and the critical condition for void coalescence was attained in these regions. The hydrostatic stresses which control the void growth were relieved in these regions as the matrix load-carrying capacity vanished, and damage in matrix was confined to these zones and did not spread immediately throughout the whole cell.

The effective composite behaviour plotted in Figure 16 reflects these mechanisms. Damage was concentrated in the matrix regions subjected to the highest plastic strains and tensile hydrostatic stresses, which contributed most to the composite flow stress. The stress relaxation due to void nucleation and growth reduced significantly the flow stress of the composite. As damage was confined to these regions and did not expand over the whole cell at once, no marked discontinuities were observed in the stress-strain curves. Eventually, a maximum was observed as damage-induced softening overcame the plastic strain hardening contribution. It is important to notice that the initiation and progress of damage was closely linked to the details of the spatial particle arrangement and to the associated local values of the stresses and strain, which can only be captured in simulations based on multiparticle unit cells.

The effect of reinforcement fracture on the tensile deformation of a sphere-reinforced metal-matrix composite was studied by Eckschlager et al. (2002) and Segurado and LLorca (2004). Spheres were assumed to fracture along predefined equatorial planes perpendicular to the loading axis in both cases. Eckschlager et al. (2002) computed the fracture probability of all the particles in the unit cell from the stresses in each Gauss point following equations (1) and (2). Particle fracture was assumed to occur when the fracture probability of a particle reached a critical value, and it was introduced using a node release technique.

Segurado and LLorca (2004) simulated the fracture of the particles along equatorial planes using cohesive elements, whose constitutive equation is given by the elastic poten-

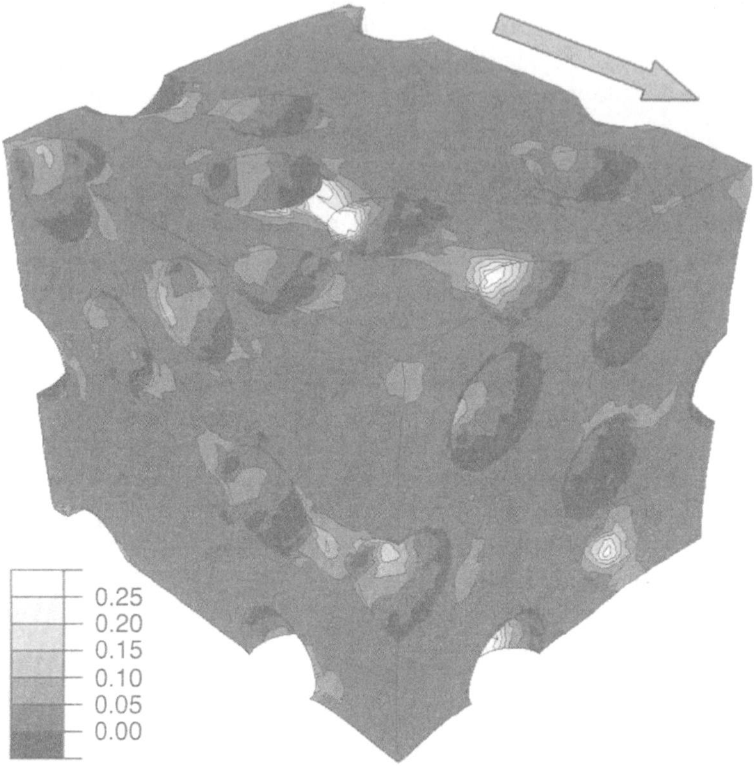


**Figure 17.** Contour plot of the void volume fraction in the matrix at a far-field axial strain of 9.5% in the arrow direction. For clarity the spherical particles are not shown.

tial of equation (4). The critical parameters controlling particle fracture were the fracture energy $\Gamma_{in}$ and maximum normal stress supported by the interface, $\sigma_{in}$. Obviously, this approach can also be used to simulate damage by decohesion by inserting the interface elements between the matrix and the particles. The tensile stress-strain curve of an Al matrix composite reinforced with 15 vol. % of elastic ceramic particles was computed using a unit cell with thirty identical spheres and periodic boundary conditions. $\Gamma_{in}$ was constant and equal to 4 J/m$^2$ in all the spheres, while $\sigma_{in}$ varied among the spheres according to a Weibull law with $m = 5$ and $\sigma_0 = 0.8$ GPa. The numerical results obtained with and without damage are plotted in Figure 18 which shows the unloadings and reloadings due to the progressive fracture of four particles. Particle fracture began at the onset of plastic deformation by those with lower strength and led to a noticeable reduction in the composite flow stress, as compared with the undamaged material. The four broken particles in the unit cell are clearly seen in Figure 19, where the displacements along the loading axis (vertical) were amplified by a factor of 3 to show the four broken particles. Evidently, the longitudinal stress at the center of the broken particles

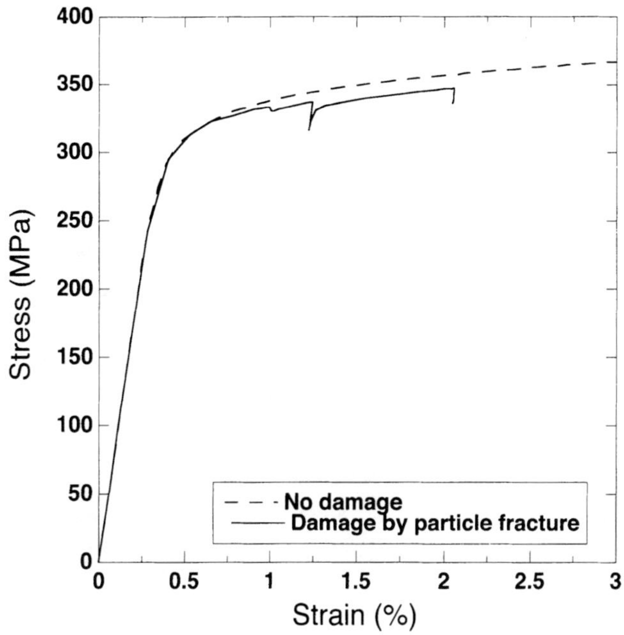


**Figure 18.** Numerical predictions of the tensile stress-strain curve of an Al matrix composite reinforced with 15 vol. % of elastic ceramics spheres with and without particle fracture.

is very close to zero but not at the surface, due to the load transfer from the matrix, which is perfectly bonded to the ceramic spheres.

## 6 Concluding remarks

Many important mechanical properties of PRC depend on the nucleation and evolution of damage in the composite microstructure during deformation. This has led to the development of simulation tools which can take into account the various damage mechanisms experimentally observed in these materials, namely reinforcement fracture, interface decohesion and matrix ductile failure. Two main approximations to the problem were pursued, and the advantages and limitations of each one to model damage are briefly commented in this section.

Models based on homogeneization techniques consider the composite as a multiphase material, where the behavior of each phase and the effective response are given by the volume-averaged properties. Damage is introduced in the model by the addition of new phases and the progression of damage is modelled as a transformation of one phase into another followed by the redistribution of the released stresses among all the phases. For instance, some models substitute the isotropic elastic inclusions by penny-shaped cracks

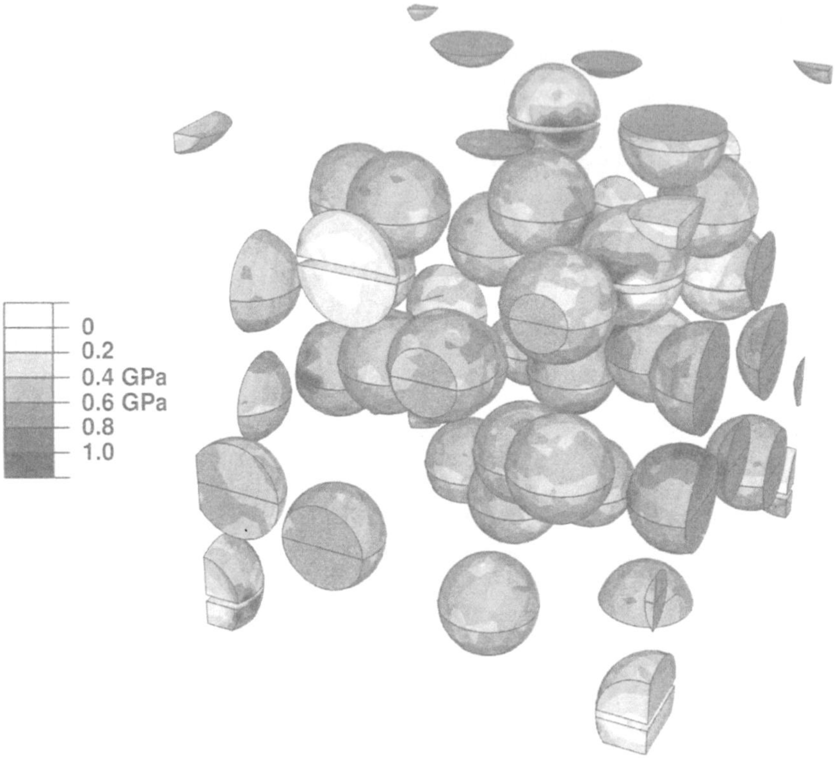


**Figure 19.** Contour plot of the stress in the loading direction (vertical) in the ceramic particles. The displacements in the vertical axis were amplified by a factor of 3 to show the cracks in the broken particles.

to model particle fracture and by anisotropic inclusions with zero stiffness perpendicular to the decohesion plane to include particle/matrix decohesion. The homogenization models are robust and rely on a well-established theoretical framework to analyze the non-linear deformation of composite materials. Although the numerical implementation of these models is sometimes complex, computing time is minimum and they can be used to provide constitutive equations for structural analyses. Limitations to the accuracy of the models come from two hypotheses. Firstly, damage is normally triggered by the extreme values and not by the average ones. Secondly, damage is nucleated at the local level and progresses inhomogeneously throughout the microstructure, leading to a strain localization which cannot be captured by the homogeneization techniques.

The first of these limitations is neatly solved by the models based on the numerical simulation of multiparticle unit cells, which provide the local stress and strain values within the microstructure. Damage nucleation and progression (for instance, by the nucleation and growth of interface cracks) is inherently provided by the model, and the effect of the various inhomogeneities in the microstructure (reinforcement shape, spatial distribution, etc.) can be easily included in the geometrical representation of the composite. Thus, the simulation of multiparticle cells presents a clear advantage over the homogeneization techniques to analyze the details of damage nucleation and growth at the microscopic level. However, the localization of damage increases significantly the minimum size of the representative volume element to obtain an effective behavior independent of the multiparticle unit cell size, and thus computing time and memory allocation rapidly reach unreasonable values which hinder the practical application of this technique in some cases. On the contrary, the effective composite properties can be computed with enough accuracy for many practical purposes (and at a significant lower cost) with the homogeneization techniques.

## Bibliography


L. Babout, E. Maire, J. Y. Buffiere, and R. Fougeres. Characterization by x-ray computed tomography of decohesion, porosity growth and coalescence in model metal matrix composites. *Acta Materialia*, 49:2055–2063, 2001.

G. Bao, J. W. Hutchinson, and R. M. McMeeking. Particle reinforcement of ductile matrices against plastic flow and creep. *Acta Metallurgica et Materialia*, 39:1871–1882, 1991.

Y. Benveniste. A new approach to the application of Mori- Tanaka's theory in composite materials. *Mechanics of Materials*, 6:147–157, 1987.

M. Berveiller and A. Zaoui. An extension of the self-consistent scheme to plastically-flowing polycrystals. *Journal of the Mechanics and Physics of Solids*, 26:325–344, 1979.

H. Böhm and W. Han. Comparisons between three-dimensional and two-dimensional statistics-based unit cell models for particle- reinforced MMCs. *Modeling and Simulation in Materials Science and Engineering*, 9:47–65, 2001.

N. Bourgeois. *Caractérisation et modélisation micromécanique du comportement et de l'endommagement d'un composite à matrice métallique: Al/SiC$_p$*. PhD Thesis, Ecole Centrale des Arts et Manufactures, Paris, 1994.

Y. Brechet, J. D. Embury, S. Tao, and L. Luo. Damage initiation in metal matrix composites. *Acta Metallurgica et Materialia*, 39:1781–1786, 1991.

L. M. Brown and J. D. Embury. The initiation and growth of voids at second phase particles. In *Proceedings of the Third International Conference on the Strength of Metals and Alloys*, pages 164–168. Institute of Metals, London, 1973.

J. Y. Buffiere, E. Maire, P. Cloetnes, G. Lormand, and R. Fougeres. Characterization of internal damage in a MMCp using X-ray synchotron phase contrast microtomography. *Acta Materialia*, 47:1613–1625, 1999.

W. J. Cantwell and A. C. Roulin-Moloney. Fractography and failure mechanisms of unfilled and particulate filled epoxy resins. In A.C. Roulin-Moloney, editor, *Fractog-*

*raphy and Failure Mechanisms of Polymers and Composites*, pages 234–289. Elsevier, London, 1989.

P. Ponte Castaneda and P. Suquet. Nonlinear composites. *Advances in Applied Mechanics*, 34:171–301, 1998.

T. Christman, A. Needleman, and S. Suresh. An experimental and numerical study of deformation in metal-ceramic composites. *Acta Metallurgica*, 37:3029–3050, 1989.

I. Doghri and A. Ouaar. Homogenization of two-phase elasto-plastic composite materials and structures: Study of tangent operators, cyclic plasticity and numerical algorithms. *International Journal of Solids and Structures*, 40:1681–1712, 2003.

T. J. Downes and J. E. King. The effect of SiC particle size on the fracture toughness of a MMC. In N. Hansen et al., editors, *Metal-matrix Composites: Processing, Microstructure and Properties*, pages 305–310, Risø National Laboratory, Roskilde, 1991.

W. J. Drugan. Micromechanics-based variational estimates for a higher-order non-local constitutive equation and optimal choice of effective moduli of elastic composites. *Journal of the Mechanics and Physics of Solids*, 48:1359–1387, 2000.

W. J. Drugan and J. R. Willis. A micromechanics-based nonlocal constitutive equation and estimates of the representative volume element size for elastic composites. *Journal of the Mechanics and Physics of Solids*, 44:497–524, 1996.

A. Eckschlager, W. Han, and H. J. Böhm. A unit cell model for brittle fracture of particles embedded in a ductile matrix. *Computational Materials Science*, 25:85–91, 2002.

J. D. Eshelby. The determination of the elastic field of an ellipsoidal inclusion and related problems. *Proceedings of the Royal Society of London*, A252:376–396, 1957.

R. Estevez, E. Maire, P. Franciosi, and D. S. Wilkinson. Effect of particle clustering on the strengthening versus damage rivalry in particulate reinforced elastic plastic materials: A 3-D analysis from a self-consistent modeling. *European Journal of Mechanics A/Solids*, 18:785–804, 1999.

K. Friedrich and J. Karger-Kocsis. Fractography and failure mechanisms of unfilled and short fiber reinforced semi-crystalline thermoplastics. In A.C. Roulin-Moloney, editor, *Fractography and Failure Mechanisms of Polymers and Composites*, pages 437–494. Elsevier, London, 1989.

J. Gammage. *Damage in Heterogeneous Aluminium Alloys.* PhD Thesis, McMaster University, Hamilton, 2002.

C. González and J. LLorca. Prediction of the tensile stress-strain curve and ductility in Al/SiC composites. *Scripta Materialia*, 35:91–97, 1996.

C. González and J. LLorca. A self-consistent approach to the elasto-plastic behavior of two-phase materials including damage. *Journal of the Mechanics and Physics of Solids*, 48:675–692, 2000.

C. González and J. LLorca. An analysis of the effect of hydrostatic pressure on the tensile deformation of aluminum-matrix composites. *Materials Science and Engineering A*, 341:256–263, 2002.

S. H. Goods and L. M. Brown. The nucleation of cavities by plastic deformation. *Acta Metallurgica*, 27:1–15, 1979.

A. L. Gurson. Continuum theory of ductile rupture by continuum void nucleation and growth. part I: yield criteria and flow rules for porous ductile media. *Journal of Engineering Materials and Technology*, 99:2–15, 1977.

A. A. Gusev. Representative volume element size for elastic composites: a numerical study. *Journal of the Mechanics and Physics of Solids*, 45:1449–1459, 1997.

M. J. Hadianfard, J. Healy, and Y.-W. Mai. Temperature effect on fracture behaviour of an alumina particulate-reinforced 6061 aluminium composite. *Advanced Composite Materials*, 1:93–113, 1994.

R. Hill. A self-consistent mechanics of composite materials. *Journal of the Mechanics and Physics of Solids*, 13:213–222, 1965.

W. H. Hunt, J. R. Brockenbrough, and P. E. Magnusen. An Al-Si-Mg composite model system: microstructural effects on deformation and damage evolution. *Scripta Metallurgica et Materialia*, 25:15–20, 1991.

J. W. Hutchinson. Elastic-plastic behavior of polycrystalline metals and composites. *Proceedings of the Royal Society of London*, A319:247–272, 1970.

J. W. Ju and L. Z. Sun. Effective elastoplastic behaviour of metal-matrix composites containing randomly located aligned spheroidal inhomogeneities. *International Journal of Solids and Structures*, 38:183–225, 2001.

M. T. Kiser, F. W. Zok, and D. S. Wilkinson. Plastic flow and fracture of a particulate metal-matrix composite. *Acta Materialia*, 44:3465–3476, 1996.

T. F. Klimowicz and K. S. Vecchio. The influence of the aging condition on the fracture toughness of alumina-reinforced aluminium composites. In P.K. Liaw and M.Gungor, editors, *Fundamental Relationships Between Microstructure and Mechanical Properties of Metal-Matrix Composites*, pages 255–267. TMS, Warrendale, 1990.

E. Kröner. Berechnung der elastischen Konstanten des Vielkristalls aus den Konstanten der Einkristalle. *Z. Physik*, 151:504–518, 1958.

N. Laws and R. McLaughlin. The effect of fibre length on the overall moduli of composite materials. *Journal of the Mechanics and Physics of Solids*, 27:1–13, 1979.

J. J. Lewandowski and P. Lowhaphandu. Effects of hydrostatic pressure on mechanical behaviour and deformation processing of materials. *International Materials Review*, 43:145–187, 1998.

C. A. Lewis and P. J. Withers. Weibull modelling of particle cracking in metal matrix composites. *Acta Metallurgica et Materialia*, 43:3685–3699, 1995.

D. S. Liu and J. J. Lewandowski. The effect of superposed hydrostatic pressure on deformation and fracture. part II: particulate reinforced 6061 Al composite. *Metallurgical Transactions*, 24A:609–615, 1993.

D. S. Liu, M. Manoharan, and J. J. Lewandowski. Effects of microstructure on the behavior of an Al alloy and an Al matrix composite tested under low levels of superimposed hydrostatic pressure. *Metallurgical Transactions A*, 20A:2409–2417, 1989.

J. LLorca. An analysis of the influence of reinforcement fracture on the strength of discontinuously-reinforced metal-matrix composite. *Acta Metallurgica et Materialia*, 43:181–192, 1995.

J. LLorca and C. González. Microstructural factors controlling the strength and ductility of particle-reinforced metal-matrix composites. *Journal of the Mechanics and Physics of Solids*, 46:1–5, 1998.

J. LLorca, A. Martín, J. Ruiz, and M. Elices. Particulate fracture during deformation in a spray-formed metal-matrix composite. *Metallurgical Transactions A*, 24:1575–1588, 1993.

J. LLorca, A. Needleman, and S. Suresh. An analysis of the effects of matrix void growth on deformation and ductility in metal-ceramic composites. *Acta Metallurgica et Materialia*, 39:2317–2335, 1991.

J. LLorca and P. Poza. Influence of matrix strength on reinforcement fracture and ductility in Al/$Al_2O_3$ composites. *Materials Science and Engineering*, A185:25–37, 1993.

J. LLorca and J. Segurado. Three-dimensional multiparticle cell simulations of deformation and damage in sphere-reinforced composites. *Materials Science and Engineering A*, 365:267–274, 2004.

D. J. LLoyd. Aspects of fracture in particulate reinforced metal matrix composites. *Acta Metallurgica et Materialia*, 39:59–71, 1991.

E. Maire, C. Verdu, G. Lormand, and R. Fougeres. Study of the damage mechanisms in an osprey Al alloy-$SiC_p$ composite by means of SEM in situ tensile tests. *Materials Science and Engineering A*, A196:135–144, 1995.

E. Maire, D. S. Wilkinson, J. D. Embury, and R. Fougeres. Role of damage on the flow and fracture of particulate reinforced alloys and metal matrix composites. *Acta Materialia*, 45:5261–5274, 1997.

M. Manoharan and J. J. Lewandowski. Crack initiation and growth toughness of an aluminum metal-matrix composite. *Acta Metallurgica et Materialia*, 38:489–496, 1991.

J. C. Michel, H. Moulinec, and P. Suquet. Effective properties of composite materials with periodic microstructure: a computational approach. *Computational Methods in Applied Mechanical Engineering*, 172:109–143, 1999.

T. Mochida, M. Taya, and D. J. Lloyd. Fracture of particles in a particle/metal matrix composite under plastic straining and its effect on the Young's modulus of the composite. *Materials Transactions JIM*, 32:931–942, 1991.

T. Mochida, M. Taya, and M. Obata. Effect of damaged particles on the stiffness of a particle/metal matrix composite. *JSME International Journal*, 34:187–193, 1996.

A. C. Moloney, H. H. Kausch, T. Kaiser, and H. R. Beer. Parameters determining the strength and toughness of particulate filled epoxide resins. *Journal of Materials Science*, 22:381–393, 1987.

T. Mori and K. Tanaka. Average stress in the matrix and average elastic energy of materials with misfitting inclusions. *Acta Metallurgica et Materialia*, 21:571–574, 1973.

P. M. Mummery, P. Anderson, G. Davies, B. Derby, and J. Elliott. Damage assessment in particle-reinforced metal-matrix composites using x-ray microtomography. *Scripta Metallurgica et Materialia*, 29:1457–1462, 1993.

T. Mura. *Micromechanics of Defects in Solids.* Martinus Nijhoff Publishers, Dordrecht, 1987.

A. Needleman. A continuum model for void nucleation by inclusion debonding. *Journal of Applied Mechanics*, 54:525–531, 1987.

S. R. Nutt and J. M. Duva. A failure mechanism in Al-Sic composites. *Scripta Metallurgica et Materialia*, 20:1055–1058, 1986.

S. R. Nutt and A. Needleman. Void nucleation at fiber ends in Al-SiC composites. *Scripta Metallurgica et Materialia*, 21:705–710, 1987.

H. E. Petermann, A. F. Planskensteiner, H. J. Böhm, and F. G. Rammerstorfer. A termo-elasto-plastic constitutive law for inhomogeneous materials based on an incremental mori.tanaka approach. *Computers and Structures*, 71:197–214, 2002.

P. Poza and J. LLorca. Mechanical behaviour and failure micromechanisms of $Al/Al_2O_3$ composites under cyclic deformation. *Metallurgical and Materials Transactions A*, 26: 3131–3141, 1995.

P. Poza and J. LLorca. Fracture toughness and fracture mechanisms of $Al/Al_2O_3$ composites at cryogenic and elevated temperature. *Materials Science and Engineering*, A206:183–193, 1996.

P. Poza and J. LLorca. Mechanical behavior of Al-Li/SiC composites. part III: micromechanical modeling. *Metallurgical and Materials Transactions A*, 30:869–878, 1999.

J. Segurado, C. González, and J. LLorca. A numerical investigation of the effect of particle clustering on the mechanical properties of composites. *Acta Materialia*, 51: 2355–2369, 2003.

J. Segurado and J. LLorca. A numerical approximation to the elastic properties of sphere-reinforced composites. *Journal of the Mechanics and Physics of Solids*, 50:2107–2121, 2002.

J. Segurado and J. LLorca. A new three-dimensional interface finite element to simulate fracture in composites. *International Journal of Solids and Structures*, 41:2977–2993, 2004.

J. Segurado, J. LLorca, and C. González. On the accuracy of mean-field approaches to simulate the plastic deformation of composites. *Scripta Materialia*, 46:525–529, 2002.

P. M. Singh and J. J. Lewandowski. Effects of heat treatment and reinforcement size on reinforcement fracture during tension testing of a $SiC_p$ discontinuously reinforced aluminum alloy. *Metallurgical and Materials Transactions A*, 24A:2531–2543, 1993.

L. Z. Sun, J. W. Ju, and H. T. Liu. Elastoplastic modeling of metal matrix composites with evolutionary particle debonding. *Mechanics of Materials*, 35:559–569, 2003.

L. Z. Sun, H. T. Liu, and J. W. Ju. Effect of particle cracking on elastoplastic behaviour of metal matrix composites. *International Journal for Numerical Methods in Engineering*, 56:2183–2198, 2003.

P. Suquet. Effective properties of nonlinear composites. In P. Suquet, editor, *Continuum Micromechanics*, pages 197–264, Springer–Verlag, Vienna, 1997.

G. P. Tandon and G. J. Weng. A theory of particle-reinforced plasticity. *Journal of Applied Mechanics*, 55:126–135, 1988.

P. F. Thomason. *Ductile Fracture of Metals*. Pergamon Press, 1990.

K. Tohgo and T. W. Chou. *Incremental theory of particulate-reinforced composite taking damage process into consideration, CCM report 91-45*. University of Delaware, 1991.

V. Tvergaard. Influence of voids on shear band instabilities under plane strain conditions. *International Journal of Fracture*, 17:389–407, 1981.

V. Tvergaard. On localization in ductile materials containing spherical voids. *International Journal of Fracture*, 18:237–252, 1982.

V. Tvergaard. Effect of fibre debonding in a whisker-reinforced metal. *Materials Science and Engineering A*, A125:203–213, 1990.

V. Tvergaard. Effects of ductile matrix failure in three dimensional analysis of metal matrix composites. *Acta Materialia*, 46:3637–3648, 1998.

V. Tvergaard and A. Needleman. Analysis of the cup-cone fracture in a round tensile bar. *Acta Metallurgica*, 32:157–169, 1984.

R. H. Van Stone, T. B. Cox, J. R. Low, and J. A. Psioda. Microstructural aspects of fracture by dimple rupture. *International Metals Reviews*, 30:157–179, 1985.

K. Wallin, T. Saario, and K. Törrönen. Fracture of brittle particles in a ductile matrix. *International Journal of Fracture*, 32:201–209, 1987.

W. Weibull. A statistical distribution function of wide applicability. *Journal of Applied Mechanics*, 18:293–299, 1951.

A. F. Whitehouse and T. W. Clyne. Cavity formation during tensile straining of particulate and short fiber metal matrix composites. *Acta Metallurgica et Materialia*, 41: 1701–1711, 1993.

D. Zhao, F. R. Tuler, and D. J. Lloyd. Fracture at elevated temperatures in a particle reinforced composite. *Acta Metallurgica et Materialia*, 42:2525–2533, 1994.

Y. H. Zhao and G. J. Weng. Plasticity of a two-phase composite with partially debonded inclusions. *International Journal of Plasticity*, 12:781–804, 1996.

# Introduction to Crystal Plasticity Theory

Peter E. McHugh

Department of Mechanical and Biomedical Engineering and National Centre for Biomedical Engineering Science, National University of Ireland, Galway, Ireland.

**Abstract.** In this chapter an introduction to crystal plasticity theory is presented. The chapter includes brief reviews of continuum mechanics and the finite element method, necessary for the understanding of the theory. The theory itself is presented in detail and its implementation in the finite element method for solving problems is outlined. The chapter closes with a review of some case study applications.

## 1 Review of Continuum Mechanics

This section presents a quick review of continuum mechanics in the context of understanding the subsequent discussion on crystal plasticity theory.

Vectors (e.g., displacement and velocity) are represented using three components in 3D. The structure presented in equation (1.1) below applies when we are considering components referred to an orthonormal set of basis vectors, $\mathbf{e}_1$, $\mathbf{e}_2$, $\mathbf{e}_3$. Note also the definition of the summation convention, identified by repeated indices

$$\mathbf{v} = v_i\mathbf{e}_i = v_1\mathbf{e}_1 + v_2\mathbf{e}_2 + v_3\mathbf{e}_3 \tag{1.1}$$

The dot product of two vectors is defined as follows

$$\mathbf{v} \cdot \mathbf{u} = v_i u_i = v_1 u_1 + v_2 u_2 + v_3 u_3 \tag{1.2}$$

Second order tensors (e.g., stress and strain) are represented using nine components in 3D. The components of a tensor A with respect to an orthonormal basis are given by the following two equivalent expressions

$$\begin{aligned} A_{ij} &= \mathbf{e}_i \cdot \mathbf{A} \cdot \mathbf{e}_j \\ \mathbf{A} &= A_{ij}\mathbf{e}_i\mathbf{e}_j \end{aligned} \tag{1.3}$$

The second of equation (1.3) illustrates the definition of a dyadic or tensor product of two vectors, $\mathbf{e}_i$ and $\mathbf{e}_j$ in this case. In general second order tensors are linear vector functions of vectors, e.g.

$$\begin{aligned} \mathbf{v} &= \mathbf{A} \cdot \mathbf{u} \\ v_i &= A_{ij}u_j \end{aligned} \tag{1.4}$$

Fourth order tensors represent linear tensor functions of second order tensors, e.g. the tensor of linear elasticities, **L**, relating strain, $\boldsymbol{\varepsilon}$, to stress, $\boldsymbol{\sigma}$.

$$\begin{aligned} \boldsymbol{\sigma} &= \mathbf{L} : \boldsymbol{\varepsilon} \\ \sigma_{ij} &= L_{ijkl}\varepsilon_{lk} = L_{ijkl}\varepsilon_{kl} \end{aligned} \tag{1.5}$$

For a fourth order tensor, major symmetry means that $L_{ijkl} = L_{klij}$ and minor symmetry means that $L_{ijkl} = L_{jikl}$ and $L_{ijkl} = L_{ijlk}$.

### 1.1 Large Deformation Kinematics

Large (finite) deformation kinematics is usually addressed in terms of considering the deformation of a reference configuration into a current (or deformed) configuration. The position vector of a material point in the reference configuration is given by the vector **x** and the position of this point in the current configuration by the vector **y(x)**. The displacement vector for the point is **u(x)** = **y(x)** – **x**. The velocity vector of the material point is **v(x)** = $d\mathbf{u}/dt$ and the surface traction distribution on the current configuration is given by **t**, as shown in Figure 1 below.

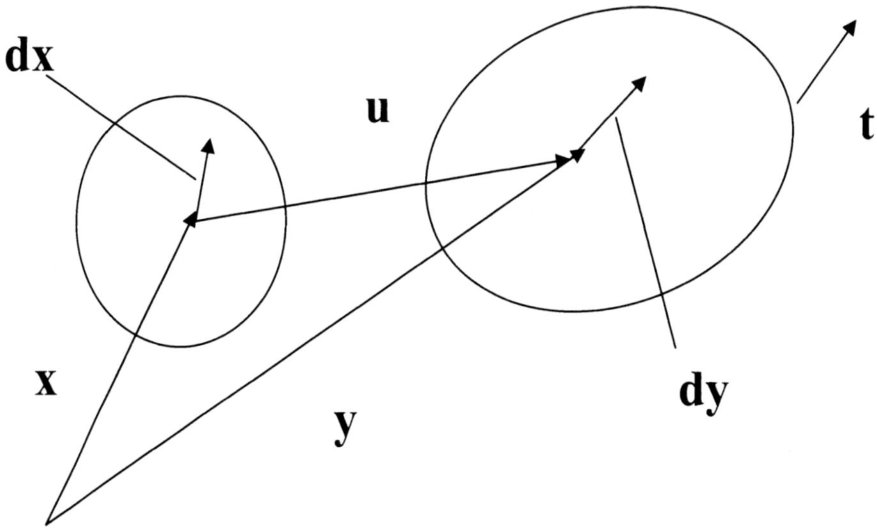


**Figure 1.** Illustration of finite deformation kinematics.

Also in Figure 1, an infinitesimal material "fibre" is described by **dx** in the reference configuration and by **dy** in the current configuration. The deformation gradient tensor **F** relates the two

$$\mathbf{dy} = \mathbf{F} \cdot \mathbf{dx} \Leftrightarrow \mathbf{F} = \frac{\partial \mathbf{y}}{\partial \mathbf{x}} = \nabla \mathbf{y} \Leftrightarrow F_{ij} = \frac{\partial y_i}{\partial x_j} = y_{i,j} \tag{1.6}$$

The right-most term in equation (1.6) presents the "shorthand" method of representing gradients in index notation. Also we can say

$$\mathbf{F} = \frac{\partial \mathbf{y}}{\partial \mathbf{x}} = \mathbf{I} + \frac{\partial \mathbf{u}}{\partial \mathbf{x}} = \mathbf{I} + \nabla \mathbf{u} \Leftrightarrow F_{ij} = \delta_{ij} + u_{i,j} \tag{1.7}$$

where **I** is the identity tensor, whose components are $\delta_{ij}$, the Kronecker delta function. A quantity that is usually of considerable interest is the spatial velocity gradient $\widetilde{\mathbf{L}}$, defined as

$$\widetilde{\mathbf{L}} = \frac{\partial \mathbf{v}}{\partial \mathbf{y}} = \dot{\mathbf{F}} \cdot \mathbf{F}^{-1} \tag{1.8}$$

## 1.2 Strain Measures

Consider the stretching of the material fibre $\mathbf{dx}$ from the point of view of change in magnitude. Consider, in particular, the square of the fibre length as follows

$$\begin{aligned}|\mathbf{dy}|^2 - |\mathbf{dx}|^2 &= (\mathbf{F}\cdot\mathbf{dx})\cdot(\mathbf{F}\cdot\mathbf{dx}) - \mathbf{dx}\cdot\mathbf{dx}\\ &= \mathbf{dx}\cdot(\mathbf{F}^{\mathrm{T}}\cdot\mathbf{F})\cdot\mathbf{dx} - \mathbf{dx}\cdot\mathbf{dx}\\ &= \mathbf{dx}\cdot(\mathbf{F}^{\mathrm{T}}\cdot\mathbf{F} - \mathbf{I})\cdot\mathbf{dx}\\ &= 2\mathbf{dx}\cdot\mathbf{E}\cdot\mathbf{dx}\end{aligned} \tag{1.9}$$

which allows the Lagrangian Strain tensor, $\mathbf{E}$, to be defined

$$\mathbf{E} = \tfrac{1}{2}(\mathbf{F}^{\mathrm{T}}\cdot\mathbf{F} - \mathbf{I}) \tag{1.10}$$

This is very clearly a non-linear relationship between strain and displacement gradient. The rate of Lagrangian strain is given by

$$\dot{\mathbf{E}} = \tfrac{1}{2}(\dot{\mathbf{F}}^{\mathrm{T}}\cdot\mathbf{F} + \mathbf{F}^{\mathrm{T}}\cdot\dot{\mathbf{F}}) \tag{1.11}$$

Another very important strain rate measure, normally associated with the current configuration, is the Rate of Deformation tensor, $\mathbf{D}$, derived from the spatial velocity gradient as

$$\mathbf{D} = \mathrm{sym}(\tilde{\mathbf{L}}) = \tfrac{1}{2}(\tilde{\mathbf{L}}^{\mathrm{T}} + \tilde{\mathbf{L}}) \tag{1.12}$$

Both $\mathbf{E}$ and $\mathbf{D}$ are symmetric tensors, i.e., $\mathbf{E} = \mathbf{E}^{\mathrm{T}}$, and $\mathbf{D} = \mathbf{D}^{\mathrm{T}}$. Under the assumption of infinitesimal (small) deformations and starting with the definition of Lagrangian strain, the strain-displacement gradient relationship expands as follows

$$\begin{aligned}\mathbf{E} &= \tfrac{1}{2}(\mathbf{F}^{\mathrm{T}}\cdot\mathbf{F} - \mathbf{I})\\ &= \tfrac{1}{2}((\mathbf{I} + \nabla\mathbf{u})^{\mathrm{T}}\cdot(\mathbf{I} + \nabla\mathbf{u}) - \mathbf{I})\\ &= \tfrac{1}{2}(\nabla\mathbf{u} + \nabla\mathbf{u}^{\mathrm{T}} + \nabla\mathbf{u}\cdot\nabla\mathbf{u}^{\mathrm{T}})\end{aligned} \tag{1.13}$$

Assuming that the product of infinitesimals is negligible we can say

$$\mathbf{E} \cong \tfrac{1}{2}(\nabla\mathbf{u} + \nabla\mathbf{u}^{\mathrm{T}}) = \boldsymbol{\varepsilon} \tag{1.14}$$

and we are left with the definition of the infinitesimal (small) strain tensor, $\boldsymbol{\varepsilon}$, which represents a linear relationship between strain and displacement gradient. Allied to this is the definition of the infinitesimal spin tensor, $\boldsymbol{\omega}$

$$\boldsymbol{\omega} = \nabla\mathbf{u} - \boldsymbol{\varepsilon} = \tfrac{1}{2}(\nabla\mathbf{u} - \nabla\mathbf{u}^{\mathrm{T}}) \tag{1.15}$$

## 1.3 Stress Measures

Cauchy (true) Stress is given the symbol $\boldsymbol{\sigma}$ and describes the force per unit area on the current configuration. It is a symmetric tensor and is related to the traction $\mathbf{t}$ on a surface (internal or external) in the current configuration and a unit normal vector to the surface $\tilde{\mathbf{n}}$,

$$\begin{aligned}\mathbf{t} &= \boldsymbol{\sigma}\cdot\tilde{\mathbf{n}}\\ t_i &= \sigma_{ij}\tilde{n}_j\end{aligned} \tag{1.16}$$

Other important stress tensors are the following: *Kirchhoff Stress* (symmetric), $\boldsymbol{\tau} = J\boldsymbol{\sigma}$, where $J = dV(\mathbf{x})/dV_0(\mathbf{x})$, is the Jacobian of the deformation, i.e., the ratio of local infinitesimal volumes in the neighbourhood of a material point $\mathbf{x}$, between the current configuration ($dV(\mathbf{x})$) and the reference configuration ($dV_0(\mathbf{x})$); *Nominal Stress* (unsymmetric), $\mathbf{n} = \mathbf{F}^{-1} \cdot \boldsymbol{\tau}$, which is the force per unit area on reference configuration; Second *Piola Kirchhoff Stress* (symmetric), $\mathbf{S} = \mathbf{F}^{-1} \cdot \boldsymbol{\tau} \cdot \mathbf{F}^{-T}$.

These different quantities satisfy work conjugacy rules; $\boldsymbol{\sigma} : \mathbf{D} = \sigma_{ij} D_{ij} = \dot{W}_C$ is the rate of work per unit current volume and $\boldsymbol{\tau} : \mathbf{D} = \mathbf{S} : \dot{\mathbf{E}} = \dot{W}_R$ is the rate of work per unit reference volume. In the small deformation case we can assume that deviations from the reference configuration are negligible, which means that $\mathbf{F}$ is approximated by $\mathbf{I}$, and all definitions of stress collapse onto one, usually denoted by $\boldsymbol{\sigma}$.

## 1.4 Equilibrium

The basic concept in static equilibrium is that the external forces and moments on a body all sum to zero. It is useful to have point-wise version of this concept, expressing equilibrium at (in the neighbourhood of) a material point $\mathbf{x}$. Such an expression can be written in terms of the divergence of the stress, for the case of small deformations as

$$\frac{\partial \sigma_{ij}}{\partial x_j} = \sigma_{ij,j} = 0 \qquad (1.17)$$

$$\nabla \cdot \boldsymbol{\sigma} = \mathbf{0}$$

For large deformation, we can say, in terms of quantities in the current configuration,

$$\frac{\partial \sigma_{ij}}{\partial y_j} = \sigma_{ij,j} = 0 \qquad (1.18)$$

$$\nabla \cdot \boldsymbol{\sigma} = \mathbf{0}$$

where $\boldsymbol{\sigma}$ is the Cauchy stress in this case and gradients are taken to be spatial gradients. If the body is subject to a body force distribution, e.g., as might come from gravity or a magnetic field, we can say

$$\frac{\partial \sigma_{ij}}{\partial y_j} + f_i = 0 \qquad (1.19)$$

where $f_i$ is known as the body force density. For dynamic situations, an inertia force density can be included

$$\frac{\partial \sigma_{ij}}{\partial y_j} + f_i - \rho \ddot{u}_i = 0 \qquad (1.20)$$

Equilibrium at a point on the surface of body can be expressed in terms of the relationship between stress and surface traction using equation (1.16).

## 1.5 Boundary Conditions

For the purposes of dealing with boundary conditions (BCs) on the surface of a body or region, the surface $S$ can be divided into the displacement boundary $S_1$ and the traction boundary $S_2$ as indicated in Figure 2 below.

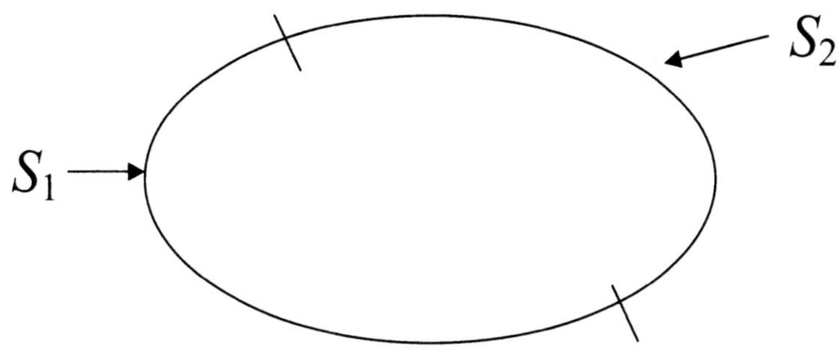


**Figure 2.** Specification of displacement and traction boundaries.

The surface $S$ is given by $S = S_1 \cup S_2$, where $S_1$ is the displacement boundary, where the kinematic boundary conditions, i.e., displacements and velocities, are prescribed, and $S_2$ is the traction boundary, where the traction boundary conditions, i.e., the surface forces are prescribed.

### 1.6 Boundary Value Problems

Problems in solid mechanics usually require that for a given body one can determine the complete mechanical state for any point in time. This means that (assuming small deformation terminology for convenience) we can determine $\boldsymbol{\varepsilon}(\mathbf{x})$, $\boldsymbol{\sigma}(\mathbf{x})$, $\mathbf{u}(\mathbf{x})$ and $\mathbf{t}(\mathbf{x})$, plus rates of these quantities for rate dependent or dynamic problems, plus other state variables (e.g., a strain hardening parameter for a plastic material) and their rates, for any time $t$. To do this we are usually required to solve the following

- Equilibrium equation (conditions on $\boldsymbol{\sigma}$)
- Constitutive equation ($\boldsymbol{\sigma} - \boldsymbol{\varepsilon}$ relationship)
- Compatibility equation ($\boldsymbol{\varepsilon} - \mathbf{u}$ relationship)

Together, these form a set of partial differential equations (PDEs). We must also consider the boundary conditions. Combining the PDEs and BCs, we are presented with a Boundary Value Problem (BVP) to solve. This concept readily extended to heat flow, diffusion problems, etc.

### 1.7 Boundary Value Problems

The Principle of Virtual Work (PVW), or in rate form the Principle of Virtual Power, can be written on either the reference or current configuration. It is an expression of the conservation of energy. Below, for example, we write it in terms of power on the current configuration with volume $V$ and surface $S$ (assuming zero body force $f_i$); the left hand side represents internal virtual power and the right hand side represents external virtual power.

$$\int_V \boldsymbol{\sigma} : \delta\mathbf{D} dV = \int_S \mathbf{t} \cdot \delta\mathbf{v} dS \qquad (1.21)$$

Here, $\boldsymbol{\sigma}$ and $\mathbf{t}$ are the real stress and surface traction, respectively, and $\delta\mathbf{D}$ and $\delta\mathbf{v}$ are the virtual rate of deformation and surface velocity, respectively. The fundamental point is that the relationship is true even when the virtual quantities (deformation rate and velocity) are not due to the real quantities (stress and traction). The virtual velocities (or displacements) must satisfy the kinematic (velocity or displacement) boundary conditions. Kinematic boundary conditions are also known as essential boundary conditions. A commonly used form of the PVW is the one used in small deformation analysis

$$\int_V \boldsymbol{\sigma} : \delta\boldsymbol{\varepsilon} dV = \int_S \mathbf{t} \cdot \delta\mathbf{u} dS \tag{1.22}$$

# 2 Crystal Plasticity Theory

## 2.1 Plasticity in Metals

The main physical mechanism of plastic deformation at ambient temperatures is the flow of dislocations along crystal slip systems. Dislocation motion results in no net compression or expansion of the crystal lattice, therefore plastic deformation is volume preserving. The lattice is undistorted when the dislocation has passed through

Slip systems are *close packed* atomic directions on *close packed* atomic planes. This yields the smallest dislocation Burgers Vector magnitude and hence the most favourable direction for slip. Some basic crystal structures and major slip systems are (indicated by their Miller Indices) as follows:

- FCC – Al, Cu, Ni, γ-Fe – <1 1 1>{1 1 0}
- BCC – α-Fe, δ-Fe, Cr, Mo – <1 1 0>{1 1 1}
- HCP – Mg, Zn – <0 0 0 1>{1 1 –1 0}

Plastic strain in a metal is due to flow of many millions of dislocations. When many dislocations come to the surface one gets a step, which results in a slip trace. Examples of this can be clearly seen for the single crystal and polycrystal cases for an Al alloy in the paper of Harren et al. (1988).

A very important side effect of crystallographic slip is crystal lattice rotation, which happens due to the finite deformation nature of massive dislocation flow. This is illustrated in Figure 3. In the figure an idealised single crystal specimen is presented, which has one active slip system, and which is loaded in tension. Due to the finite deformation nature of slip on the slip system an out of balance moment is generated. To re-establish equilibrium, the specimen and hence the lattice, must rotate. This generates bending stresses at the supports because of the gripping constraints. Generalising this, one can say that lattice rotation is geometrical in nature and also that rotation does not have to be uniform throughout a crystal; it can be constrained in certain locations, which can result in the generation of stresses.

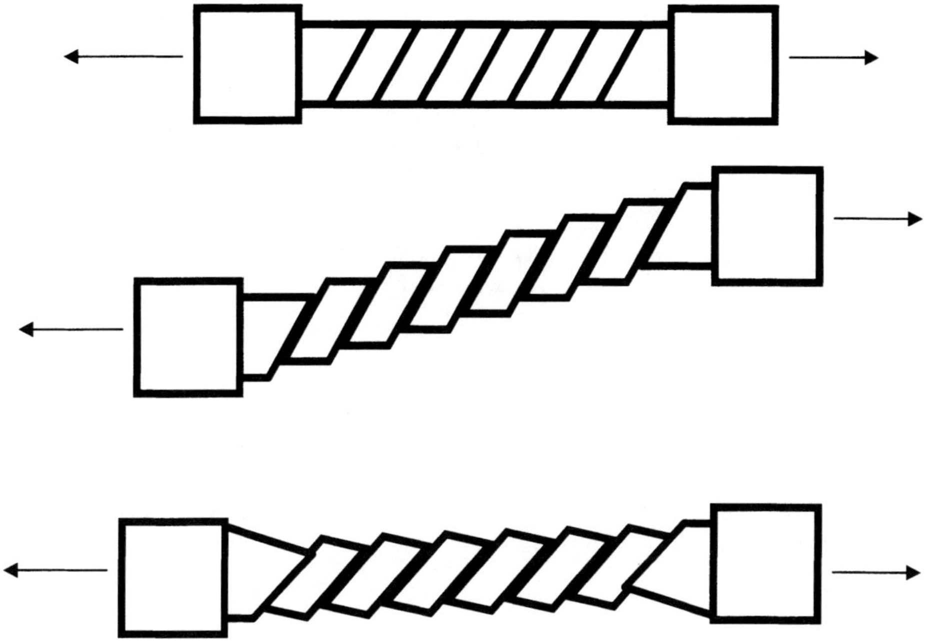

**Figure 3.** Single crystal tensile specimen, illustrating lattice rotation.

## 2.2 Kinematics of Crystal Plasticity

We are interested in representing dislocation motion and the associated plastic deformation in the context of continuum mechanics. This issue has been addressed by a large number of researchers (including Taylor, Hill, Rice, Asaro, Needleman, Teodosiu, Mandel, Berveiller and Cailletaud), all of whom have made vital contributions to the development of crystal plasticity theory. The reader should consult the reference list for examples of specific papers by these authors. In crystal plasticity theory, we represent (massive) dislocation flow on a slip system in a continuum sense as a plastic shear strain $\gamma$, as illustrated in Figure 4, where the slip plane normal is designated by the unit vector **m** and the slip direction by the unit vector **s**. The implications of adopting this continuum viewpoint are that we lose the representation of individual atoms and inter-atomic plane spacing, but on the other hand we keep the notion of stress and a direction of slip (slip system geometry). For more than one slip system active at a point, the plastic strain is a superposition of plastic shear strain in each slip direction. Crystal plasticity is considered to be a Physically Based theory, since it is based on the microstructural behaviour of materials. This is in contrast to Phenomenological theories, such as $J_2$ flow theory, which are based on either macroscopic or effective material behaviour. For crystal plasticity, to make this kinematic concept of shearing in slip directions useful, we need to relate the shear strains to a tensorial representation of deformation and/or strain.

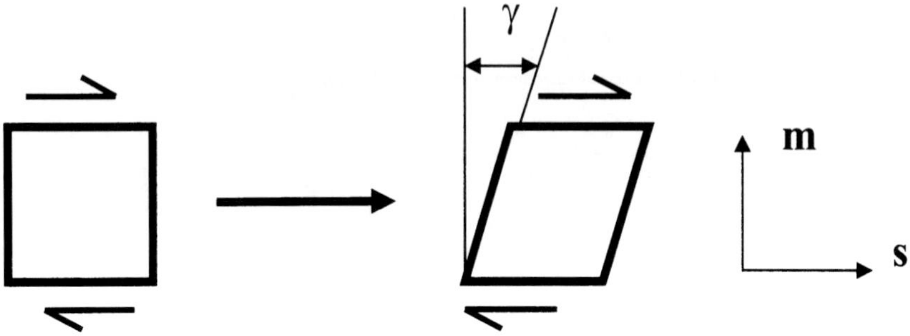


**Figure 4.** Plastic shear strain on a slip system.

In order to do this, consider the following situation (illustrated in Figure 5), under the assumption of small strain kinematics, where we have a material point **x** within a crystal and a material fibre **dx**, in the reference configuration.

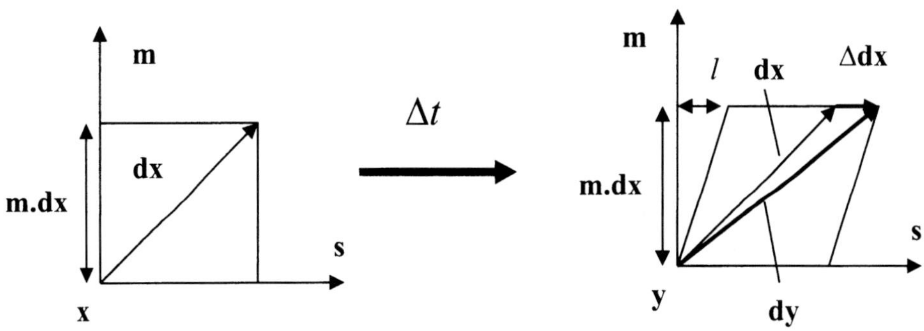


**Figure 5.** Kinematics of crystal deformation.

An increment in deformation, caused by shearing through an amount $\Delta\gamma$ on one slip system, occurs in a time $\Delta t$. The material point at **x** moves to **y** and the fibre is now described by **dy**. The stretching and rotation of the fibre is described by $\Delta\mathbf{dx}$. Consider the following, with reference to Figure 5,

$$\frac{l}{\mathbf{m}\cdot\mathbf{dx}} = \Delta\gamma = \dot{\gamma}\Delta t \tag{2.1}$$

$$l = \dot{\gamma}\Delta t\mathbf{m}\cdot\mathbf{dx} \tag{2.2}$$

but,

$$l = |\Delta\mathbf{dx}| \tag{2.3}$$

$$|\Delta\mathbf{dx}| = \dot{\gamma}\Delta t\mathbf{m}\cdot\mathbf{dx} \tag{2.4}$$

Now,

$$\Delta\mathbf{dx} = |\Delta\mathbf{dx}|\mathbf{s} \tag{2.5}$$

$$\Delta\mathbf{dx} = \dot{\gamma}\Delta t\mathbf{s}(\mathbf{m}\cdot\mathbf{dx}) \tag{2.6}$$

$$\frac{\Delta\mathbf{dx}}{\Delta t} = \dot{\gamma}\mathbf{s}(\mathbf{m}\cdot\mathbf{dx})$$
$$\lim_{\Delta t\to 0}\frac{\Delta\mathbf{dx}}{\Delta t} = \frac{d\mathbf{dx}}{dt} = d\left(\frac{\mathbf{dx}}{dt}\right) = d\dot{\mathbf{x}} \tag{2.7}$$

So,

$$d\dot{\mathbf{x}} = \dot{\gamma}\mathbf{s}(\mathbf{m}\cdot\mathbf{dx}) \tag{2.8}$$

In index notation, this can be written as

$$d\dot{x}_i = \dot{\gamma}s_i m_j dx_j \tag{2.9}$$

Also, from the chain rule for differentiation,

$$d\dot{x}_i = \frac{\partial \dot{x}_i}{\partial x_j}dx_j \tag{2.10}$$

So,

$$\frac{\partial \dot{x}_i}{\partial x_j}dx_j = \dot{\gamma}s_i m_j dx_j \tag{2.11}$$

For arbitrary $dx_j$ values we have

$$\frac{\partial \dot{x}_i}{\partial x_j} = \dot{\gamma}s_i m_j \tag{2.12}$$

However, the left hand side of this is the velocity gradient $\tilde{\mathbf{L}}$ (definition relevant for small strain kinematics, i.e., defined w.r.t. the reference configuration) so,

$$\frac{\partial \dot{x}_i}{\partial x_j} = \tilde{L}_{ij} = \dot{\gamma}s_i m_j$$
$$\tilde{\mathbf{L}} = \dot{\gamma}\mathbf{sm} \tag{2.13}$$

where **sm** is the dyadic or tensor product of the vectors **s** and **m**. The velocity gradient describes both the stretching rate and the spin rate. For $n$ slip systems, $\alpha$, active at a point we have

$$\tilde{\mathbf{L}} = \sum_{\alpha=1}^{n}\dot{\gamma}_\alpha \mathbf{s}_\alpha \mathbf{m}_\alpha \tag{2.14}$$

For the finite deformation case, the same framework can be used to describe the deformation of the crystal in going from a deformed configuration at time $t$ to a more incrementally deformed configuration at time $t$+$\Delta t$. In the case **x** and **y** in the above are replaced with $\mathbf{y}_t$ and $\mathbf{y}_{t+\Delta t}$. Following the same reasoning one can arrive at the following:

$$d\dot{\mathbf{y}} = \dot{\gamma}\mathbf{s}(\mathbf{m}\cdot\mathbf{dy}) \tag{2.15}$$

$$d\dot{y}_i = \dot{\gamma}s_i m_j dy_j \tag{2.16}$$

$$\frac{\partial \dot{y}_i}{\partial y_j} = \tilde{L}_{ij} = \dot{\gamma}s_i m_j \tag{2.17}$$

$$\tilde{\mathbf{L}} = \dot{\gamma}\mathbf{sm} \tag{2.18}$$

Here, $\tilde{\mathbf{L}}$ is the spatial velocity gradient (defined w.r.t. the current configuration). For $n$ slip systems, $\alpha$, active at a point we have

$$\tilde{\mathbf{L}} = \sum_{\alpha=1}^{n} \dot{\gamma}_\alpha \mathbf{s}_\alpha \mathbf{m}_\alpha \tag{2.19}$$

This result is due to Rice (1971). As we know from section 1.1, the deformation gradient $\mathbf{F}$ is given by

$$\mathbf{F} = \frac{\partial \mathbf{y}}{\partial \mathbf{x}} \tag{2.20}$$

and this is related to $\tilde{\mathbf{L}}$ through

$$\tilde{\mathbf{L}} = \frac{\partial \dot{\mathbf{y}}}{\partial \mathbf{y}} = \frac{\partial \dot{\mathbf{y}}}{\partial \mathbf{x}} \cdot \frac{\partial \mathbf{x}}{\partial \mathbf{y}} = \dot{\mathbf{F}} \cdot \mathbf{F}^{-1} \tag{2.21}$$

Decomposing this into Stretching (symmetric) and Spin (anti-symmetric) parts we get the following

$$\begin{aligned} \mathbf{D} &= \frac{1}{2}\left(\tilde{\mathbf{L}} + \tilde{\mathbf{L}}^{\mathrm{T}}\right) \\ \mathbf{\Omega} &= \frac{1}{2}\left(\tilde{\mathbf{L}} - \tilde{\mathbf{L}}^{\mathrm{T}}\right) \end{aligned} \tag{2.22}$$

$\mathbf{D}$ is the rate of deformation tensor ("strain" rate, defined by equation 1.12) and $\mathbf{\Omega}$ is the spin tensor (plastic spin in this case). For example if we take single slip (as per Figure 4, and identify the $x_1$ direction with $\mathbf{s}$ and the $x_2$ direction with $\mathbf{m}$) we can say

$$\mathbf{F} = \begin{bmatrix} 1 & \gamma \\ 0 & 1 \end{bmatrix} \Rightarrow \tilde{\mathbf{L}} = \begin{bmatrix} 0 & \dot{\gamma} \\ 0 & 0 \end{bmatrix} \Rightarrow \mathbf{D} = \begin{bmatrix} 0 & \frac{\dot{\gamma}}{2} \\ \frac{\dot{\gamma}}{2} & 0 \end{bmatrix}, \mathbf{\Omega} = \begin{bmatrix} 0 & \frac{\dot{\gamma}}{2} \\ -\frac{\dot{\gamma}}{2} & 0 \end{bmatrix} \tag{2.23}$$

This breakdown is illustrated in the following figure (Figure 6). Note that plastic deformation is through the lattice, there is no deformation of the lattice itself.

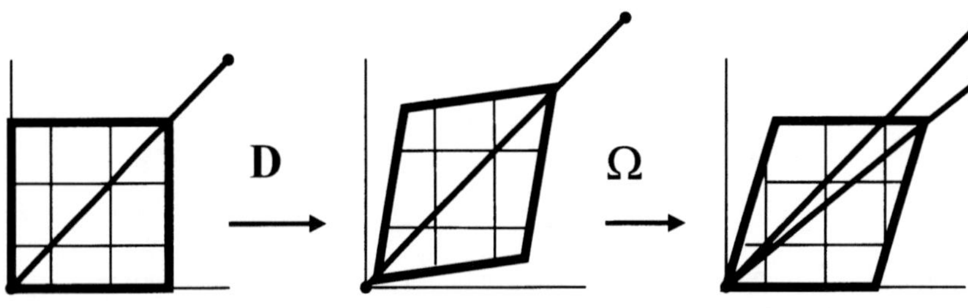


**Figure 6.** Kinematics of crystal deformation; strain and spin.

If there is lattice rotation then this will contribute an additional spin to the velocity gradient

$$\tilde{\mathbf{L}} = \mathbf{D} + \mathbf{\Omega}_{\text{plastic}} + \mathbf{\Omega}_{\text{lattice}} \tag{2.24}$$

An important feature of the kinematics of crystal plasticity theory is that, due to its physical basis, both the stretching and spin due to plasticity are explicitly given, i.e.,

$$\widetilde{\mathbf{L}} = \mathbf{D} + \mathbf{\Omega}_{\text{plastic}} = \sum_{\alpha=1}^{n} \dot{\gamma}_\alpha \mathbf{s}_\alpha \mathbf{m}_\alpha \tag{2.25}$$

This is in contrast to phenomenological theories where extra conditions must be invoked to determine the plastic spin, or where the plastic spin is assumed to vanish.

The next step is to extend the kinematical framework to include elastic deformation and rigid body rotation, i.e., to combine plastic deformation with elastic and rigid body deformation. This combination can be achieved in terms of a multiplicative decomposition of the deformation gradient **F** that is given as follows

$$\mathbf{F} = \mathbf{F}^* \cdot \mathbf{F}^p \tag{2.26}$$

$\mathbf{F}^p$ is the plastic part of deformation gradient and $\mathbf{F}^*$ is the elastic/rigid body part. With this we consider that, starting with the reference configuration, deformation occurs by plastic flow through the lattice, without deformation of the lattice, until an intermediate configuration is reached. This is followed by elastic deformation and rigid body motion of the lattice and its embedded material, until the final or current configuration is reached. In reality of course both types of deformation occur at the same time. However, this decomposition generates a convenient mathematical framework for the construction of a constitutive theory. We can define reference configuration, intermediate configuration and current configuration quantities. On the reference configuration we can (echoing the discussion in sections 1.2 and 1.3) define **S**, the Second Piola Kirchhoff stress (symmetric), and **E**, the Lagrangian strain tensor

$$\mathbf{S} = \mathbf{F}^{-1} \cdot \boldsymbol{\tau} \cdot \mathbf{F}^{-\mathrm{T}} \qquad \mathbf{E} = \tfrac{1}{2}\left(\mathbf{F}^{\mathrm{T}} \cdot \mathbf{F} - \mathbf{I}\right) \tag{2.27}$$

On the intermediate configuration we can define $\mathbf{S}^*$, the Second Piola Krichhoff stress on the intermediate configuration (symmetric), and $\mathbf{E}^*$, the Lagrangian strain tensor of the lattice, which represents elastic strain

$$\mathbf{S}^* = \mathbf{F}^{*-1} \cdot \boldsymbol{\tau} \cdot \mathbf{F}^{*-\mathrm{T}} \qquad \mathbf{E}^* = \tfrac{1}{2}\left(\mathbf{F}^{*\mathrm{T}} \cdot \mathbf{F}^* - \mathbf{I}\right) \tag{2.28}$$

On the deformed configuration we can speak in terms of the Cauchy or true stress $\boldsymbol{\sigma}$, the rate of deformation tensor **D** and the Kirchhoff stress $\boldsymbol{\tau}$, which is given by $\boldsymbol{\tau} = J\boldsymbol{\sigma}$, where $J$ is the Jacobian of the deformation. These quantities are work conjugates; the rate of internal work per unit reference volume is given by $\mathbf{S} : \dot{\mathbf{E}} = \boldsymbol{\tau} : \mathbf{D}$ and the rate of elastic lattice work per unit reference volume is given by $\mathbf{S}^* : \dot{\mathbf{E}}^*$.

## 2.3 Plastic Deformation

On this kinematical framework we can build a constitutive law. Let us first concentrate on plasticity, which is crystallographic. The basis of the description of the plasticity is that the slip system shear strain is driven by slip system resolved shear stress: $\tau_\alpha \Rightarrow \dot{\gamma}_\alpha$. Slip systems move (convect) with the crystal lattice, whose deformation is described by $\mathbf{F}^*$. We can say that $\mathbf{m}_\alpha$ and $\mathbf{s}_\alpha$ convect to $\mathbf{m}^*_\alpha$ and $\mathbf{s}^*_\alpha$ in the current configuration, where $\mathbf{s}^*_\alpha = \mathbf{F}^* \mathbf{s}_\alpha$ and $\mathbf{m}^*_\alpha = \mathbf{F}^{*\text{-T}} \mathbf{m}_\alpha$ (the latter ensures that $\mathbf{s}^*_\alpha \perp \mathbf{m}^*_\alpha$). The resolved shear stress is given by $\tau_\alpha = \mathbf{m}^*_\alpha \cdot \boldsymbol{\tau} \cdot \mathbf{s}^*_\alpha$.

The constitutive law to be developed can be either rate independent or rate dependent. The rate independent formulation of the constitutive law is based on identifying either active or inactive slip systems. For the active slip systems the shearing rate is non-zero ($\dot{\gamma}_\alpha \neq 0$) and this means that $|\tau_\alpha| = g_\alpha$, where $g_\alpha$ is the current slip system strain hardness. For inactive slip systems $|\tau_\alpha| < g_\alpha$ which means that the shearing rate is zero $\dot{\gamma}_\alpha = 0$. For perfect plasticity, $g_\alpha = g_0 =$ constant. For strain hardening plasticity $g_\alpha$ grows with strain.

For the strain hardening case $g_\alpha$ can be determined by the path dependent integration of the following evolution equation

$$\dot{g}_\alpha = \sum_{\beta=1}^{n} h_{\alpha\beta} |\dot{\gamma}_\beta| \tag{2.29}$$

$$g_\alpha = g_0, t = 0$$

In the above, $n$ is the number of slip systems and $h_{\alpha\beta}$ is a matrix of hardening moduli. The terms in this matrix relate the rate of hardening of slip system $\alpha$ to slip on $\alpha$ (self-hardening) and slip on all other slip systems $\beta$ (latent-hardening). So, for inactive slip systems $\delta$

$$\dot{\gamma}_\delta = 0 \Rightarrow \dot{\tau}_\delta < \dot{g}_\delta = \sum_{\beta=1}^{n} h_{\delta\beta} |\dot{\gamma}_\beta| \tag{2.30}$$

and for active slip systems $\alpha$

$$\dot{\gamma}_\alpha \neq 0 \Rightarrow \dot{\tau}_\alpha = \dot{g}_\alpha = \sum_{\beta=1}^{n} h_{\alpha\beta} |\dot{\gamma}_\beta| \tag{2.31}$$

Given the stress rate, the hardening moduli and a knowledge of the slip system geometry (**s** and **m** vectors) we would like to invert equations (2.31) (in fact modified versions) to yield the active slip system shear strain rates.

There is however a problem here. We have seen in the previous equations that the shear strain rates can be related to the macroscopic deformation measures (e.g., velocity gradient). If the active slip systems are *linearly dependent* then there is redundancy, i.e., different sets of slip system strain rates can be chosen to yield the same macroscopic strain rate. This means that there is non-uniqueness of solution for the active slip system strain rates and it is not possible to invert the equations. Quite a number of researchers have worked to overcome this non-uniqueness problem, e.g., Miehe (1996) and Anand and Kothari (1996), who developed iterative algorithms based on approximate non-singular version of the above equations (2.31). The problem is aggravated the more slip systems there are, e.g., more than 2 in 2D and more than 5 in 3D. Note that FCC has 12 slip systems.

In the rate dependent formulation the non-uniqueness problem is removed by effectively introducing an extra constraint equation for each slip system that establishes a direct and unique relationship between the resolved shear stress and the shear strain rate for that slip system. The most commonly used relationship was presented by Hutchinson (1976) and Pierce et al. (1983) and takes the form of a viscoplastic, power-law expression; it involves a reference strain rate $a_\alpha$ and a rate sensitivity exponent $m$, that are material properties.

$$\dot{\gamma}_\alpha = a_\alpha \,\mathrm{sgn}(\tau_\alpha) \left\{ \left| \frac{\tau_\alpha}{g_\alpha} \right| \right\}^{\frac{1}{m}} \tag{2.32}$$

Each slip system strain rate is a function of the resolved shear stress and the current slip system strain hardness. One is free to choose $m$ to capture strain rate dependent nature of material and this is useful at elevated temperatures and for creep. It is important to note that in this formulation, the resolved shear stress can be greater than current slip system hardness, depending on the strain rate and the value of $m$, e.g., if we assume that all quantities are positive and that $m$ = 1 (which would mean extreme rate sensitivity, usually $m$ <<1), we get

$$\frac{\dot{\gamma}_\alpha}{a_\alpha} = \frac{\tau_\alpha}{g_\alpha} \tag{2.33}$$

The rate independent limit is achieved as $m$ gets smaller. In this case, the resolved shear stress and the slip system strain hardness approach one another $m \to 0 \Leftrightarrow \tau_\alpha \cong g_\alpha$. This is particularly useful for effectively rate independent conditions (e.g., deformation of metals at ambient temperatures). Note that there is no explicit yield in this formulation. This means that plastic deformation happens from onset of loading on all slip systems and technically all slip systems are active. This is not a big disadvantage, e.g., for metals at ambient conditions $m$ <<<1 and so for stress levels below the current strain hardness, the shear strain rate is very small: $\tau_\alpha < g_\alpha \Rightarrow \dot{\gamma}_\alpha \cong 0$. In any event, it is not difficult to imagine formulations where both explicit yield and rate dependent plasticity are combined, e.g., Cuitiño and Ortiz (1996).

We have described crystallographic plasticity in terms of the slip system shear strain rate and the slip system resolved shear stress (as defined above). These two descriptors are not arbitrary but have been chosen to possess two important properties. (i) Work Conjugacy: $\tau_\alpha \dot{\gamma}_\alpha$ = rate of work due to slip on slip system $\alpha$ per unit reference volume, and (ii) Normality: when the descriptors are used as defined then the formulations (rate independent and rate dependent) can be shown to obey normality rules (Hill and Rice 1973, Asaro 1983).

## 2.4 Strain Hardening

This is a complex aspect of crystal plasticity theory and has led to much investigation. Here we look at approach of Asaro and co-workers. Recall equation (2.29) and expand it for the example of two slip systems: 1 and 2

$$\begin{aligned} \dot{g}_1 &= h_{11}|\dot{\gamma}_1| + h_{12}|\dot{\gamma}_2| \\ \dot{g}_2 &= h_{21}|\dot{\gamma}_1| + h_{22}|\dot{\gamma}_2| \end{aligned} \tag{2.34}$$

$h_{11}$ and $h_{22}$ are the self-hardening moduli, and $h_{12}$ and $h_{21}$ are the latent-hardening moduli. Assuming that both slip systems have identical properties, then $h_{11} = h_{22} = h$ and $h_{12} = h_{21} = h_{lh}$. We need to determine $h$ and $h_{lh}$. $l_h$, the latent hardening ratio, is the ratio of latent-hardening modulus to self-hardening modulus

$$l_h = \frac{h_{lh}}{h} \tag{2.35}$$

can be (and usually is) non-zero. This is because of dislocation interactions on intersecting slip systems. $l_h$ is not necessarily equal to 1 because dislocation flow on one slip system may generate different amounts of strain hardening on that slip system in comparison to intersecting slip systems. To determine $l_h$ many experimental investigations have been performed, a good number based on the "conjugate slip overshoot" method, outlined as follows. Take single crystal samples oriented for double slip, where only two planes are active (effective slip systems) with one slip system (primary slip system) more favourably oriented for slip (closer to 45°) than the other (conjugate slip system). Following initial yielding the primary slip system will slip first (single slip). The lattice will rotate to bring the primary slip system closer to the loading direction. This brings the conjugate slip system into a more favourable orientation for slip. At some point the conjugate slip system begins to slip. From measurements, the yield stress of the conjugate system and the current flow stress of the primary slip system are determined. The ratio of the two (conjugate yield stress/primary flow stress) produces the latent hardening ratio (see equation 2.35 above). If the slip systems are symmetrically oriented about the loading axis at this point, the latent hardening ratio is equal to 1. If not then the resolved shear stresses are not the same and latent hardening ratio is less than or greater than 1. Typical results for most metals range from the symmetric condition to an "overshoot" where the conjugate yield stress is higher than the primary flow stress: 1 < latent hardening ratio < 1.4. Based on such work, typical values of $l_h$ that have been used in simulations using crystal plasticity theory are: $1 < l_h < 1.4$. For $l_h = 1$ one has "Taylor Isotropic Hardening" which is very commonly used in simulations (e.g., in all simulations of McHugh and co-workers). With this structure

$$h_{\alpha\beta} = hq_{\alpha\beta} \tag{2.36}$$

In the 2 slip systems example,

$$h_{\alpha\beta} = h\begin{bmatrix} 1 & l_h \\ l_h & 1 \end{bmatrix} \qquad q_{\alpha\beta} = \begin{bmatrix} 1 & l_h \\ l_h & 1 \end{bmatrix} \tag{2.37}$$

To describe self hardening we must determine the self-hardening modulus, $h$. To deal with this, first consider single slip, a single crystal sample loaded so that only one slip system is active (we can drop the subscripts for convenience). $h$ is the derivative of the slip system resolved shear stress ($\tau = g$, assuming very low loading rate/rate independent conditions) – shear strain ($\gamma$) curve for the sample

$$\dot{g} = h\dot{\gamma} \Rightarrow \frac{dg}{d\gamma} = h \tag{2.38}$$

For different metals experiments can be done to determine the $g(\gamma)$ and hence the $h(\gamma)$ curves and functional forms can be fitted to them. For example (Asaro and co-workers) have used a function of the following type that has been found suitable for Al-Cu alloys (amongst others), which involves four material parameters.

$$g(\gamma) = g_0 + h_\infty\gamma + (g_\infty - g_0)\tanh\left\{\frac{h_0 - h_\infty}{g_\infty - g_0}\gamma\right\} \tag{2.39}$$

Also, power-law versions are common (e.g., Becker and co-corkers, McHugh and co-workers) which involve three material parameters, e.g.,

$$g(\gamma) = g_0\left(1 + \frac{\gamma}{\gamma_0}\right)^N \tag{2.40}$$

Going beyond single slip, to the more general case of multiple slip with many slip systems active, the single slip system strain $\gamma$ used in the above expressions to determine $g$ and hence $h$ (by differentiation) is of no use since each slip system will have a different shear strain. To deal with this, we can introduce the accumulated slip $\gamma_a$, which is a measure of total plastic shearing at a point

$$\gamma_a = \int_0^t \sum_{\alpha=1}^{n} |\dot{\gamma}_\alpha| dt \tag{2.41}$$

This can be used in place of $\gamma$ in the expressions for $g$ and $h$ yielding, for example

$$g(\gamma_a) = g_0 + h_\infty \gamma_a + (g_\infty\text{-}g_0) \tanh\left\{\frac{h_0\text{-}h_\infty}{g_\infty\text{-}g_0}\gamma_a\right\} \tag{2.42}$$

and,

$$h(\gamma_a) = \frac{dg(\gamma_a)}{d\gamma_a} = h_\infty + (h_0\text{-}h_\infty)\,\mathrm{sech}^2\left\{\frac{h_0\text{-}h_\infty}{g_\infty\text{-}g_0}\gamma_a\right\} \tag{2.43}$$

The replacement of $\gamma$ with $\gamma_a$ is analogous to the use of effective plastic strain, in place of tensile plastic strain, in simple tension stress-strain curves to describe multi-axial strain hardening in $J_2$ flow theory.

This description of hardening is essentially phenomenological and it attempts to account for the whole range of dislocation interaction behaviour using the self-hardening function $h$ and the single constant latent hardening ratio. In general this works quite well. However it misses out on physical details and subtle effects such as the following.

- It does not predict complete crystal orientation dependence.
- It does not allow description of transitions from Stage I (easy glide) to Stage II (strain hardening) to Stage III (saturation) hardening.
- Latent hardening ratios are not constant with strain and can be quite high (> 1.4) at small strains.
- It does not explicitly account for different effects of different types of dislocation interactions (Hirth lock, Co-planar junction, etc.).

A discussion on alternative and more sophisticated approaches is given later.

## 2.5 Elasticity

Having dealt with the plasticity, let us now consider elasticity. The fundamental issue here is that stress in the crystal depends only on elastic deformation of the crystal. In our finite deformation kinematical framework, we can phrase the elasticity in terms of quantities referred back to the intermediate configuration (equation 2.28). In rate form, we have

$$\dot{\mathbf{S}}^* = \mathbf{K} : \dot{\mathbf{E}}^*$$

$$K_{ijkl} = \frac{\partial^2 \Phi}{\partial E^*_{ij} \partial E^*_{kl}} \tag{2.44}$$

**K** is a tensor of elastic moduli and $\Phi$ is the Helmholtz free energy of the lattice per unit reference volume. In many cases we are interested in situations where plastic strains are much greater than elastic strains. In this event, it is sufficient to use the linear elastic constants for $\Phi$ and K. Both anisotropic and isotropic forms exist. In the isotropic case we have

$$\Phi = \tfrac{1}{2}\lambda\left(E^*_{11} + E^*_{22}\right)^2 + \mu\left(E^{*2}_{11} + E^{*2}_{22}\right) + \tfrac{1}{2}\mu\left(E^*_{12} + E^*_{21}\right)^2 \tag{2.45}$$

where $\lambda$ and $\mu$ are the Lamé elastic constants.

### 2.6 Overall Constitutive Law

The overall constitutive law for the elastic plastic material can be phrased in terms of quantities on the reference configuration (e.g., Harren et al. 1988, McHugh et al. 1993a) or the deformed configuration (e.g., Asaro and co-workers pre 1988, Needleman and co-workers). Both are essentially equivalent. In the following presentation we use the reference configuration. One of the implications of this is that the formulation does not involve quantities, and time derivatives of those quantities, on axes that are rotating.

Recalling the expressions for the strains from equations (2.27) and (2.28), differentiating and combining yields

$$\dot{\mathbf{E}}^* = \mathbf{F}^{p\text{-T}} \cdot \dot{\mathbf{E}} \cdot \mathbf{F}^{p\text{-1}} - \sum_{\alpha=1}^{n} \dot{\gamma}_\alpha \mathbf{A}_\alpha \tag{2.46}$$

$$\mathbf{A}_\alpha = \text{sym}\left\{\mathbf{F}^{*\text{T}} \cdot \mathbf{F}^* \cdot \mathbf{s}_\alpha \mathbf{m}_\alpha\right\}$$

Looking at the stresses and combining yields

$$\mathbf{S}^* = \mathbf{F}^p \cdot \mathbf{S} \cdot \mathbf{F}^{p\text{T}} \tag{2.47}$$

Differentiation yields

$$\dot{\mathbf{S}}^* = \mathbf{F}^p \cdot \dot{\mathbf{S}} \cdot \mathbf{F}^{p\text{T}} + 2\sum_{\alpha=1}^{n} \dot{\gamma}_\alpha \mathbf{B}_\alpha \tag{2.48}$$

$$\mathbf{B}_\alpha = \text{sym}\left\{\mathbf{s}_\alpha \mathbf{m}_\alpha \cdot \mathbf{S}^*\right\}$$

Substituting the stress and strain rates into the elastic rate equation yields the final form of the governing rate equation:

$$\begin{aligned} \dot{\mathbf{S}} &= \mathbf{L} : \dot{\mathbf{E}} - \sum_{\alpha=1}^{n} \dot{\gamma}_\alpha \mathbf{X}_\alpha \\ \mathbf{X}_\alpha &= \mathbf{F}^{p\text{-1}} \cdot \mathbf{R}_\alpha \cdot \mathbf{F}^{p\text{-T}} \\ \mathbf{R}_\alpha &= \mathbf{K} : \mathbf{A}_\alpha + 2\mathbf{B}_\alpha \\ L_{ijrn} &= F_{ik}^{p\text{-1}} F_{jl}^{p\text{-1}} K_{klpq} F_{rp}^{p\text{-1}} F_{nq}^{p\text{-1}} \end{aligned} \tag{2.49}$$

**L** here is a tensor of modified elastic moduli that has both the same major and minor symmetries as the tensor **K**. For an isotropic material, both major and minor symmetry will exit for both **L** and **K**. To appreciate the structure of this expression, consider a simple small strain analogy in 1D where strain ($\varepsilon$) can be decomposed additively into elastic ($\varepsilon^e$) and plastic ($\varepsilon^p$) parts and where the stress $\sigma$ is related to the elastic strain by $E$, the Young's modulus. The structure of the last line in equation (2.50) is the same as that of the first line in equation (2.49).

$$\begin{aligned} \varepsilon &= \varepsilon^e + \varepsilon^p \Rightarrow \dot{\varepsilon} = \dot{\varepsilon}^e + \dot{\varepsilon}^p \\ \sigma &= E\varepsilon^e \Rightarrow \dot{\sigma} = E\dot{\varepsilon}^e \\ \dot{\sigma} &= E(\dot{\varepsilon} - \dot{\varepsilon}^p) \\ \dot{\sigma} &= E\dot{\varepsilon} - E\dot{\varepsilon}^p \end{aligned} \tag{2.50}$$

# 3 Numerical Implementation

To solve problems with the theory presented in the previous section, the rate equation must be integrated with respect to time; we have stress and strain rates and we must generate stress and strain. As the governing rate equations are highly non-linear, integration is usually achieved incrementally, i.e., going from time $t$ to time $t$ plus increment in $t$: $t \rightarrow t+\Delta t$. Usually, very many increments/steps are required to achieve an accurate solution. For a single increment, stress at time $t$ changes to stress at $t+\Delta t$: $\mathbf{S}^t \rightarrow \mathbf{S}^{t+\Delta t} = \mathbf{S}^t + \Delta\mathbf{S}$. So, for incremental integration, the increment in stress is required. This is generated by a *Stress Update Algorithm* which evaluates the increment in stress and updates the stress (and also does this for all other relevant state variables). Various stress update algorithms have been used in the literature, for example the Rate Tangent Method of Pierce et al. (1983). This is an explicit update algorithm, i.e., it uses the information at time $t$ to get to time $t+\Delta t$ directly (as distinct from iteratively).

Methods of this type are concerned with converting the rate constitutive equation into a time discrete form in terms of average stress and average strain rates for the increment, ensuring an accurate relationship between the average stress and average strain rates. Equivalently, (when multiplied by $\Delta t$) they ensure an accurate relationship between the stress and strain increments.

Consider the following very simplified case to illustrate. Suppose we have a constitutive law in the following form

$$\dot{\sigma} = C\dot{\varepsilon} - d \tag{3.1}$$

Using the Simple/Forward Euler method, the stress increment would be calculated using $C$ and $d$ at time $t$,

$$\Delta\sigma_{\text{simple-euler}} = C\Delta\varepsilon - d\Delta t \tag{3.2}$$

Using the rate tangent method, the stress increment would be calculated using modified versions of $C$ and $d$, which can be considered as time discrete versions of these quantities,

$$\Delta\sigma_{\text{RT}} = C_{\text{RT}}\Delta\varepsilon - d_{\text{RT}}\Delta t \tag{3.3}$$

This is illustrated below in Figure 7, and the greater accuracy achieved by the rate tangent method is evident.

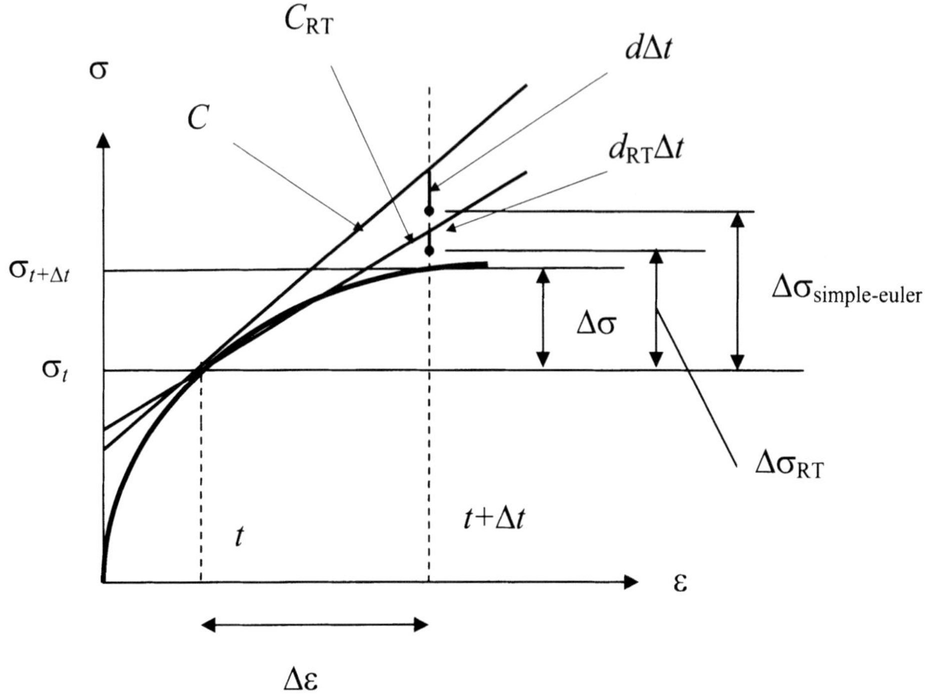


**Figure 7.** Illustration of different integration schemes.

### 3.1 Summary of Rate Tangent Method for Crystal Plasticity

In the context of crystal plasticity theory, the rate tangent algorithm is based on estimating the change in shear strain rates during the time increment. The slip system shear strain increment is given as follows

$$\Delta\gamma_\alpha = \gamma_\alpha^{t+\Delta t} - \gamma_\alpha^t \tag{3.3}$$

Linear interpolation is applied within increment, to estimate $\Delta\gamma_\alpha$, where $\Theta$ is a differencing parameter in the range $0 \le \Theta \le 1$,

$$\Delta\gamma_\alpha = \left[(1-\Theta)\dot{\gamma}_\alpha^t + \Theta\dot{\gamma}_\alpha^{t+\Delta t}\right]\Delta t \tag{3.4}$$

$\Theta = 0$ corresponds to Simple Euler integration and $0.5 \le \Theta \le 1$ is commonly used. Assuming that $\dot{\gamma}_\alpha$ is a function of $\tau_\alpha$ and $g_\alpha$, a first order Taylor expansion yields

$$\dot{\gamma}_\alpha^{t+\Delta t} = \dot{\gamma}_\alpha^t + \left.\frac{\partial\dot{\gamma}_\alpha}{\partial\tau_\alpha}\right|_t \Delta\tau_\alpha + \left.\frac{\partial\dot{\gamma}_\alpha}{\partial g_\alpha}\right|_t \Delta g_\alpha \tag{3.5}$$

with $\Delta\tau_\alpha = \dot{\tau}_\alpha\Delta t$ and $\Delta g_\alpha = \dot{g}_\alpha\Delta t$. Substituting in the power-law expression for $\dot{\gamma}_\alpha$ (equation 2.32) yields

$$\dot{\gamma}_\alpha^{t+\Delta t} = \operatorname{sgn}\{\tau_\alpha\}\left\{\left|\frac{\tau_\alpha}{g_\alpha}\right|\right\}^{\frac{1}{m}}\left\{1+\frac{1}{m}\left(\frac{\Delta\tau_\alpha}{\tau_\alpha}-\frac{\Delta g_\alpha}{g_\alpha}\right)\right\} \tag{3.6}$$

To make use of this expression in equation (3.4) we need expressions relating $\dot{\tau}_\alpha$ and $\dot{g}_\alpha$ to $\dot{\gamma}_\alpha$. It can be shown that

$$\tau_\alpha = \mathbf{s}_\alpha \cdot \mathbf{C}^* \cdot \mathbf{S}^* \cdot \mathbf{m}_\alpha \tag{3.7}$$

where $\mathbf{C}^* = \mathbf{F}^{*\mathrm{T}} \cdot \mathbf{F}^*$ is the right Cauchy-Green strain of elastic deformation. Differentiating this and inserting the previous expressions for rates of stress and strain yields

$$\dot{\tau}_\alpha = \mathbf{X}_\alpha : \dot{\mathbf{E}} - \mathbf{R}_\alpha : \sum_{\beta=1}^{n} \dot{\gamma}_\beta \mathbf{A}_\beta \tag{3.8}$$

The $\dot{g}_\alpha$ to $\dot{\gamma}_\alpha$ relationship is given by the strain hardening law (equation 2.29). Combining all of this, substituting equation (3.6) into equation (3.4) and manipulating, yields a linear system of equations for the $\Delta\gamma_\beta$, the increments in slip system shear strain

$$\sum_{\beta=1}^{n} N_{\alpha\beta}\Delta\gamma_\beta = \dot{\gamma}_\alpha^t \Delta t\left[1+\frac{\Theta\Delta t}{m\tau_\alpha}\mathbf{X}_\alpha : \dot{\mathbf{E}}\right] \tag{3.9}$$

These equations involve the following matrix

$$N_{\alpha\beta} = \delta_{\alpha\beta} + \frac{\Theta\dot{\gamma}_\alpha^t\Delta t}{m}\left\{\frac{\mathbf{R}_\alpha : \mathbf{A}_\beta}{\tau_\alpha} + \operatorname{sgn}\{\tau_\beta\}\frac{h_{\alpha\beta}}{g_\alpha}\right\} \tag{3.10}$$

For fixed $\Theta$ and $m$, $N_{\alpha\beta}$ is invertible if $\Delta t$ is sufficiently small. This is a limitation of the rate tangent explicit method and puts a constraint on the maximum time step that can be used. If we let $M_{\alpha\beta} = N_{\alpha\beta}^{-1}$ then we can write down the time discrete shear strain rate (denoted by the superscript $^d$) as

$$\dot{\gamma}_\alpha^d = \frac{\Delta\gamma_\alpha}{\Delta t} = \dot{f}_\alpha + \mathbf{F}_\alpha : \dot{\mathbf{E}} \tag{3.11}$$

with

$$\dot{f}_\alpha = \sum_{\beta=1}^{n} M_{\alpha\beta}\dot{\gamma}_\beta^t\left[1-\frac{\Theta\Delta t}{m}\right] \quad \text{and} \quad \mathbf{F}_\alpha = \sum_{\beta=1}^{n} M_{\alpha\beta}\frac{\Theta\dot{\gamma}_\beta^t\Delta t}{m\tau_\beta}\mathbf{X}_\beta \tag{3.12}$$

If we now recall the governing rate equations (2.49) and if we replace the instantaneous rate of shear strain with the time discrete rate of shear strain from equations (3.11) and (3.12) we can, following manipulation, write down the time discrete form of the governing rate equations

$$\dot{\mathbf{S}}^d = \mathbf{C}_{\mathrm{RT}} : \dot{\mathbf{E}}^d - \boldsymbol{\rho}_{\mathrm{RT}}$$

$$\mathbf{C}_{\mathrm{RT}} = \mathbf{L} - \sum_{\alpha=1}^{\mathrm{n}} \mathbf{X}_\alpha \cdot \mathbf{F}_\alpha \tag{3.13}$$

$$\boldsymbol{\rho}_{\mathrm{RT}} = \sum_{\alpha=1}^{\mathrm{n}} \mathbf{X}_\alpha \dot{f}_\alpha$$

or by multiplying by $\Delta t$, we can write down the time discrete incremental form

$$\Delta\mathbf{S} = \mathbf{C}_{\mathrm{RT}} : \Delta\mathbf{E} - \boldsymbol{\rho}_{\mathrm{RT}}\Delta t \tag{3.14}$$

$\mathbf{C}_{RT}$ is the rate tangent modulus tensor. It is a fourth order tensor and in general, because of its structure (equation 3.13), it may not have major symmetry, even when the tensor $\mathbf{L}$ does have major symmetry.

### 3.2 Solving Problems Using Crystal Plasticity

In using crystal plasticity theory to solve problems in solid mechanics, two main areas emerge.

1. Use with the finite element method to study the details of deformation in single crystals and polycrystals at the microscale. This involves the solution of BVPs.
2. Use to develop "macroscale" constitutive models for polycrystalline aggregates. Here, the theory is commonly used to predict deformation "textures", i.e., the alignment of crystal lattices, for initially randomly oriented grains, into well defined patterns during large scale deformation (rolling, forging, etc.). This kind of deformation results in material anisotropy at the macroscale. Self-consistent methods and "Taylor-type" modelling methods are commonly used to generate the polycrystalline aggregate constitutive models. These models can me implemented in finite element programs to study macroscale behaviour of polycrystalline bodies. There has been extensive research done in this area, for example by the research groups of McDowell, Dawson, Neale, Becker, Raabe, Asaro and Needleman. The reader should consult the reference list for specific papers by these authors.

Our main interest here is area (1). BVPs involving crystal plasticity theory can be solved using an appropriate stress update algorithm (to provide the temporal solution) and the finite element method (to provide the spatial solution). The usual starting point is the PVW (essentially equivalent to the Galerkin method). In the following we look at an example of an explicit finite element formulation that uses the explicit rate tangent integration algorithm discussed before. What is presented is a summary and the details are given in Dève et al. (1988).

To enforce equilibrium at the end of a time step, the principle of virtual work is written on the reference configuration $V_0$, whose surface is $S_0$.

$$\int_{V_0} \mathbf{n}^{t+\Delta t} : \delta\mathbf{F} dV = \int_{S_0} \mathbf{t}^{t+\Delta t} \cdot \delta\mathbf{u} dS \tag{3.15}$$

where $\mathbf{t} = \mathbf{N} \cdot \mathbf{n}$ is the nominal traction vector on the reference configuration ($\mathbf{N}$ is the outward unit normal to $S_0$), $\mathbf{n} = \mathbf{F}^{-1} \cdot \boldsymbol{\tau} = \mathbf{S} \cdot \mathbf{F}^{\mathrm{T}}$ is the nominal stress tensor, $\delta\mathbf{u}$ is the variational displacement field and $\delta\mathbf{F}$ is the variational deformation gradient. To obtain the evolution of the solution through time, the time discrete rate of nominal stress is used

$$\mathbf{n}^{t+\Delta t} = \mathbf{n}^{t} + \dot{\mathbf{n}}^{d} \Delta t \tag{3.16}$$

Substitution of this into equation (3.15) and manipulation yields

$$\int_{V_0} \dot{\mathbf{n}}^{d} : \delta\mathbf{F} dV = \frac{1}{\Delta t} \left\{ \int_{S_0} \mathbf{t}^{t+\Delta t} \cdot \delta\mathbf{u} dS - \int_{V_0} \mathbf{n}^{t+\Delta t} : \delta\mathbf{F} dV \right\} \tag{3.17}$$

Consider now the rate tangent time discrete form of the governing rate equations (3.13). The main quantities in (3.13) and (3.17) can be related (see Dève et al. 1988 for details). Essentially, relationships between the following pairs of quantities can be established: $\mathbf{n}^{t} \Leftrightarrow \mathbf{S}^{t}$, $\dot{\mathbf{n}}^{d} \Leftrightarrow \dot{\mathbf{S}}^{d}$, $\dot{\mathbf{F}}^{\mathrm{d}} \Leftrightarrow \dot{\mathbf{E}}^{d}$ and $\delta\mathbf{F} \Leftrightarrow \delta\mathbf{E}$. Using these relationships and equations (3.13), equation (3.17) can be transformed into the following form

$$\int_{V_0}\left[\delta\mathbf{E}:\mathbf{C}_{\mathrm{RT}}:\dot{\mathbf{E}}^d+\mathbf{S}^t\cdot\left(\dot{\mathbf{F}}^d\right)^{\mathrm{T}}:\delta\mathbf{F}\right]dV=\int_{V_0}\left\{\boldsymbol{\rho}_{\mathrm{RT}}-\frac{1}{\Delta t}\mathbf{S}^t\right\}:\delta\mathbf{E}dV+\frac{1}{\Delta t}\int_{S_0}\mathbf{t}^{t+\Delta t}\cdot\delta\mathbf{u}dS \quad (3.18)$$

This is a highly non-linear differential equation but it can be transformed into a set of linear algebraic equations by insertion of finite element interpolation functions. Consider the set of finite element interpolation/shape functions for a finite element "e" expressed in matrix form, $\mathbf{N}_{\mathrm{e}}$. Using the standard finite element approximation approach we can say the following

$$\mathbf{u}=\mathbf{N}_{\mathrm{e}}\,\mathbf{u}_{\mathrm{e}} \quad (3.19)$$

where $\mathbf{u}$ is the continuum displacement vector and $\mathbf{u}_{\mathrm{e}}$ is the vector of nodal displacements,

$$\mathbf{v}=\mathbf{N}_{\mathrm{e}}\,\mathbf{v}_{\mathrm{e}} \quad (3.20)$$

where $\mathbf{v}$ is the continuum velocity vector and $\mathbf{v}_{\mathrm{e}}$ is the vector of nodal velocities. Furthermore, by evaluating spatial gradients, we can say that

$$\mathbf{F}=\mathbf{I}+\nabla\mathbf{u}=\mathbf{I}+\nabla\,\mathbf{N}_{\mathrm{e}}\,\mathbf{u}_{\mathrm{e}} \quad (3.21)$$

This is of course the critical step in the finite element approach whereby gradients in the primary solution variable are replaced by a product of gradients in the known shape functions, and (approximate) nodal values of the primary solution variable itself. This of course leads to the situation where a set of partial differential equations (PDEs) such as those embedded in equation (3.18) can be converted into a set of algebraic equations. Going further, based on the finite element interpolation (equations 3.19 to 3.21), relationships between the following pairs of quantities can be readily established: $\mathbf{E}\Leftrightarrow\mathbf{u}_{\mathrm{e}}$, $\delta\mathbf{F}\Leftrightarrow\delta\mathbf{u}_{\mathrm{e}}$, $\delta\mathbf{E}\Leftrightarrow\delta\mathbf{u}_{\mathrm{e}}$, $\dot{\mathbf{E}}^d\Leftrightarrow\mathbf{v}_{\mathrm{e}}$ and $\dot{\mathbf{F}}^d\Leftrightarrow\mathbf{v}_{\mathrm{e}}$. Substitution of these relationships into equation (3.18), and elimination of the arbitrary variational quantities $\delta\mathbf{u}_{\mathrm{e}}$, yields a set of linear algebraic equations. These equations can be expressed in the standard linear finite element format, where the left hand side of equation (3.18) produces the stiffness matrix $\mathbf{K}_{\mathrm{e}}$ and nodal velocity vector $\mathbf{v}_{\mathrm{e}}$ (the direct solution variables in this case), and the right hand side produces the nodal force rate vector $\mathbf{P}_{\mathrm{e}}$

$$\mathbf{K}_{\mathrm{e}}\mathbf{v}_{\mathrm{e}}=\mathbf{P}_{\mathrm{e}} \quad (3.22)$$

Inversion of $\mathbf{K}_{\mathrm{e}}$ and multiplication by $\mathbf{P}_{\mathrm{e}}$ yields the solution $\mathbf{v}_{\mathrm{e}}$. In this case, since the equations were generated using the time discrete rate form of the governing equations, $\mathbf{v}_{\mathrm{e}}$ is the vector of average nodal velocities. From this the incremental nodal displacements can be determined by multiplying by $\Delta t$: $\Delta\mathbf{u}_{\mathrm{e}}=\mathbf{v}_{\mathrm{e}}\Delta t$. The rate tangent algorithm is then used to

- calculate increments in all relevant quantities for the increment (including state variables): $\Delta\mathbf{E}$, $\Delta\mathbf{F}$, $\Delta\gamma_\alpha$, $\Delta\gamma_\alpha$, $\Delta\tau_\alpha$, $\Delta g_\alpha$ and finally, $\Delta\mathbf{S}$
- update all quantities: $\mathbf{S}^{t+\Delta t}=\mathbf{S}^t+\Delta\mathbf{S}$, etc.

Following this, we then proceed to the next increment.

In summary, what has been presented in the above is an explicit finite element formulation based on an explicit stress update algorithm. The method is very accurate when small time steps are used. The time step size itself can be determined by empirical relationships or trial and error.

$\mathbf{K}_{\mathrm{e}}$ is usually an unsymmetric matrix, due to both the lack of major symmetry in $\mathbf{C}_{\mathrm{RT}}$ and the finite deformation kinematics, as can be inferred from the structure of equation (3.18). This means that in a computational implementation, an unsymmetric solver must be used.

# 4 Conventional Finite Element Formulation

The finite element formulation of crystal plasticity theory presented above does not immediately lend itself to implementation in a commercial finite element code. Therefore it is useful to consider crystal plasticity theory in the context of a "conventional" finite element formulation. A very good reference on finite element formulations of this nature is Zienkiewicz and Taylor (1991).

Consider a BVP, assuming small deformation kinematics, posed in terms of the PVW, where $\mathbf{t}$ is the surface traction vector and the $\delta$ refers to virtual quantities

$$\int_V \boldsymbol{\sigma} : \delta\boldsymbol{\varepsilon} dV = \int_S \mathbf{t} \cdot \delta\mathbf{u} dS \tag{4.1}$$

Rewrite this in matrix/vector form

$$\int_V \delta\boldsymbol{\varepsilon}^{\mathrm{T}} \boldsymbol{\sigma} dV = \int_S \delta\mathbf{u}^{\mathrm{T}} \mathbf{t} dS \tag{4.2}$$

For example in 2D we would have:

$$\underset{\delta\boldsymbol{\varepsilon}}{\begin{pmatrix} \delta\varepsilon_{11} \\ \delta\varepsilon_{22} \\ 2\delta\varepsilon_{12} \end{pmatrix}} \quad \underset{\boldsymbol{\sigma}}{\begin{pmatrix} \sigma_{11} \\ \sigma_{22} \\ \sigma_{12} \end{pmatrix}} \quad \underset{\delta\mathbf{u}}{\begin{pmatrix} \delta u_1 \\ \delta u_2 \end{pmatrix}} \quad \underset{\mathbf{t}}{\begin{pmatrix} t_1 \\ t_2 \end{pmatrix}} \tag{4.3}$$

Using the definition of the infinitesimal strain tensor and inserting the finite element interpolation from equation (3.19) we can say

$$\boldsymbol{\varepsilon} = \tfrac{1}{2}(\nabla\mathbf{u} + \nabla\mathbf{u}^{\mathrm{T}}) = \mathbf{B}_{\mathrm{e}}\mathbf{u}_{\mathrm{e}} \text{ and } \delta\boldsymbol{\varepsilon} = \mathbf{B}_{\mathrm{e}}\delta\mathbf{u}_{\mathrm{e}} \tag{4.4}$$

Here $\mathbf{B}_{\mathrm{e}}$ is the matrix of shape function gradients. For a material with a non-linear constitutive response we can say

$$\boldsymbol{\sigma} = \boldsymbol{\sigma}(\boldsymbol{\varepsilon}) \rightarrow \boldsymbol{\sigma} = \boldsymbol{\sigma}(\mathbf{u}_{\mathrm{e}}) \tag{4.5}$$

Substitution into the PVW and summation over all the elements "e" in a finite element mesh, each with a volume $V_{\mathrm{e}}$ and a surface $S_{\mathrm{e}}$, yields

$$\sum_{\mathrm{e}} \int_{V_{\mathrm{e}}} \delta\mathbf{u}_{\mathrm{e}}^{\mathrm{T}} \mathbf{B}_{\mathrm{e}}^{\mathrm{T}} \boldsymbol{\sigma}(\mathbf{u}_{\mathrm{e}}) dV = \sum_{\mathrm{e}} \int_{S_{\mathrm{e}}} \delta\mathbf{u}_{\mathrm{e}}^{\mathrm{T}} \mathbf{N}_{\mathrm{e}}^{\mathrm{T}} \mathbf{t} dS \tag{4.6}$$

If we drop the summation notation and the subscripts for convenience and eliminate or "cancel" the arbitrary virtual displacements we get

$$\int_V \mathbf{B}^{\mathrm{T}} \boldsymbol{\sigma}(\mathbf{u}) dV = \int_S \mathbf{N}^{\mathrm{T}} \mathbf{t} dS \tag{4.7}$$

If we rearrange this we get

$$\mathbf{G}(\mathbf{u}) = \int_V \mathbf{B}^{\mathrm{T}} \boldsymbol{\sigma}(\mathbf{u}) dV - \int_S \mathbf{N}^{\mathrm{T}} \mathbf{t} dS = \mathbf{0} \tag{4.8}$$

$\mathbf{G}$ is a set of non-linear equations in $\mathbf{u}$, where $\mathbf{u}$ now means the vector of nodal displacements and $\mathbf{G}$ is the residual force vector. The solution of this set of equations is usually achieved by incremental methods, i.e., applying load in increments/steps and stepping to final time $t_{final}$ in time steps $\Delta t$. Numerous algorithms have been developed to solve the equations for the increment and these can be generally classified under the headings of Implicit or Explicit methods.

Considering an increment in time, $t \rightarrow t+\Delta t$, implicit methods involve determining the state at $t+\Delta t$ based on information at time $t$ and $t+\Delta t$, whereas explicit methods involve determining the state at $t+\Delta t$ based on information at time $t$ alone. An example of an explicit method is the previously presented finite element formulation for crystal plasticity based around the rate tangent method. In implicit methods iteration is usually required to obtain the solution for an increment. Generally, for reasonable accuracy, $\Delta t$ values for implicit methods can be an order of magnitude greater than $\Delta t$ values for explicit methods. In the following, we consider a very common implicit method, the Newton-Raphson method.

### 4.1 Newton-Raphson Method

Assume that we have solved for state at time $t$ and we wish to update state to $t+\Delta t$. The $\mathbf{u}^{t+\Delta t}$ are considered as main solution variables and we must solve the following non-linear set of equations for $\mathbf{u}^{t+\Delta t}$

$$\mathbf{G}\left(\mathbf{u}^{t+\Delta t}\right)=\mathbf{0} \tag{4.9}$$

The Newton-Raphson (NR) method can be applied to determine the $\mathbf{u}^{t+\Delta t}$ by iteration. Let's first look at a 1D analogy. Suppose we wish to solve $f(x) = 0$ by NR. If we have a guess at the root $x_i$, a better guess $x_{i+1}$ given by the NR formula

$$x_{i+1}=x_i-\left[\frac{df}{dx}\right]_{x_i}^{-1}\cdot f(x_i) \tag{4.10}$$

The method is applied iteratively, i.e., $x_{i+1}$ is substituted for $x_i$ and the NR formula is reapplied. This process is continued until a *convergence criterion* is satisfied

$$\begin{gathered}\left|x_{i+1}-x_i\right|<Tolerance\\ \left|f(x_{i+1})\right|<Tolerance\end{gathered} \tag{4.11}$$

If we apply the same concept to our non-linear finite element equations (4.9) we can say for the $i^{\text{th}}$ iteration

$$\mathbf{u}_{i+1}^{t+\Delta t}=\mathbf{u}_i^{t+\Delta t}-\left[\frac{\partial\mathbf{G}\left(\mathbf{u}_i^{t+\Delta t}\right)}{\partial\mathbf{u}}\right]^{-1}\mathbf{G}\left(\mathbf{u}_i^{t+\Delta t}\right) \tag{4.12}$$

If we reorganise this in terms of the change in incremental displacement we get

$$\delta\mathbf{u}_{i+1}=\mathbf{u}_{i+1}^{t+\Delta t}-\mathbf{u}_i^{t+\Delta t}=-\left[\frac{\partial\mathbf{G}\left(\mathbf{u}_i^{t+\Delta t}\right)}{\partial\mathbf{u}}\right]^{-1}\mathbf{G}\left(\mathbf{u}_i^{t+\Delta t}\right) \tag{4.13}$$

The partial derivative term on the right hand side of equation (4.13) is the Jacobian of the governing equations. We can identify this as the Tangent Stiffness matrix, $\mathbf{K}$. Using this we can say

$$\delta\mathbf{u}_{i+1}=-\mathbf{K}\left(\mathbf{u}_i^{t+\Delta t}\right)^{-1}\mathbf{G}\left(\mathbf{u}_i^{t+\Delta t}\right) \tag{4.14}$$

and hence by inversion

$$\mathbf{K}\left(\mathbf{u}_i^{t+\Delta t}\right)\delta\mathbf{u}_{i+1}=-\mathbf{G}\left(\mathbf{u}_i^{t+\Delta t}\right) \tag{4.15}$$

The matrix equation (4.15) represents a set of linear algebraic equations in the $\delta\mathbf{u}_{i+1}$. It has the same form as for a linear problem ($\mathbf{Ku} = \mathbf{P}$). It must be solved, for each iteration, for the change in incremental displacements. The matrix **K** and vector **G** (the residual force vector) are different for each iteration. At any particular iteration, the current increment in displacements is given by

$$\mathbf{u}_i^{t+\Delta t} - \mathbf{u}^t = \Delta\mathbf{u}_i = \sum_{k=1}^{i} \delta\mathbf{u}_k \tag{4.16}$$

The convergence criterion can be expressed in terms of **G** as follows

$$\left|\mathbf{G}\left(\mathbf{u}_{i+1}^{t+\Delta t}\right)\right| < Tolerance \tag{4.17}$$

It is important to note that the method requires accurate evaluation of $\mathbf{K}\left(\mathbf{u}_i^{t+\Delta t}\right)$ and $\mathbf{G}\left(\mathbf{u}_i^{t+\Delta t}\right)$ for each iteration *i*. **G** requires accurate evaluation of the stress $\boldsymbol{\sigma}$ at the current estimate of $\mathbf{u}^{t+\Delta t}$. This can be seen from the form of **G**,

$$\mathbf{G}\left(\mathbf{u}_i^{t+\Delta t}\right) = \int_V \mathbf{B}^{\mathrm{T}} \boldsymbol{\sigma}(\mathbf{u}_i^{t+\Delta t}) dV - \int_S \mathbf{N}^{\mathrm{T}} \mathbf{t}^{t+\Delta t} dS \tag{4.18}$$

which in turn means that a stress update algorithm is required. Next, look at structure of **K**.

$$\mathbf{K}\left(\mathbf{u}_i^{t+\Delta t}\right) = \frac{\partial \mathbf{G}\left(\mathbf{u}_i^{t+\Delta t}\right)}{\partial \mathbf{u}} = \int_V \mathbf{B}^{\mathrm{T}} \left.\frac{\partial \boldsymbol{\sigma}}{\partial \mathbf{u}}\right|_{\left(\mathbf{u}_i^{t+\Delta t}\right)} dV = \int_V \mathbf{B}^{\mathrm{T}} \left.\frac{\partial \boldsymbol{\sigma}}{\partial \boldsymbol{\varepsilon}}\right|_{\left(\mathbf{u}_i^{t+\Delta t}\right)} \frac{\partial \boldsymbol{\varepsilon}}{\partial \mathbf{u}} dV = \int_V \mathbf{B}^{\mathrm{T}} \left.\frac{\partial \boldsymbol{\sigma}}{\partial \boldsymbol{\varepsilon}}\right|_{\left(\mathbf{u}_i^{t+\Delta t}\right)} \mathbf{B} dV \tag{4.19}$$

The partial derivative in the right-most term in equation (4.19) is the Jacobian of the constitutive law and can be identified as the Tangent matrix, $\mathbf{D}^{tan}$.

$$\mathbf{D}^{tan} = \left.\frac{\partial \boldsymbol{\sigma}}{\partial \boldsymbol{\varepsilon}}\right|_{\left(\mathbf{u}_i^{t+\Delta t}\right)} \Rightarrow \mathbf{K}\left(\mathbf{u}_i^{t+\Delta t}\right) = \int_V \mathbf{B}^{\mathrm{T}} \mathbf{D}^{tan} \mathbf{B} dV \tag{4.20}$$

The tangent stiffness matrix **K** has the same form as for linear problems, i.e., the classic "$\mathbf{B}^{\mathrm{T}}\mathbf{D}\mathbf{B}$" structure. However here **K** is different for each iteration. This can lead to problems since the tangent matrix can be difficult to evaluate for very non-linear materials.

The NR method is accurate and displays rapid convergence. However sometimes it is modified, e.g.,

- use of a constant **K**, from the first iteration in the increment (or even the first increment in the loading history – initial stress method) which means we avoid having to recalculate **K** for each iteration
- use of a symmetric **K** if the problem generates an unsymmetric **K**, otherwise we must use unsymmetric solvers
- more complex versions of the algorithm: BFGS method (Broyden-Fletcher-Goldfarb-Shanno).

These and other methods can work well. They sometimes result in slower convergence, but they can actually increase the radius of convergence.

It is important to note that it is acceptable to modify **K** but we always must calculate **G** correctly and since **G** depends directly on the stress this means that we must have a stress update algorithm that is accurate:

$$\mathbf{u}_i^{t+\Delta t} = \mathbf{u}^t + \Delta\mathbf{u}_i$$

$$\mathbf{G}\left(\mathbf{u}_i^{t+\Delta t}\right) = \int_V \mathbf{B}^{\mathrm{T}} \boldsymbol{\sigma}(\mathbf{u}_i^{t+\Delta t}) dV - \int_S \mathbf{N}^{\mathrm{T}} \mathbf{t}^{t+\Delta t} dS \tag{4.21}$$

$$\boldsymbol{\sigma}(\mathbf{u}_i^{t+\Delta t}) = \boldsymbol{\sigma}(\mathbf{u}^t + \Delta\mathbf{u}_i) = \boldsymbol{\sigma}(\mathbf{u}^t) + \Delta\boldsymbol{\sigma}_i$$

From equation (4.21) we must determine $\Delta\boldsymbol{\sigma}_i$ accurately, given $\Delta\mathbf{u}_i$. This can be quite difficult for complex non-linear constitutive laws.

### 4.2 Explicit Stress Update Algorithms

Explicit algorithms use information at time $t$ to get to time $t+\Delta t$ directly. Perhaps the simplest explicit method is the Simple/Forward Euler method which is based on a linearisation of the constitutive law about time $t$. It is usually easy to implement but it is not very accurate unless very small time steps are used. Another explicit method is the Runge Kutta method. This is a multi-step explicit method based on subdivision of the time step. The rate tangent method, introduced above, is also an explicit method. For the general case of a rate dependent elastic-plastic constitutive law, the method is based on linearising the non-linear expression for plastic strain rate, using a Taylor expansion about time $t$ to estimate the plastic strain rate at $t+\Delta t$. Since it involves this estimation of the state at $t+\Delta t$, it is also sometimes called a "semi-implicit" method.

### 4.3 Implicit Stress Update Algorithms

For an implicit algorithm the state at time $t+\Delta t$ is determined from information at time $t$ and $t+\Delta t$. An example of a "fully" implicit method is the Backward Euler method which is based on a linearisation of the constitutive law about time $t+\Delta t$. Since the state at $t+\Delta t$ is unknown, the method usually produces a system of coupled non-linear equations in the stress/state variables at $t+\Delta t$ that must be solved to perform the stress/state variable update. The equations can usually be solved using a NR iteration. The Backward Euler method is very accurate.

There are many other forms of explicit and implicit stress update algorithms presented in the literature. Implicit algorithms usually allow bigger time increments to be taken than explicit formulations, from the point of view of accuracy and stability, due to the iteration and the shooting for convergence at the end of the time step. Factors of at least 10 in time step size difference between implicit and explicit are possible.

Remember that we are still discussing stress update algorithms within the context of the implicit NR (or similar) solution of the overall system equations $\mathbf{G(u)} = \mathbf{0}$. The implication of this is that we can have explicit or implicit stress update algorithm operating within each NR iteration. For example, if using Backward Euler method we would have a series of iterations to update stresses and state variables within each element, and in fact a series of iterations at each integration point within each element, and this is done for each iteration of the NR solution of the governing equations for each increment.

It is interesting to note that people commonly write user material subroutines (UMATs) for the implicit commercial finite element code ABAQUS/Standard. Usually these routines implement a stress update algorithm. It should be realised that it is up to the user to write a UMAT that produces a sufficiently accurate stress update based on the time and strain increments provided by ABAQUS. For a given iteration within a given increment, ABAQUS will provide the UMAT with a number of quantities, including the time step and the current estimate of the strain increment (ABAQUS → UMAT: $\Delta t$, $\Delta\varepsilon$,...). Based on these, the UMAT increments and updates the stress and the other state variables and calculates the Jacobian of the constitutive law (UMAT → ABAQUS: $\Delta\sigma$, $d\sigma/d\varepsilon$,...). The important issue to consider is whether the stress increment/updated stress is accurate based on the ABAQUS $\Delta t$. ABAQUS assumes that it is and if the overall NR iteration converges then it will assume that equilibrium has been achieved for $t+\Delta t$. But just because convergence has been reached doesn't mean that the solution is correct. It will only be correct if the stress increment/updated stress provided by the UMAT is correct. Therefore, care must be taken in writing UMATs, and for a given $\Delta t$, in order to ensure accuracy, it might be necessary to modify or reduce $\Delta t$ within the UMAT. Fortunately, this can readily be done.

### 4.4 Dynamical Solution Methods

Since we have used the word explicit above, it is important to distinguish the above approach from the dynamic explicit approach that is also commonly (and therefore confusingly) referred to as explicit.

First of all, consider including inertia and formulating the BVP dynamically as expressed by the following

$$\mathbf{M}\ddot{\mathbf{u}} + \mathbf{G}(\mathbf{u},\mathbf{v}) = \mathbf{0} \tag{4.22}$$

Here, **M** is a mass matrix, $\ddot{\mathbf{u}}$ is the nodal acceleration vector, and **G** is now a function of the nodal displacements **u** and the nodal velocities **v**. This is a general form valid for non-linear materials, finite deformation kinematics, etc. Both implicit and explicit methods have been developed to solve such equations. In the dynamic context the terms implicit and explicit usually have the following meanings.

- Explicit: no matrix inversion is required and there is usually no need to form a tangent stiffness matrix **K** from **G**. A popular explicit method is the Central Difference Method.
- Implicit: matrix inversion/equation solution is required, and this can be iterative. A popular implicit method is Newmark's Algorithm.

Let us consider the central difference method. A good description of the method in the finite element context is given in the ABAQUS Theory manual (HKS 1997a). The central difference method is a fully explicit method, involving no matrix inversion and no formation of a stiffness matrix, **K**. **M** must be in diagonal or "lumped" mass form. The method works in terms of "half" increments: $i$-1/2, $i$, $i$+1/2, $i$+1, etc. The algorithm works in terms of first evaluating accelerations by solving the first of equation (4.23), which is trivial since **M** is diagonal, and then solving for velocities and displacements.

$$\begin{aligned} \ddot{\mathbf{u}}_i &= -\mathbf{M}^{-1}\mathbf{G}_i \\ \mathbf{v}_{i+\frac{1}{2}} &= \mathbf{v}_{i-\frac{1}{2}} + \frac{\Delta t_{i+1} + \Delta t_i}{2}\ddot{\mathbf{u}}_i \\ \mathbf{u}_{i+1} &= \mathbf{u}_i + \Delta t_{i+1}\mathbf{v}_{i+\frac{1}{2}} \end{aligned} \tag{4.23}$$

This is a very computationally efficient method for updating the state variables. It integrates through time using many small times increments for accuracy. It is conditionally stable and the time step must be kept within stability limit. Algorithms have been developed to evaluate this limit. Explicit methods are becoming more popular since there are no iteration and no convergence difficulties. They are suited to large deformation problems and highly non-linear problems that might arise due to material constitutive law complexity and surface contact, for example. The main down side it the necessity to use a very small time step size that can slow down computation. Time step size can be increased using mass scaling, i.e., artificially factoring-up the mass. However this has to be carefully monitored to avoid the introduction of artificially high kinetic energy effects.

## 4.5 Quasi-static Formulation: Implicit Solution for Finite Deformation Kinematics

Let us now return to considering the quasi-static formulation of the BVP expressed in equation (4.8), and introduce finite deformation kinematics. Let us look at the complexities this can introduce. A good discussion on this is also presented in the ABAQUS Theory Manual (HKS 1997a). Consider now the PVW expressed in rate form on the current configuration with volume $V$ and surface $S$. In this case $\mathbf{t}$ is the surface traction vector in the deformed configuration

$$\int_V \delta\mathbf{D}^{\mathrm{T}}\boldsymbol{\sigma}\,dV = \int_S \delta\mathbf{v}^{\mathrm{T}}\mathbf{t}\,dS \tag{4.24}$$

For small elastic strains and incompressible plasticity, the Jacobian of the mapping $J$ is approximately equal to 1, and hence the Cauchy stress is approximately equal to the Kirchhoff stress. Referring the PVW back to the reference configuration, $V_0$ and $S_0$, we can say

$$\int_{V_0} \delta\mathbf{D}^{\mathrm{T}}\boldsymbol{\sigma}\,dV = \int_{S_0} \delta\mathbf{v}^{\mathrm{T}}\mathbf{t}\Re\,dS \tag{4.25}$$

Here, $\Re$ is the infinitesimal surface area ratio between the current and the reference configurations. Introducing the finite element interpolation as before we can say $\mathbf{u} = \mathbf{N}_{\mathrm{e}}\mathbf{u}_{\mathrm{e}}$, $\delta\mathbf{u} = \mathbf{N}_{\mathrm{e}}\delta\mathbf{u}_{\mathrm{e}}$, $\mathbf{v} = \mathbf{N}_{\mathrm{e}}\mathbf{v}_{\mathrm{e}}$ and $\delta\mathbf{v} = \mathbf{N}_{\mathrm{e}}\delta\mathbf{v}_{\mathrm{e}}$. The rate of deformation tensor $\mathbf{D}$ is linearly related to $\mathbf{v}_{\mathrm{e}}$ at any instant in time through the matrix of shape function spatial gradients, $\mathbf{B}_{\mathrm{e}}$, as follows

$$\mathbf{D} = \tfrac{1}{2}\left(\tilde{\mathbf{L}} + \tilde{\mathbf{L}}^{\mathrm{T}}\right) = \tfrac{1}{2}\left(\frac{\partial \mathbf{v}}{\partial \mathbf{y}} + \frac{\partial \mathbf{v}}{\partial \mathbf{y}}^{\mathrm{T}}\right) = \mathbf{B}_{\mathrm{e}}\mathbf{v}_{\mathrm{e}} \tag{4.26}$$

In contrast to the small deformation case, here the shape functions and hence the shape function gradients can be functions of the deformation $\mathbf{y}$ and the displacements $\mathbf{u}$. This means that $\mathbf{B}_{\mathrm{e}}$ can be different at each instant in time. The linear relationship is only true instantaneously and is not true in a time integrated form, i.e., the strain and displacement are not linearly related. In an incremental form over a time step $\Delta t$ the linear relationship is approximately true and we can say the following

$$\Delta\boldsymbol{\varepsilon} = \mathbf{D}\Delta t = \mathbf{B}_{\mathrm{e}}\mathbf{v}_{\mathrm{e}}\Delta t = \mathbf{B}_{\mathrm{e}}\Delta\mathbf{u}_{\mathrm{e}} \tag{4.27}$$

In this case, $\Delta\boldsymbol{\varepsilon}$ is the increment in logarithmic strain. Substituting into the PVW, equation (4.25), summing over all the elements "e" in a finite element mesh and dropping the summation notation and the subscripts for convenience we get

$$\int_{V_0} \delta\mathbf{v}^{\mathrm{T}}\mathbf{B}(\mathbf{u})^{\mathrm{T}}\boldsymbol{\sigma}(\mathbf{u})\,dV = \int_{S_0} \delta\mathbf{v}^{\mathrm{T}}\mathbf{N}^{\mathrm{T}}\mathbf{t}\Re\,dS \tag{4.28}$$

Eliminating or "cancelling" the arbitrary virtual velocities we get

$$\int_{V_0} \mathbf{B}(\mathbf{u})^{\mathrm{T}}\boldsymbol{\sigma}(\mathbf{u})\,dV = \int_{S_0} \mathbf{N}^{\mathrm{T}}\mathbf{t}\Re\,dS \tag{4.29}$$

This has a similar structure to equation (4.7) except for complexity of $\mathbf{B}$. Finally, we can define $\mathbf{G}$ as before

$$\mathbf{G}(\mathbf{u}) = \int_{V_0} \mathbf{B}(\mathbf{u})^{\mathrm{T}}\boldsymbol{\sigma}(\mathbf{u})\,dV - \int_{S_0} \mathbf{N}^{\mathrm{T}}\mathbf{t}\Re\,dS = \mathbf{0} \tag{4.30}$$

For solution of these equations using and implicit NR based method an accurate stress update algorithm required, as before. Let us consider the form of the Jacobian of governing equilibrium equations (tangent stiffness matrix) for an NR solution scheme in this case. Referring back to equation (4.19) we can say

$$\mathbf{K}\left(\mathbf{u}_i^{t+\Delta t}\right)=\frac{\partial \mathbf{G}\left(\mathbf{u}_i^{t+\Delta t}\right)}{\partial \mathbf{u}}=\int_{V_0}\left.\frac{\partial \mathbf{B}^{\mathrm{T}}}{\partial \mathbf{u}}\right|_{\left(\mathbf{u}_i^{t+\Delta t}\right)}\boldsymbol{\sigma}dV+\int_{V_0}\mathbf{B}^{\mathrm{T}}\left.\frac{\partial \boldsymbol{\sigma}}{\partial \mathbf{u}}\right|_{\left(\mathbf{u}_i^{t+\Delta t}\right)}dV$$
$$-\int_{S_0}\mathbf{N}^{\mathrm{T}}\left.\frac{\partial \mathbf{t}}{\partial \mathbf{u}}\right|_{\left(\mathbf{u}_i^{t+\Delta t}\right)}\Re dS-\int_{S_0}\mathbf{N}^{\mathrm{T}}\mathbf{t}\left.\frac{\partial \Re}{\partial \mathbf{u}}\right|_{\left(\mathbf{u}_i^{t+\Delta t}\right)}dS \tag{4.31}$$

However, as before,

$$\left.\frac{\partial \boldsymbol{\sigma}}{\partial \mathbf{u}}\right|_{\left(\mathbf{u}_i^{t+\Delta t}\right)}=\left.\frac{\partial \boldsymbol{\sigma}}{\partial \boldsymbol{\varepsilon}}\frac{\partial \boldsymbol{\varepsilon}}{\partial \mathbf{u}}\right|_{\left(\mathbf{u}_i^{t+\Delta t}\right)}=\mathbf{D}^{tan}\mathbf{B} \tag{4.32}$$

so that

$$\mathbf{K}\left(\mathbf{u}_i^{t+\Delta t}\right)=\int_{V_0}\left.\frac{\partial \mathbf{B}^{\mathrm{T}}}{\partial \mathbf{u}}\right|_{\left(\mathbf{u}_i^{t+\Delta t}\right)}\boldsymbol{\sigma}dV+\int_{V_0}\mathbf{B}^{\mathrm{T}}\mathbf{D}^{tan}\mathbf{B}dV-\int_{S_0}\mathbf{N}^{\mathrm{T}}\left.\frac{\partial \mathbf{t}}{\partial \mathbf{u}}\right|_{\left(\mathbf{u}_i^{t+\Delta t}\right)}\Re dS-\int_{S_0}\mathbf{N}^{\mathrm{T}}\mathbf{t}\left.\frac{\partial \Re}{\partial \mathbf{u}}\right|_{\left(\mathbf{u}_i^{t+\Delta t}\right)}dS \tag{4.33}$$

Here we see that finite deformation kinematics have introduced considerable complexities. In the first instance **B** will be different at each point in time in the evaluation of **G**, from equation (4.30). More significantly however, **K** is quite complicated, being composed of the familiar term involving the material tangent matrix (second term on right in equation 4.33) and other terms coming from the variation in **B**, **t** and $\Re$, due to large deformations. A finite element programme written to implement this method must keep track of all these variations to generate an accurate **G** and an accurate **K**. As part of this, the stress update algorithm is specifically required to generate an accurate $\boldsymbol{\sigma}$ and an accurate $\mathbf{D}^{tan}$. The complexity of equations (4.30) and (4.33) is analogous to that seen for the explicit finite element finite deformation formulation for crystal plasticity theory presented in equation (3.18). Echoing the discussion in section 3.2, **K** will in general be an unsymmetric matrix, requiring the use of unsymmetric solvers when implemented in a computer programme.

## 4.6 Crystal Plasticity Implementation

Now that we have seen the details of a more conventional implicit NR based finite element formulation in the general sense, let us relate this directly to finite deformation crystal plasticity theory, focusing in particular on the stress update algorithm. There is, of course, a choice of appropriate algorithms. One obvious choice, that we will examine here, is the rate tangent method. We have seen this before expressed in terms of **S** and $\dot{\mathbf{E}}$ on the reference configuration (equation 3.13). It can, by mapping forward to the current configuration, be expressed in terms of $\boldsymbol{\tau}$ and **D**, and $\boldsymbol{\sigma}$ and **D**. A very commonly used version of this is presented by Huang (1991). It is written in terms of $\boldsymbol{\sigma}$ and **D** and it has been coded as UMAT for ABAQUS/Standard. An outline of the formulation is given below. The governing rate equation written in tensor notation in terms of the Jaumann rate of Cauchy stress, $\overset{\nabla}{\boldsymbol{\sigma}}$, which is an objective stress rate, is as follows; it is equivalent to that presented in equation (2.49)

$$\overset{\nabla}{\boldsymbol{\sigma}} = \mathbf{K} : \mathbf{D} - \boldsymbol{\sigma}(\mathbf{I} : \mathbf{D}) - \sum_{\alpha} \dot{\gamma}_\alpha \mathbf{Y}_\alpha$$

$$\mathbf{Y}_\alpha = \mathbf{K} : \boldsymbol{\mu}_\alpha + \boldsymbol{\omega}_\alpha \cdot \boldsymbol{\sigma} - \boldsymbol{\sigma} \cdot \boldsymbol{\omega}_\alpha \tag{4.34}$$

$$\boldsymbol{\mu}_\alpha = \tfrac{1}{2}\left(\mathbf{s}^*_\alpha \mathbf{m}^*_\alpha + \mathbf{m}^*_\alpha \mathbf{s}^*_\alpha\right)$$

$$\boldsymbol{\omega}_\alpha = \tfrac{1}{2}\left(\mathbf{s}^*_\alpha \mathbf{m}^*_\alpha - \mathbf{m}^*_\alpha \mathbf{s}^*_\alpha\right)$$

where $\mathbf{K}$ here represents the tensor of linear elastic moduli (from equation 2.44). In incremental terms we can say

$$\Delta \boldsymbol{\sigma} = \mathbf{K} : \Delta \boldsymbol{\varepsilon} - \boldsymbol{\sigma}(\mathbf{I} : \Delta \boldsymbol{\varepsilon}) - \sum_{\alpha} \Delta \gamma_\alpha \mathbf{Y}_\alpha \tag{4.35}$$

The rate tangent method is employed to determine the $\Delta\gamma_\alpha$ through equation (3.4) repeated here

$$\Delta\gamma_\alpha = \left[(1-\Theta)\dot{\gamma}^t_\alpha + \Theta\dot{\gamma}^{t+\Delta t}_\alpha\right]\Delta t \tag{4.36}$$

As before, linearisation of the expression for $\dot{\gamma}^{t+\Delta t}_\alpha$ about time $t$ (a Taylor expansion of equation 2.32) leads to the following linear set of equations for the increments in slip system shear strain $\Delta\gamma_\beta$,

$$\sum_{\beta=1}^{n} \widetilde{N}_{\alpha\beta} \Delta\gamma_\beta = \dot{\gamma}^t_\alpha \Delta t + \mathbf{Z}_\alpha : \Delta \boldsymbol{\varepsilon} \tag{4.37}$$

In this expression, $\widetilde{N}_{\alpha\beta}$ is a matrix similar to $N_{\alpha\beta}$ used before (equation 3.10) and $\mathbf{Z}_\alpha$ is a tensor involving $\mathbf{K}$, $\boldsymbol{\mu}_\alpha$, $\boldsymbol{\omega}_\alpha$, etc. The structure of equation (4.37) is equivalent to that of equation (3.9).

Inverting $\widetilde{N}_{\alpha\beta}$ and substituting for the $\Delta\gamma_\beta$ in equation (4.35) we can establish the following linear relationship between $\Delta\boldsymbol{\sigma}$ and $\Delta\boldsymbol{\varepsilon}$

$$\Delta\boldsymbol{\sigma} = \widetilde{\mathbf{C}}_{\mathrm{RT}} : \Delta\boldsymbol{\varepsilon} - \widetilde{\boldsymbol{\rho}}_{\mathrm{RT}} \Delta t \tag{4.38}$$

In this equation, $\widetilde{\mathbf{C}}_{\mathrm{RT}}$ the rate tangent modulus tensor, which is similar to $\mathbf{C}_{\mathrm{RT}}$ seen before in equation (3.13). Both are unsymmetric in general. $\widetilde{\boldsymbol{\rho}}_{\mathrm{RT}}$ is similar to $\boldsymbol{\rho}_{\mathrm{RT}}$ also seen before in equation (3.13). For the NR method, the stress increment to allow an update to $t+\Delta t$ for the current iteration is achieved using equation (4.38). The NR method requires also the Jacobian of the constitutive law at $t+\Delta t$ for the current iteration, $\left.\dfrac{\partial\boldsymbol{\sigma}}{\partial\boldsymbol{\varepsilon}}\right|_{\left(\mathbf{u}_i^{t+\Delta t}\right)}$. This can be written as $\left.\dfrac{\partial\Delta\boldsymbol{\sigma}}{\partial\Delta\boldsymbol{\varepsilon}}\right|_{\left(\mathbf{u}_i^{t+\Delta t}\right)}$ since $\boldsymbol{\sigma}^t$ and $\boldsymbol{\varepsilon}^t$ are fixed, having been determined as the solution to the previous increment, and $\boldsymbol{\sigma}^{t+\Delta t} = \boldsymbol{\sigma}^t + \Delta\boldsymbol{\sigma}$ and $\boldsymbol{\varepsilon}^{t+\Delta t} = \boldsymbol{\varepsilon}^t + \Delta\boldsymbol{\varepsilon}$. This notation is commonly used (see for example the ABAQUS Users Manuals, HKS 1997b). Since the constitutive law is now in a linear form for the increment (equation 4.38), the Jacobian is easily written down as $\widetilde{\mathbf{C}}_{\mathrm{RT}}$ and when converted to matrix form becomes $\mathbf{D}^{tan}$, which is an unsymmetric matrix in general and will yield an unsymmetric stiffness matrix through equation (4.33)

$$\frac{\partial\Delta\boldsymbol{\sigma}}{\partial\Delta\boldsymbol{\varepsilon}} = \widetilde{\mathbf{C}}_{\mathrm{RT}} \Rightarrow \mathbf{D}^{tan} \tag{4.39}$$

The numerical performance of the method can be improved by not linearising the expression for $\dot{\gamma}_\alpha^{t+\Delta t}$, i.e., not using the Taylor expansion, as presented in Huang (1991). As before, start with equation (4.36) and substitute the non-linearlised power-law form of $\dot{\gamma}_\alpha^{t+\Delta t}$ coming from equation (2.32)

$$\Delta\gamma_\alpha = (1-\Theta)\Delta t\dot{\gamma}_\alpha^t + \Theta a_\alpha \operatorname{sgn}(\tau_\alpha)\left\{\left|\frac{\tau_\alpha + \Delta\tau_\alpha}{g_\alpha + \Delta g_\alpha}\right|\right\}^{\frac{1}{m}} \Delta t \tag{4.40}$$

This is a non-linear equation for the $\Delta\gamma_\alpha$ since $\Delta\tau_\alpha$ and $\Delta g_\alpha$ are functions of $\Delta\gamma_\alpha$. Since it is non-linear, a straightforward linear form of the constitutive law for the increment (like equation 4.38) will not result. A "local" NR iterative procedure can be used to solve equation (4.40) for the $\Delta\gamma_\alpha$. This procedure also yields the increments in the stress, $\Delta\boldsymbol{\sigma}$, and other state variables, allowing an update to be performed. As such, this is then essentially an implicit stress update algorithm. It is interesting to note that in this approach, the required Jacobian of the consitutitve law is very well approximated by the $\widetilde{\mathbf{C}}_{\mathrm{RT}}$, as required in equation (4.38), evaluated at the end of the local NR iteration process. More discussion on the evaluation of the Jacobian is given in Huang (1991).

### 4.7 General Comments

Numerical integration algorithms (stress update algorithms) of various forms (in addition to the rate tangent method) have been developed and used by a number of groups to deal with the complexities of crystal plasticity theory. In addition, algorithms have been implemented in commercial finite element codes, e.g., ABAQUS, ANSYS , etc. For example:

- Teodosiu, Raphanel and co-workers have used explicit algorithms.
- Meric, Cailletaud and co-workers (Meric et al. 1991, Meric and Cailletaud 1991) developed an implicit algorithm and implemented it in the code Zébulon. Cailletaud and co-workers more recently report an implicit method, based on a NR/BFGS approach, incorporating a Runge Kutta algorithm to deal with large numbers of variables, implemented in a parallel processing environment.
- Rashid and Nemat-Nasser (1992) developed an explicit algorithm based on a partition method, a weighted residual method. More recently Nemat-Nasser and co-workers developed predictor-corrector iterative methods that show improved performance over the rate tangent method.
- Implicit algorithms have been developed by,

  Ortiz and co-workers (implemented in the FEAP programme)
  Busso, O'Dowd and co-workers (implemented in ABAQUS)
  Steinmann and Stein (1996)
  Grujicic and Zhang (1999) (implemented in ABAQUS)

  These authors present full Backward Euler implicit NR formulations and significant attention is given to *explicit* derivation of an exact Jacobian for the constitutive law, what is also called the "consistent" tangent matrix. As can be appreciated from the derivations presented above, this matrix can be particularly difficult to derive due to the complexity of the crystal plasticity constitutive equations, in the context of finite deformation kinematics.

- Zikry has developed an explicit-implicit algorithm combining, and allowing, interchange between Runge Kutta (explicit) components and BFGS/Backward Euler (implicit) components.
- Implementing various stress update algorithms in UMATs for ABAQUS is quite common, for example (in addition to Huang 1991) there are the following,
  - Harder (1999)
  - Gall et al. (2000)
  - Kysar (2001a,b)
  - Choi (2003).
- Most of the emphasis has been on rate dependent algorithms but rate independent algorithms have also been developed, for example
  - Anand and co-workers, Staroselsky and Anand (2003) (ABAQUS/Explicit)
  - Potirniche and Daniewicz (2003) (ANSYS)
  - Miehe and co-workers (FEAP)
  - Bertram and co-workers.

The reader should consult the reference list for examples of papers by these and other authors in this area. An interesting comparison of different stress update algorithms is presented in Miehe and Schröder (2001). Current trends in finite element implementations of crystal plasticity are as follows:

- Both implicit and dynamic explicit finite element codes are in continuous development.
- Implicit codes are used for their accuracy and stability.
- Dynamic explicit codes are used for their robustness (for highly non-linear problems, severe localisation, crack growth, etc.).
- There is a strong move to parallel processing.

# 5 Enhancements of the Basic Theory

## 5.1 Strain Hardening

Recall that strain hardening was discussed in section 2 and that a phenomenological model was presented. This model works quite well but it has its limitations and these were listed. Much detailed analytical and experimental work has been done to study strain hardening in crystals (e.g., Franciosi and co-workers). Based on work of this type, models that are more physically based have been proposed. These models address some of the limitations of the previously discussed phenomenological approach.

Consider the approach of Bassani and Wu (1989). They proposed the following structure for determining the hardening moduli.

$$h_{\alpha\alpha} = \left[ (h_0 - h_s)\,\mathrm{sech}^2\left\{ \frac{h_0 - h_s}{\tau_1 - \tau_0} \gamma_\alpha \right\} + h_s \right] \times \left[ 1 + \sum_{\beta \neq \alpha} f_{\alpha\beta} \tanh\left( \frac{\gamma_\beta}{\gamma_0} \right) \right] \tag{5.1}$$

$$h_{\alpha\beta} = q h_{\alpha\alpha}$$

In this model there are a series of material constants that must be fitted from experiment: $\tau_1, \tau_0, h_0, h_s, q$. $f_{\alpha\beta}$ is a matrix of amplitude factors for different types of dislocation interactions, which are divided into five classifications. Tests show that that model captures physical effects very well (Bassani 1990). However it doesn't deal with the Bauschinger effect. Dislocation mechanics has been used to generate other models, for example those proposed by

- Cuitiño, Ortiz and co-workers
- Zikry and co-workers
- Ohashi (1994)
- Teodosiu, Raphanel and co-workers
- Harder (1999)
- Acharya (2001)
- Arsenlis and Parks (2002)
- Choi (2003).

These models explicitly account for such things as quantification of dislocation density evolution and kinetics of dislocation interaction. These types of models involve many parameters and coefficients that must be determined from experiments. The reader should consult the reference list for examples of specific papers describing these models. Hardening description can be important in stress and strain prediction. See, for example, Povirk et al. (1995), on the investigation of single-crystal thin films using different models, and also Kysar (2001a,b).

Note that some very interesting work has been done recently comparing the predictions of microstructural deformation generated by continuum crystal plasticity models with those generated by discrete dislocation based models (e.g., van der Giessen, Needleman and co-workers) and discrete atomistic models (Farrissey et al. 2000). These have shown that the continuum mechanics approach is still lacking in representing certain details of microstructural deformation picked up by the discrete models. There is still tremendous scope here for research into linking the two approaches and for improving continuum mechanics formulations to give greater physical realism.

### 5.2 Cyclic Loading

To deal with reversed loading and cyclic loading, a "Backstress" can be incorporated into the modelling framework to represent the Bauschinger effect. This introduces a kinematic hardening component into the hardening model. Meric, Cailletaud and co-workers have presented a viscoplastic, power-law crystallographic plasticity model based on both isotropic and kinematic hardening (e.g., Meric et al. 1991), as is illustrated by the following expressions

$$\dot{\gamma}_\alpha = \dot{v}_\alpha \operatorname{sgn}(\tau_\alpha - x_\alpha)$$

$$\dot{v}_\alpha = \left( \frac{|\tau_\alpha - x_\alpha| - r_\alpha}{K} \right)^n \qquad (5.2)$$

where $x_\alpha$ is a kinematic hardening variable and $r_\alpha$ is an isotropic hardening variable. $K$ and $n$ are material parameters. Voyiadjis and Huang (1996) and Fedelich (2002) have also presented models incorporating kinematic hardening. In addition, the Bauschinger effect can be addressed by using some dislocation mechanics based models, e.g., Harder (1999).

### 5.3 Size Scale Representation

The representation of size scale effects is one of the most important current and future issues for mechanics and is hugely relevant in the development of crystal plasticity theory. Experiments have shown that when one is considering physical size scales going down from tens of microns to fractions of microns, an increase in stress for a given strain is observed. This is relevant in the context of nano-indentation testing, the mechanics of sub-micron particle reinforced materials and when considering the mechanical effects of varying grain size in polycrystals (as has traditionally been thought of in terms of the Hall-Petch effect).

This size effect in crystalline materials has been explained in terms of the presence of Geometrically Necessary Dislocations (GNDs). The GND can be visualised in terms of dislocations necessary to cause plastic distortion of crystal, as is illustrated in Figure 8 below (Fleck and Hutchinson 1997). They can be viewed as "wedges" required to keep atomic planes misaligned in the plastically deformed state.

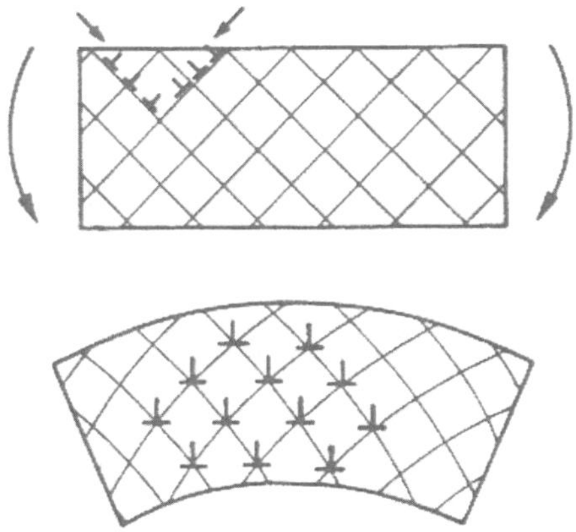

**Figure 8.** Illustration of geometrically necessary dislocations. Reprinted from Fleck, N.A. and Hutchinson, J.W., Strain Gradient Plasticity, *Advances in Applied Mechanics*, Vol. 33, pp. 295-361, © 1997, with permission from Elsevier.

One can consider the total dislocation population in a crystal being composed of two types of dislocations, Statistically Stored Dislocations (SSD) and Geometrically Necessary Dislocations (GND). The SSD are related to plastic strain and the GND are related to plastic strain gradients. As argued in Fleck and Hutchinson (1997), plastic strain gradients (and hence the GND) are dependent on the physical size scale over which deformation is occurring, and the smaller the size scale the greater the GND density relative to the SSD density. For deformation happening over large physical size scales, the GND density relative to the SSD density is small; consequently GND densities and local plastic strain gradients have not traditionally formed a part of classical continuum mechanics plasticity formulations. In recent years new plasticity formulations have emerged that take the GND density into account and hence purport to represent a size scale effect.

In the context of crystal plasticity theory, numerous research groups have addressed this issue, for example Hutchinson, Fleck, Shu, Busso, Bassani, Cermelli, Gurtin, Svendsen, Acharya, Baumann and Steinmann, and their co-workers. Crystal plasticity theories have been developed that represent a size scale effect and that could be called "strain gradient crystal plasticity" theories. The reader should consult the reference list for examples of papers by these authors and their co-workers in this area. Such theories sometimes referred to as non-local theories. A Local theory is one where σ at a material point is related to ε at the point. A Non-local theory is one where σ at a point is related to ε at the point *and in the neighbourhood of* the point, and ε in the neighbourhood of the point is related to the spatial gradient in ε at the point.

# 6 Examples of Applications

## 6.1 Thermal Deformations

In McHugh et al. (1993a) the kinematical framework is augmented to include a deformation gradient, $\mathbf{F}^{\theta}$, that represents thermal expansion

$$\mathbf{F} = \mathbf{F}^{*} \cdot \mathbf{F}^{\theta} \cdot \mathbf{F}^{p} \quad (6.1)$$

The thermal velocity gradient is given by

$$\dot{\mathbf{F}}^{\theta} \cdot \mathbf{F}^{\theta-1} = \dot{\theta}\boldsymbol{\alpha} \quad (6.2)$$

$\dot{\theta}$ is the rate of temperature change and $\boldsymbol{\alpha}$ is a tensor of thermal expansion coefficients, that can be anisotropic. This kinematical framework very flexible, for example $\mathbf{F}^{\theta}$ could represent deformation due to a phase change or a chemical reaction.

## 6.2 Deformation in Metal Matrix Composites (MMCs)

Crystal plasticity theory has been used to study the microscale deformation and strengthening mechanisms in Al-SiC MMCs (e.g., McHugh et al. 1993a,b,c,d, Needleman and Tvergaard 1993, Bruzzi et al. 2001). Periodic unit cell models containing single and multiple particles have been used and the matrix has been represented as either a single crystal or a polycrystal. During deformation the crystal plasticity models show greater levels of microscale plastic strain non-uniformity and localisation in comparison to models with the same unit cell geometry and loading but where the matrix is represented using isotropic $J_2$ flow theory. This has significant implications for the prediction of microscale damage in the context of critical local stress or strain criteria. As the grain size in the polycrystal models is made smaller, relative to the particle size, the macroscopic stress strain curve prediction approaches that produced by the $J_2$ flow theory matrix model (Bruzzi et al. 2001).

## 6.3 Deformation in Intermetallic Alloys

Fischer and co-workers have represented twinning in TiAl (see reference list). They used an extended crystal plasticity formulation where twinning is represented as a shear strain within a twinned volume and is driven by a critical stress. Staroselsky and Anand (2003) studied twinning in Mg alloy using a rate independent model.

Dao, Asaro and co-workers studied non-Schmid effects in crystals. Here the shear strain is driven by loading parameter derived from Schmid resolved shear stress. Also Brockmann (2003) performed investigations in this area.

### 6.4 Deformation in Ni base Superalloys

Ni base "γ-γ' superalloys" take the form of two-phase single crystals and are used in high temperature turbine blade applications due to low creep rates. The material is FCC composed of γ' cuboids (typically 0.5μm along edge) embedded in a γ matrix, as illustrated in Figure 9 below. The γ' volume fraction is typically 70%. The creep resistance comes mainly from the presence of the γ-γ' phase boundaries that act as barriers to dislocation flow. A number of approaches have been taken to modelling these materials, and predicting creep rates, based on single crystal plasticity theory, for example,

- Cailletaud, Nouailhas and co-workers
- Busso et al. (2000)
- Feng et al. (2002)
- McHugh and Mohrmann (1997)
- MacLachlan et al. (2002).

Finite element models are based on either homogenised "effective" single phase crystal properties or two phase unit cell micromechanical models.

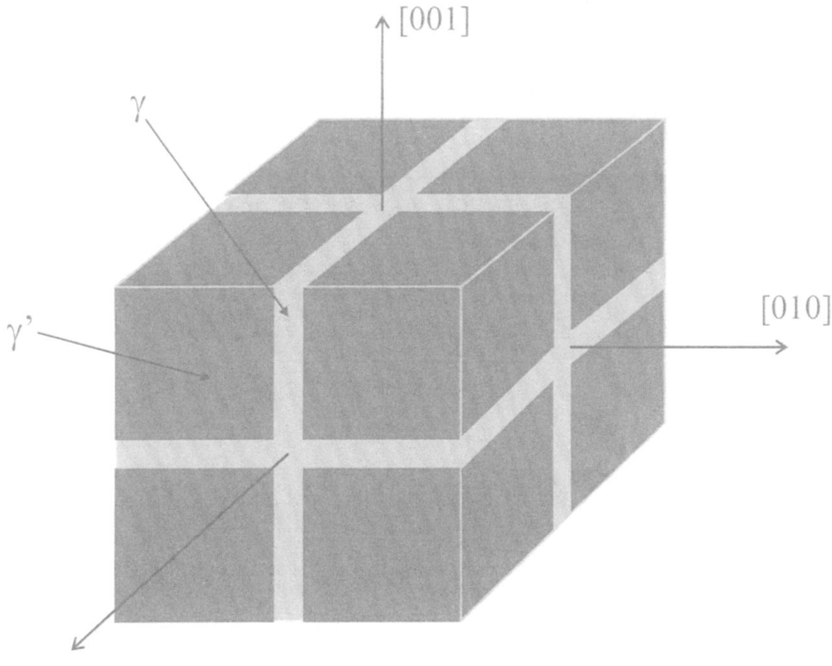


**Figure 9.** Schematic diagram of the Ni base γ-γ'superalloy microstructure. Reprinted from McHugh, P.E. and Mohrmann, R., Modelling of Creep in Ni Base Superalloy using a Single Crystal Plasticity Model, *Computational Materials Science,* Vol. 9, pp. 134-140, © 1997, with permission from Elsevier.

### 6.5 Link with Damage Modelling

Crystal plasticity theory has been linked with some representation of microstructural damage in metals, e.g., crack growth, void growth, etc. An example of this is the simulation of interface debonding using the Needleman (1987) debonding model, which is based on a debonding potential. Xu and Needleman (1993) used it for the study of reinforcement/matrix interface debonding in MMCs where the matrix was represented by crystal plasticity theory. Kysar (2001a,b) modelled copper/sapphire interface debonding, using the Needleman model implemented as a UEL for ABAQUS. A second example is the use of continuum damage models to represent crack growth through a metallic crystal. Feng et al. (2002) presented a formulation based on crystal plasticity coupled with anisotropic continuum damage mechanics (CMD) and applied it to superalloys. McHugh and Connolly (2003) investigated fracture in the cobalt binder layer of WC-Co hard metal and used an isotropic Rice and Tracey damage model and element removal to represent crack growth through a single Co crystal. Idealised geometry was used here, as shown in Figure 10, where a rectangular single crystal Co ligament is enclosed in an elastic WC region that is pre-cracked on either side of the ligament. Loading is applied by embedding the "cut-out" in a homogenous region containing a surface crack. Figure 11 presents contour plots of the Rice and Tracey damage parameter (indicating element that have failed/been removed) in the upper half of the ligament, and crack growth is clearly evident.

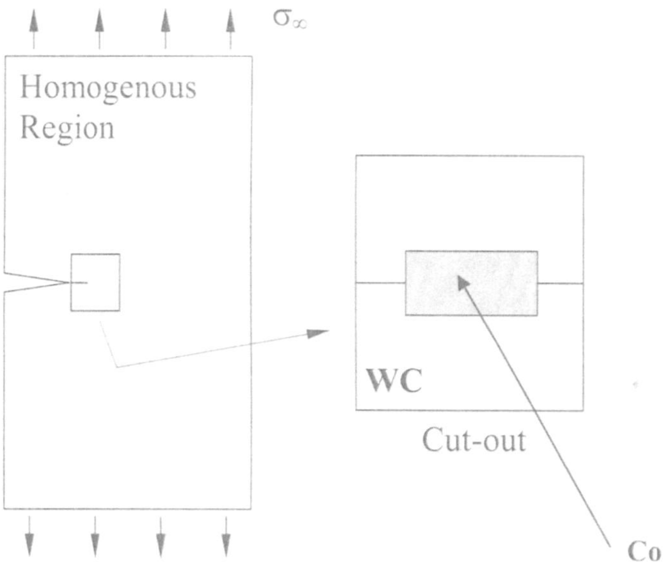


**Figure 10.** Idealised microstructural geometry for WC-Co material. Reprinted from McHugh, P.E. and Connolly, P.J., Micromechanical Modelling of Ductile Crack Growth in the Binder Phase of WC-Co, *Computational Materials Science*, Vol. 27, pp. 423-436, © 2003, with permission from Elsevier.

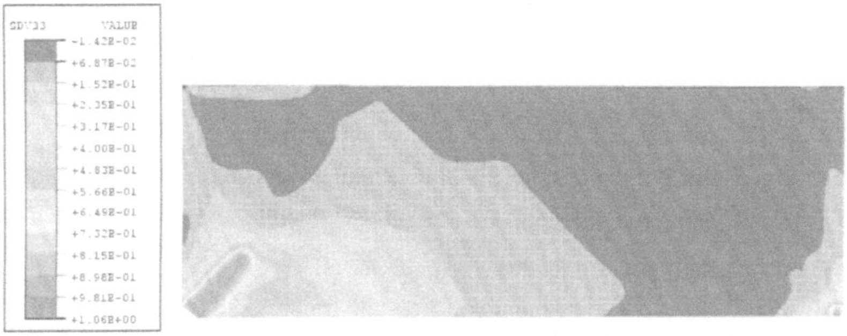


(a)

(b)

**Figure 11.** Contour plots of ductile damage parameter in the Co ligament. Reprinted from McHugh, P.E. and Connolly, P.J., Micromechanical Modelling of Ductile Crack Growth in the Binder Phase of WC-Co, *Computational Materials Science*, Vol. 27, pp. 423-436, © 2003, with permission from Elsevier.

### 6.6 Biomedical Applications

Crystal plasticity theory has been used to study size effects in small scale metallic medical devices, in particular vascular stents for the treatment if coronary heart disease (Savage et al. 2004). Stents, when manufactured, typically take the form of networks of thin stainless steel struts. The structure undergoes considerable plastic deformation when the stent is being deployed in a blocked artery; it is expanded into position in the artery by the inflation of a balloon upon which it is mounted. Once deployed, the balloon is deflated and removed. Since stent struts are quite small, on the order of 75 μm x 75 μm in cross-section, and may only have a few grains across the thickness (on the order of 10), the issue of size effects on material properties arises. In particular, Murphy et al. (2003) studied the influence of strut size on ductility, finding that thin strut tensile ductility was only on the order of 50% of the ductility of bulk material. In Savage et al. (2004), crystal plasticity polycrystalline models were used to study this effect in 2D generalised plane strain deformation. Grains were assumed to be regular hexagons in the plane (see Figure 12), with random crystal lattice orientations (in 3D). The properties of 316L austenitic (FCC) stainless steel were assumed.

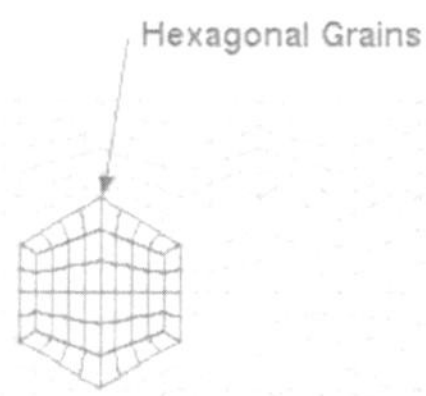


**Figure 12.** Example of 2D stent strut model with hexagonal grains. Reprinted from Savage, P., O'Donnell, B.P., McHugh, P.E. Murphy, B.P. and Quinn, D.F., Coronary stent strut size dependent stress-strain response investigated using micromechanical finite element models, *Annals of Biomedical Engineering*, Vol. 32, No. 2, pp. 202–211, 2004 © Biomedical Engineering Society.

Due to the inhomogeneous lattice orientations, plastic strain localisation occurred, leading to necking and failure. An example of this is shown in Figure 13 for a model with nine grains through the thickness. By considering models of different thicknesses, where the grain size was kept constant, a clear relationship between failure strain and the thickness was predicted, following the trend seen in the experiments. This is shown in Figure 14, where experimental and computed macroscopic failure strains are compared. Failure strain was taken as the strain at the UTS point from the macroscopic tensile stress-strain diagram in both cases. The results show that the models predict the observed trend but that further refinement is necessary to obtain full quantitative agreement.

**Figure 13.** Contour plot of accumulated slip for a stent strut model with nine grains across the thickness. Reprinted from Savage, P., O'Donnell, B.P., McHugh, P.E. Murphy, B.P. and Quinn, D.F., Coronary stent strut size dependent stress-strain response investigated using micromechanical finite element models, *Annals of Biomedical Engineering*, Vol. 32, No. 2, pp. 202–211, 2004 © Biomedical Engineering Society.

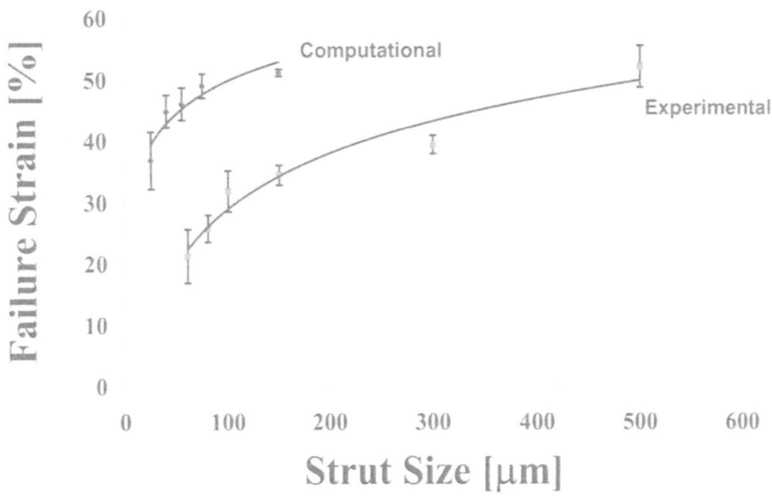


**Figure 14.** Comparison of experimental and computational failure strain for different stent strut thicknesses. Reprinted from Savage, P., O'Donnell, B.P., McHugh, P.E. Murphy, B.P. and Quinn, D.F., Coronary stent strut size dependent stress-strain response investigated using micromechanical finite element models, *Annals of Biomedical Engineering*, Vol. 32, No. 2, pp. 202–211, 2004 © Biomedical Engineering Society.

# 7 Organisation of References

References have been provided to allow the reader to access more detail on the discussion presented here. To help the reader find further information mode quickly, the references have been organised under the following headings

- General Formulations
- Rate Independent Methods
- Strain Hardening
- Developments and General Applications
- Applications to Intermetallics
- Applications to Superalloys
- Strain Gradients
- Related References

The last section includes references on links between continuum mechanics and discrete entity modelling methods, e.g., atomistics and dislocation mechanics. It also contains some selected references on macroscale polycrystal modelling.

# References


### General Formulations

A. Acharya. A model of crystal plasticity based on the theory of continuously distributed dislocations. *J. Mech. Phys. Solids,* 49:761–784, 2001.

A. Arsenlis and D.M. Parks. Modelling the evolution of crystallographic dislocation density in crystal plasticity. *J. Mech. Phys. Solids,* 50:1979–2009, 2002.

R.J. Asaro and J.R. Rice. Strain localisation in ductile single crystals. *J. Mech. Phys. Solids,* 25:309–338, 1977.

R.J. Asaro. Crystal plasticity, *J. Appl. Mech.*, 50:921–934, 1983.

R.J. Asaro and A. Needleman. Texture development and strain hardening in rate dependent polycrystals, *Acta metall.*, 33:923–953, 1985.

F. Barbe, L. Decker, D. Jeulin, and G. Cailletaud. Intergranular behaviour of polycrystaline aggregates. Part1: F.E. model. *Int. J. Plasticity,* 17:513–536, 2001.

F. Barbe, L. Decker, D. Jeulin, and G. Cailletaud. Intergranular behaviour of polycrystaline aggregates. Part 2: Results. *Int. J. Plasticity,* 17:537–563, 2001.

A.M. Cuitiño and M. Ortiz. Computational modelling of single crystals. *Modell. Simul. Mater. Sci. Engng.*, 1:255–263, 1992.

H.E. Dève, S. Harren, C. McCullough and R.J. Asaro. Micro and macroscopic aspects of shear band formation in internally nitrided single crystals of Fe-Ti-Mn alloys. *Acta metall.*, 36:341–365, 1988.

H.E. Dève and R.J. Asaro. The development of plastic failure modes in crystalline materials: Shear Bands in fcc polycrystals. *Metall. Trans. A*, 20A:579–593, 1989.

J. Harder. A Crystallographic model for the study of local deformation processes in polycrystals. *Int. J. Plasticity*, 15:605–624, 1999.

S.V. Harren, H.E. Dève, and R.J. Asaro. Shear band formation in plane strain compression. *Acta metall.*, 36:2435–2480, 1988.

S.V. Harren and R.J. Asaro. Nonuniform deformations in polycrystals and aspects of the validity of the Taylor model. *J. Mech. Phys. Solids*, 37:191–232, 1989.

S.V. Harren, T.C. Lowe, R.J. Asaro, and A. Needleman. Analysis of large-strain shear in rate-dependant face-centred cubic polycrystals: Correlation of micro- and macromechanics. *Phil. Trans. Royal Soc. London A*, 328:443–500, 1989.

R. Hill. Generalised consititutive Relations for incremental deformation of metal crystals by multislip. *J. Mech. Phys. Solids*, 14:95–102, 1966.

R. Hill and J.R. Rice. Constitutive analysis of elastic-plastic crystals at arbitrary strain. *J. Mech. Phys. Solids*, 20:401–413, 1972.

R. Hill and J.R. Rice. Elastic potentials and the structure of inelastic constitutive laws. *SIAM J. Appl. Math.*, 25:448–461, 1973.

HKS. *ABAQUS Theory Manual Version 5.7*. Hibbitt, Karlsson and Sorensen, Inc., Pawtucket, RI, 1997a.

HKS. *ABAQUS Users Manuals Version 5.7*. Hibbitt, Karlsson and Sorensen, Inc., Pawtucket, RI, 1997b.

J.W. Hutchinson. Bounds and self-consistent estimates for creep of polycrystalline materials. *Proc. Royal Soc. London*, 348A:101, 1976.

Y.G. Huang. *A User-Material Subroutine Incorporating Single Crystal Plasticity in the ABAQUS Finite Element Program*. Harvard University Report MECH-178, 1991.

T. Kameda and M.A. Zikry. Three dimensional dislocation-based crystalline constitutive formulation for ordered intermetallics. *Scr. mater.*, 38:631–636, 1998.

J. Lemonds, R.J. Asaro and A. Needleman. A numerical study of localised deformation in bi-crystals. *Mech. Mater.*, 4:417–435, 1985.

J. Mandel. *Plasticité Classique et Viscoplasticité*. CISM Lecture Notes, Springer, Berlin, 1972.

P.E. McHugh, R.J. Asaro, and C.F. Shih. Computational modelling of metal matrix composite materials – I. Isothermal deformation patterns in ideal microstructures. *Acta metall. mater.*, 41:1461–1476, 1993a.

P.E. Hugh, R.J. Asaro and C.F. Shih. Computational modelling of metal matrix composite materials – II. Isothermal stress-strain behaviour. *Acta metall. mater.*, 41:1477–1488, 1993b.

P.E. McHugh, R.J. Asaro, and C.F. Shih. Computational modelling of metal matrix composite materials – III. Comparisons with phenomenological models. *Acta metall. mater.*, 41:1489–1499, 1993c.

P.E. McHugh, R.J. Asaro, and C.F. Shih. Computational modelling of metal matrix composite materials – IV. Thermal deformations. *Acta metall. mater.*, 41:1501–1510, 1993d.

L. Meric, P. Poubanne, and G. Cailletaud. Single crystal modelling for structural calculations: Part I – Model presentation, *J. Engng. Mater. Techn.*, 113:162, 1991.

L. Meric and G. Cailletaud. Single crystal modelling for structural calculations: Part II – Finite element implementation. *J. Engn. Mater. Techn.*, 113:171, 1991.

R. Mohan, M. Ortiz, and C.F. Shih. Mode mixity effects on crack tip deformation in ductile single crystals. *Acta metall. mater.*, 40:1907–1922, 1990.

A. Needleman. A continuum model for void nucleation by inclusion debonding. *J. Appl. Mech.*, 54:525–531, 1987.

D. Peirce, R.J. Asaro, and A. Needleman. An analysis of nonuniform and localised deformation in ductile single crystals. *Acta metall*, 30:1087–1119, 1982.

D. Peirce, R.J. Asaro, and A. Needleman. Material rate dependence and localised deformation in crystalline solids. *Acta metall*, 31:1951–1976, 1983.

M.M. Rashid and S. Nemat-Nasser. A constitutive algorithm for rate-dependant crystal plasticity. *Comput. Meth. Appl. Mech. Engng.*, 94:201–228, 1992.

J.R. Rice. Inelastic constitutive relations for solids: An internal-variable theory and its application to metal plasticity. *J. Mech. Phys. Solids*, 19:433–455, 1971.

N.J. Sørensen and B.S. Andersen. A parallel finite element method for the analysis of crystalline solids. *Comput. Meth. Appl. Mech. Engng.*, 132:345–357, 1996.

S. Suresh and A. Mortensen. Functionally graded metals and metal-ceramic composites: Part 2 Thermomechanical behaviour. *Int. Mater. Rev.*, 42:85–116, 1997.

L. Stainier, A.M. Cuitiño, and M. Ortiz. A micromechanical model of hardening, rate sensitivity and thermal softening in BCC single crystals. *J. Mech. Phys. Solids*, 50, 1511–1545, 2002.

G.I. Taylor. Plastic strain in metals. *J. Inst. Metals*, 62:307–324, 1938.

C. Teodosiu and F. Sidoroff. A finite theory of the elastoviscoplasticity of single crystals. *Int. J. Eng. Sci.*, 14:713–723, 1976.

C. Teodosiu, J.L. Raphanel, and L. Tabourot. Finite element simulation of the large elastoplastic deformation of multicrystals. In C. Teodosiu, J.L. Raphanel and F. Sidoroff, editors, *Large Plastic*

*Deformations: Fundamental Aspects and Applications to Metal Forming*, pages 153–168. Balkema, Rotterdam, 1993.

G.Z. Voyiadjis and W. Huang. A modelling of single crystal plasticity with backstress evolution. *Eur. J. Mech. A/Solids*, 15:553–573, 1996.

X.P. Xu and A. Needleman. Void nucleation by inclusion debonding in crystal matrix. *Modell. Simul. Mater. Sci. Engng.*, 1:111–132, 1993.

O.C. Zienkiewicz and R.L. Taylor. *The Finite Element Method Fourth Edition Volume 2, Solid and Fluid Mechanics Dynamics and Non-Linearity*. McGraw-Hill, London, 1991.

## Rate Independent Methods

L. Anand and M. Kothari. A computational procedure for rate-independent crystal plasticity. *J. Mech. Phys. Solids*, 44:525–558, 1996.

C. Miehe. Exponential map algorithm for stress updates in anisotropic multiplicative elastoplasticity for single crystals. *Int. J. Num. Meth. Eng.*, 39:3367–3390, 1996.

J. Schröder and C. Miehe. Aspects of computational rate-independent crystal plasticity. *Comput. Mater. Sci.*, 9:168–176, 1997.

M. Schurig and A. Bertram. A rate independent approach to crystal plasticity with a power law. *Comput. Mater. Sci.*, 26:154–158, 2003.

## Strain Hardening

A. Acharya and J.L. Bassani. Incompatible lattice deformations and crystal plasticity. In N. Ghoniem, editor, *Plastic and Fracture Instabilities*, AMD-Vol.200/MD-Vol.52, pages 75–80. ASME, New York, 1995.

J.L. Bassani and T.Y. Wu. *Latent Hardening in Single Crystals: Part II, Analytical Characterisation and Predictions*. Univ. Pennsylvanya, 1989.

J.L. Bassani. Single crystal hardening. *Appl. Mech. Rev.*, 43:S320–S327, 1990.

R. Estevez, G. Hoinard, and P. Franciosi. Hardening anisotropy of γ/γ' superalloy single crystals – II. Numerical a nalysis of heterogeneity effects. *Acta mater.*, 45:1567–1584, 1997.

P. Franciosi, M. Berveiller, and A. Zaoui. Latent hardening in copper and aluminium single crystals. *Acta metall.*, 28:273–283, 1980.

P. Franciosi and A. Zaoui. Multislip in F.C.C. crystals – A theoretical approach compared with cxternal data. *Acta metall.*, 30:1627–1637, 1982.

P. Franciosi. The concepts of latent hardening and strain hardening in metallic single crystals. *Acta metall.*, 33:1601–1612, 1985.

L. Llanes, J.L. Bassani, and C. Laird. Cyclic response of polycrystalline copper-composite-grain model. *Acta metall. mater.*, 42:1279–1288, 1994.

A. Mecif, B. Bacroix, and P. Franciosi. Temperature and orientation dependent plasticity features of Cu and Al single crystals under axial compression – I. Lattice rotation effects and true hardening stages. *Acta mater.*, 45:371–381, 1997.

M. Miller and P. Dawson. Influence of slip system hardening assumptions on modeling stress dependence of work hardening. *J. Mech. Phys. Solids*, 45:1781–1804, 1997.

M. Ortiz. Plastic yielding as a phase transition. *J. Appl. Mech.*, 66:289–298, 1999.

T.Y. Wu, J.L. Bassani, and C. Laird. *Latent Hardening in Single Crystals: Part I, Experimental Characterisation*. Univ. Pennsylvanya, 1989.

Y. Zhou, K.W. Neale and L.S. Tóth. A modified model for simulating latent hardening during the plastic deformation of rate-dependent FCC polycrystals. *Int. J. Plasticity*, 9:961–978, 1993.

**Developments and General Applications**

W.M. Ashmawi and M.A. Zikry. Single void morphological and grain-boundary effects on overall Failure in FCC polycrystalline systems. *Mater. Sci. Engng.*, A343:126–142, 2003.

R. Becker, R.E. Smelser, and S. Panchanadeeswaran. Simulations of earing in highly textured materials. *Comput. Meth. Mater. Processing*, 39:47–60, 1992.

R. Becker, R.E. Smelser, and S. Panchanadeeswaran. Simulations of earing in aluminium single crystals and polycrystals. *Modell. Simul. Mater. Sci. Engng.*, 1:203–224, 1993.

R. Becker and S. Panchanadeeswaran. Effects of grain interactions on deformation and local texture in polycrystals. *Acta metall. mater.*, 43:2701–2719, 1995.

R. Becker. Effects of strain localization on surface roughening during sheet forming. *Acta mater.*, 46:1385–1401, 1998.

M.S. Bruzzi, P.E. McHugh, F. O'Rourke, and T. Linder. Micromechanical modelling of the static and cyclic loading of an Al 2124-SiC MMC. *Int. J. Plasticity*, 17:565–599, 2001.

T.E. Buchheit, R.J. Bourcier, G.W. Wellman, and M.K. Neilsen. Capturing the influence of surface constraints in small and thin samples using polycrystalline plasticity theory. *Modell. Simul. Mater. Sci. Engng.*, 5:421–437, 1997.

S.H. Choi. Simulation of stored energy and orientation gradients in cold-rolled interstitial free steels. *Acta mater.*, 51:1775–1788, 2003.

H.H.M. Cleveringa, E. Van der Giessen, and A. Needleman. Comparison of discrete dislocation and continuum plasticity predictions for a composite material. *Acta mater.*, 45:3163–3179, 1997.

P. Connolly and P.E. McHugh. Fracture modelling of WC-Co hardmetals using crystal plasticity theory and the Gurson model. *Fatigue Fract. Engng. Mater. Struct.*, 22:77–86, 1999.

A.M. Cuitiño and M. Ortiz. Three-dimensional crack-tip fields in four-point-bending copper single-crystal specimens. *J. Mech. Phys. Solids*, 44:863–904, 1996.

M. Dao, P. Gu, A. Maewal, and R.J. Asaro. A micromechanical study of residual stresses in functionally graded materials. *Acta mater.*, 45:3265–3276, 1997.

F. Delaire, J.L. Raphanel, and C. Rey. Plastic heterogeneities of a copper multicrystal deformed in uniaxial tension: Experimental study and finite element simulations. *Acta mater.*, 48:1075–1087, 2000.

O. Diard, S. Leclercq, G. Rousselier, and G. Cailletaud. Distribution of normal stress at grain boundaries in multicrystals: Application to an intergranular damage modelling. *Comput. Mater. Sci.*, 25:77–84, 2002.

F. Feyel, S. Calloch, D. Marquis, and G. Cailletaud. FE computation of a triaxial specimen using a polycrystalline model. *Comput. Mater. Sci.*, 9:141–157, 1997.

S. Forest, B. Barbe, and G. Cailletaud. Cosserat modelling of size effects in the mechanical behaviour of polycrystals and multi-phase materials. *Int. J. Solids Struct.*, 37:7105–7126, 2000.

K. Gall, T.J. Lim, D.L. McDowell, H. Sehitoglu, and Y.I. Chumlyakov. The role of intergranular constraint on the stress-induced martensitic transformation in textured polycrystalline NiTi. *Int. J. Plasticity*, 16:1189–1214, 2000.

M. Grujicic and Y. Zhang. Crystal plasticity analysis of the effect of dispersed $\beta$-phase on deformation and fracture of laminar $\gamma + \alpha_2$ titanium aluminide. *Mater. Sci. Engng.*, A265:285–300, 1999.

J.A. Hines, K.S. Vecchio, and S. Ahzi. A model for microstructure evolution in adiabatic shear bands. *Metall. Mater. Trans. A*, 29A:191–203, 1998.

H.K. Kim and S.L. Oh. Finite element analysis of grain-by-grain deformation by crystal plasticity with couple stress. *Int. J. Plasticity*, 19:1245–1270, 2003.

J.W. Kysar. Continuum simulations of directional dependence of crack growth along a copper/sapphire bicrystal interface. Part I: Experiments and crystal plasticity background. *J. Mech. Phys. Solids*, 49:1099–1128, 2001a.

J.W. Kysar. Continuum simulations of directional dependence of crack growth along a copper/sapphire bicrystal interface. Part II: Crack tip stress/deformation analysis. *J. Mech. Phys. Solids*, 49:1129–1153, 2001b.

C.T. Lin, S.Y. Shen, R. Becker, and S. Suresh. Grain morphology, texture, and microhardness gradients in aluminum diffusion-bonded to aluminum oxide. *Acta mater.*, 47:501–515, 1999.

P.E. McHugh and P.J. Connolly. Micromechanical modelling of ductile crack growth in the binder phase of WC-Co. *Comput. Mater. Sci.*, 27:423–436, 2003.

C. Miehe and J. Schröder. A comparative study of stress update algorithms for rate-independent and rate-dependent crystal plasticity. *Int. J. Numer. Meth. Engng*, 50:273–298, 2001.

A. Needleman. and V. Tvergaard. Comparison of crystal plasticity and isotropic hardening predictions for metal-matrix composites. *J. Appl. Mech.*, 60:70–76, 1993.

S. Nemat-Nasser and T. Okinaka. A new computational approach to crystal plasticity: FCC single crystal. *Mech. Mater.*, 24:43–57, 1996.

T. Ohashi. Numerical modelling of plastic multislip in metal crystals of f.c.c. type. *Phil. Mag. A*, 70:793–803, 1994.

T. Ohashi. Modelling and evaluation of plastic slip deformation in metal microstructures. In S. Tanimura and A.S. Khan, editors, *Dynamic Plasticity and Structural Behaviours*, pages 301–304. Gordon and Breach, New York, 1995.

T.L. O'Regan, D.F. Quinn, M.A. Howe, and P.E. McHugh. Void growth simulations in single crystals. *Comput. Mech.*, 20:115–121, 1997.

V.C. Orsini and M.A. Zikry. Void growth and interaction in crystalline materials. *Int. J. Plasticity*, 17:1393–1417, 2001.

G.P. Potirniche and S.R. Daniewicz. Analysis of tip plasticity for microstructurally small cracks using crystal plasticity theory. *Engng. Frac. Mech.*, 70:1623–1643, 2003.

G.L. Povirk, R. Mohan, and S.B. Brown. Crystal plasticity simulations of thermal stresses in thin-film aluminium interconnects. *J. Appl. Phys.*, 77:598–606, 1995.

S. Quilici and G. Cailletaud. FE simulation of macro-, meso- and micro-scales in polycrystalline plasticity. *Comput. Mater. Sci.*, 16:383–390, 1999.

D.F. Quinn, P.J. Connolly, M.A. Howe, and P.E. McHugh. Simulation of void growth in WC-Co hardmetals using crystal plasticity theory. *Int. J. Mech. Sci.*, 39:173–183, 1997.

E.A. Repetto and M. Ortiz. A micromechanical model of cyclic deformation and fatigue-crack nucleation in F.C.C. single crystals. *Acta mater.*, 45:2577–2595, 1997.

J.R. Rice and M. Saeedvafa. Crack tip singular fields in ductile crystals with Taylor power-law hardening. *J. Mech Phys. Solids,* 37, 673–691, 1989.

J.R. Rice, D.E. Hawk, and R.J. Asaro. Crack tip fields in ductile crystals. *Int. J. Fracture*, 42:301–321, 1990.

M.M. Rashid, G.T. Gray, and S. Nemat-Nasser. Heterogeneous deformations in copper single crystals at high and low strain rates. *Phil. Mag. A*, 65, 707–735, 1992.

M. Saeedvafa. Domain of validity of HRR fields in ductile crystals. *Mech. Mater.*, 19:73–88, 1994.

P. Savage, B.P. O'Donnell, P.E. McHugh, B.P. Murphy, and D.F. Quinn. Coronary stent strut size dependent stress-strain response investigated using micromechanical finite element models. *Ann. Biomed. Engng.*, 32:202–211, 2004.

C. Schmitt, P. Lipinski, and M. Berveiller. Micromechanical modelling of the elastoplastic behavior of polycrystals containing precipitates — applications to hypo- and hyper-eutectoid steels. *Int. J. Plasticity*, 13:183–199, 1997.

A. Staroselsky and L. Anand. A constitutive model for HCP materials deforming by slip and twinning: application to magnesium alloy AZ31B. *Int. J. Plasticity*, 19:1843–1864, 2003.

P. Steinmann and E. Stein. On the numerical treatment and analysis of finite deformation ductile single crystal plasticity. *Comput. Methods. Appl. Mech. Engrg.*, 129:235–254, 1996.

C. Tsakmakis and A. Willuweit. A comparative study of kinematic hardening rules at finite deformations. *Int. J. Non-Linear Mech.*, 39:539–554, 2003.

P.D. Wu, K.W. Neale, and E. Van der Giessen. Simulation of the behaviour of FCC polycrystals during reversed torsion. *Int. J. Plasticity*, 12:1199–1219, 1996.

M.A. Zikry. Failure modes in cubic crystalline materials. *Fatigue Fract. Aerospace Struct. Mater.*, 36:199–214, 1993.

M.A. Zikry. An accurate and stable algorithm for high strain-rate finite strain plasticity. *Comput. Struct.*, 50:337–350, 1994.

M.A. Zikry and M. Kao. Dislocation based multiple-slip crystalline constitutive formulation for finite-strain plasticity. *Scr. mater.*, 34:1115–1121, 1996.

M.A. Zykry and M. Kao. Inelastic microstructural failure mechanisms in crystalline materials with high angle grain boundaries. *J. Mech. Phys. Solids,* 44:1765–1798, 1996.

M.A. Zikry and M. Kao. Inelastic microstructural failure modes in crystalline materials: The Σ33A and Σ11 high angle grain boundaries. *Int. J. Plasticity*, 13:403–434, 1997.

M.A. Zikry, M.R. Pothier, and J.N. Baucom. High strain-rate shear-strain localisation in FCC crystalline materials: A perturbation analysis. *Int. J. Solids Struct.*, 37:6177–6202, 2000.

### Applications to Intermetallics

R.A. Brockman. Analysis of elastic-plastic deformation in TiAl-polycrystals. *Int. J. Plasticity*, 19:1749–1772, 2003.

E.P. Busso and F.A. McClintock. A dislocation mechanics-based crystallographic model of a B2-type intermetallic alloy. *Int. J. Plasticity*, 12:1–28. 1996.

M. Dao and R.J. Asaro. Non-Schmid effects and localised plastic flow in intermetallic alloys. *Mater. Sci. Engng.*, A170:143–160, 1993.

M. Dao and R.J. Asaro. Coarse slip bands and the transition to macroscopic shear bands. *Scr. metall. mater.*, 30:791–796, 1994.

B.K. Kad, M. Dao, and R.J. Asaro. Numerical simulations of stress-strain behaviour in two-phase $\alpha_2$+$\gamma$ lamellar TiAl alloys. *Mater. Sci. Engng.,* A192–A193:97–193, 1994.

B.K. Kad, M. Dao, and R.J. Asaro. Numerical simulations of tensile deformation behaviour in two-phase $\alpha_2$+$\gamma$ lamellar TiAl alloys. *Mat. Res. Soc. Symp. Proc.*, 364:169–174, 1995.

E. Parteder, T. Siegmund, F.D. Fischer, and S. Schlögl. Numerical simulation of the plastic behaviour of polysynthetically twinned Ti-Al crystals. *Mater. Sci. Engng.*, Vol. A192/193:149–154, 1995.

S.M. Schlögl and F.D. Fischer. Micromechanical modelling of the deformation behaviour of polysynthetically twinned (PST) TiAl crystals. In Y.W. Kim, R. Wagner, and M. Yamaguchi, editors, *Gamma Titanium Aluminides*, pages 267–274. TMS, Warrendale, 1995.

S.M. Schlögl and F.D. Fischer. Micromechanical modelling of TiAl intermetallics. *Comput. Mater. Sci.*, 7:34–39, 1996.

S.M. Schlögl and F.D. Fischer. The role of slip and twinning on the deformation behaviour of polysynthetically twinned (PST) crystals of TiAl – A micromechanical model. *Phil. Mag. A*, 75:631–636, 1996.

### Applications to Superalloys

D.W. MacLachlan, G.S.K. Gunturi, and D.M. Knowles. Modelling the uniaxial creep anisotropy of nickel base single crystal superalloys CMSX-4 and RR2000 at 1023 K using a slip system based finite element approach. *Comput. Mater. Sci.,* 25:129–141, 2002.

P.E. McHugh and R. Mohrmann. Modelling of creep in Ni Base superalloy using a single crystal plasticity model. *Comput. Mater. Sci.,* 9:134–140, 1997.

B. Fedelich. A microstructural model for the monotonic and the cyclic mechanical behaviour of single crystals of superalloys at high temperatures. *Int. J. Plasticity*, 18:1–49, 2002.

L. Feng, K.S. Zhang, G. Zhang, and H.D. Yu. Anisotropic damage model under continuum slip crystal theory for single crystals. *Int. J. Solids Struct.*, 39:5279–5293, 2002.

D. Nouailhas, J.P. Culié, G. Cailletaud, and L. Méric. Finite element analysis of the stress-strain behaviour of single-crystal tubes. *Eur. J. Mech. A/Solids*, 14:137–154, 1995.

D. Nouailhas and G. Cailletaud. Tension-torsion behaviour of single-crystal superalloys : Experiment and finite element analysis. *Int. J. Plasticity*, 11:451–470, 1995.

D. Nouailhas and G. Cailletaud. Multiaxial behaviour of Ni-base single crystals. In A. Pineau, G. Cailletaud, and T.C. Lindley, editors, *Multiaxial Fatigue and Design*, ESIS 21, pages 75–82. Mechanical Engineering Publications, London, 1996.

D. Nouailhas and G. Cailletaud. Finite element analysis of the mechanical behaviour of two-phase single-crystal superalloys. *Scr. mater.*, 34:565–571, 1996.

D. Nouilhas and S. Lhuillier. On the micro-macro modelling of $\gamma/\gamma'$ single crystal behaviour. *Comput. Mater. Sci.*, 9:177–187, 1997.

### Strain Gradients

A. Acharya, J.L. Bassani, and A. Beaudoin. Geometrically necessary dislocations, hardening and a simple gradient theory of crystal plasticity. *Scr. mater.*, 48:167–172, 2003.

A. Acharya and J.L. Bassani. Lattice incompatibility and a gradient theory of crystal plasticity. *J. Mech. Phys. Solids*, 48:1565–1595, 2000.

D.J. Baumann. A model of crystal plasticity containing a natural length scale. *Mater. Sci. Engng.* A309-310:406–410, 2001.

J.L. Bassani. Incompatibility and a single gradient theory of plasticity. *J. Mech. Phys. Solids*, 49:1983–1996, 2001.

E. Bittencourt, A. Needleman, M.E. Gurtin and E. Van der Giessen. A comparison of nonlocal continuum and discrete dislocation plasticity predictions. *J. Mech. Phys. Solids*, 51:281–310, 2003.

E.P. Busso, F.T. Meissonnier and N.P. O'Dowd. Gradient-dependant deformation of two-phase single crystals. *J. Mech. Phys. Solids*, 48:2333–2361, 2000.

M.E. Gurtin. A gradient theory of single-crystal viscoplasticity that accounts for geometrically necessary dislocations. *J. Mech. Phys. Solids*, 50:5–32, 2002.

N.A. Fleck and J.W. Hutchinson. Strain gradient plasticity. *Adv. Appl. Mech.*, 33:295–361, 1997.

F.T. Meissonnier, E.P. Busso, and N.P. O'Dowd. Finite element implementation of a generalised non-local rate dependant crystallographic formulation for finite strains. *Int. J. Plasticity*, 17:601–640, 2001.

J.Y. Shu, N.A. Fleck, and W.E. King. Bi-crystals with strain gradient effects. *Mat. Res. Soc. Symp. Proc.*, 458:295–300, 1997.

J.Y. Shu and N.A. Fleck. Strain gradient crystal plasticity: Size dependant deformation of bicrystals. *J. Mech. Phys. Solids*, 47:297–324, 1999.

B. Svendsen. Continuum thermodynamic models for crystal plasticity including the effects of geometrically-necessary dislocations. *J. Mech. Phys. Solids*, 50:1297–1329, 2002.

### Related References

S. Ahzi, B.J. Lee, and R.J. Asaro. Plasticity and anisotropy evolution in crystalline polymers, *Mater. Sci. Engng.*, A189:35–44, 1994.

R. Becker and L.A. Lalli. Texture evolution in channel-die compression – Part II: Effects of grains which shear. *Textures and Microstructures*, 14:145–150, 1991.

A. Bertram and J. Olschewski. Anisotropic creep modelling of the single crystal superalloy SRR99, *Comput. Mater. Sci.*, 5:12–16, 1996.

P. Boisse, P. Ladeveze, M. Poss, and P. Rougee. A new large time increment algorithm for anisotropic plasticity, *Int. J. Plasticity*, 7:65–77, 1991.

N. Bonfoh, A. Carmasol, and P. Lipinski. Modelling of intra-crystalline hardening of materials with particles. *Int. J. Plasticity*, 19:1167–1193, 2003.

G. Cailletaud, S. Forest, D. Jeulin, F. Feyel, I. Galliet, V. Mounoury, and S. Quilici. Some elements of microstructural mechanics. *Comput. Mater. Sci.*, 27:351–374, 2003.

J.D. Clayton, and D.L. McDowell. A multiscale multiplicative decomposition for elastoplasticity of polycrystals. *Int. J. Plasticity*, 19:1401–1444, 2003.

J. Crépin, T. Bretheau, and D. Caldemaison. Cavity growth and rupture of β-treated zirconium: A crystallographic model. *Acta mater.* 44:4927–4935, 1996.

J.P. Dingli, A. Abdul-Latif, and K. Saanouni. Predictions of the complex cyclic behaviour of polycrystals using a self-consistent modelling. *Int. J. Plasticity*, 16:411–437, 2000.

L. Farrissey, M. Ludwig, P.E. McHugh and S. Schmauder. An atomistic study of void growth in single crystalline copper. *Comput. Mater. Sci.*, 18:102–117, 2000.

N.A. Fleck. A crystal plasticity view of powder compaction. *Acta metall. mater.*, 43:3177–3184, 1995.

K. Inal, P.D. Wu, and K.W. Neale. Simulation of earing in textured aluminum sheets, *Int. J. Plasticity*, 16:635–648, 2000.

H. Kitagawa, A. Nakatani, and Y. Shibutani. Molecular dynamics study of crack processes associated with dislocation nucleated at the tip. *Mater. Sci. Engng.*, A176:263–269, 1994.

J. Knap and M. Ortiz. An analysis of the quasicontinuum method. *J. Mech. Phys. Solids*, 49:1899–1923, 2001.

K. Kowalczyk and W. Gambin. Model of plastic anisotropy evolution with texture dependent yield surface. *Int. J. Plasticity*, 20:19–54, 2004.

S. Li, E. Hoferlin, E. Van Bael, P. Van Houtte, and C. Teodosiu. Finite element modelling of plastic anisotropy induced by texture and strain-path change. *Int. J. Plasticity* 19:647–674, 2003.

M. Ortiz. Computational micromechanics. *Comput. Mech.*, 18:321–338, 1996.

S. Panchanadeeswaran, R.D. Doherty and R. Becker. Direct observation of orientation change by channel die compression of polycrystalline aluminum — Use of a split sample. *Acta mater.* 44:1233–1262, 1996.

D. Raabe. Modelling of texture evolution during rolling and compression deformation of intermetallic $Ni_3Al$ and NiAl polycrystals. *Comput. Mater. Sci.*, 3:231–240, 1994.

D. Raabe. Texture simulation for hot rolling of aluminium by use of a Taylor model considering grain interactions. *Acta metall. mater.* 43:1023–1028, 1995.

D. Raabe and W. Mao. Experimental investigation and simulation of the evolution during rolling deformation of an intermetallic Fe-28% Al-2 at. % Cr polycrystal at elevated temperatures. *Phil. Mag. A*, 71: 805–813, 1995.

D. Raabe. Modelling of active slip systems, Taylor factors and grain rotations during rolling and compression deformation of polycrystalline intermetallic $L1_2$ compounds. *Acta metall. mater.*, 43:1531–1540, 1995.

J.L. Raphanel and P. Van Houtte. Simulation of the rolling textures of BCC metals by means of the relaxed Taylor theory. *Acta metall*, 33:1481–1488, 1985.

B.M. Schroter and D.L. McDowell. Measurment of deformation fields in polycrystaline OFHC copper. *Int. J. Plasticity*, 19:1355–1376, 2003.

E.B. Tadmor, R. Phillips, and M. Ortiz. Hierarchical modeling in the mechanics of materials. *Int. J. Solids Struct.*, 37:379–389, 2000.

L.S. Tóth, J.J. Jonas, P. Gilormini, and B. Bacroix. Length changes during free end torsion: A rate sensitive analysis. *Int. J. Plasticity*, 6:83–108, 1990.

E. Van Der Giessen. Micromechanical and thermodynamic aspects of the plastic spin. *Int. J. Plasticity*, 7:365–386, 1991.

E. Van der Giessen, V.S. Deshpande, H.H.M. Cleveringa, and A. Needleman. Discrete dislocation plasticity and crack tip fields in single crystals. *J. Mech. Phys. Solids*, 49:2133–2153, 2001.

A. Zaoui and J.J. Raphanel. On the nature of the intergranular accommodation in the modelling of elastoviscoplastic behaviour of polycrystalline aggregates. In C. Teodosiu, J.L. Raphanel and F. Sidoroff, editors, *Large Plastic Deformations: Fundamental Aspects and Applications to Metal Forming*. Balkema, Rotterdam, 1993.

Y. Zhou, J.J. Jonas, and K.W. Neale. Behaviour of initial texture components during the plane strain drawing of F.C.C. sheet metals. *Acta mater.*, 44:607–619, 1996.

# Microstructural Description of Composites, Statistical Methods

Ryszard Pyrz

Institute of Mechanical Engineering, Aalborg University, Aalborg, Denmark

**Abstract.** For the purpose of mechanical modeling the behaviour of composite materials it is necessary to identify the descriptors that in a best way characterize the special dispersion of fillers. Different morphological measures are discussed along with the description of relevant experimental techniques and simulation methods.

# 1 Introduction

The physical properties of a material are strongly influenced by the microstructure, which is designed during processing. Amount of constituent elements forming the material is in many cases known beforehand as for example the volume fraction of reinforcing phase in composite materials. However, the final architecture of a microstructure is controlled only to a limited extent, and on the microscale the geometrical arrangement of second-phase inclusions is a result of complex and interacting processing micromechanism rather than a design variable. This seems to create an obstacle in modeling the relation between microstructure and overall material properties.

In order to overcome this difficulty it is customarily assumed that the geometrical features of a microstructure are randomly distributed without any precise explanation of what randomness means and how it can be quantified. On the other extreme lies the assumption that the dispersion of fillers is regular. It significantly simplifies calculations, especially with the finite element method, and provides satisfactory results as far as highly non-linear phenomena are not concerned. For strongly localized effects such as initiation of microcracks and plastic zones, which subsequently propagate at different scale lengths, the regularity assumption may lead to seriously erroneous results. Therefore, it is important to describe dispersion characteristics of the microstructure in an unambiguous way by quantitative factors, which eventually could be related to physical properties of the material, Pyrz (1997b).

A multiphase material is by definition heterogeneous and its local properties vary spatially. If the material is statistically homogeneous, which means that the local material properties are constant when averaged over a representative volume element, then it is possible to replace the real disordered material by a homogeneous one where the local material properties are the averages over the representative volume element in the real material. Estimation of those averages presents a fundamental issue for different effective medium theories, Nemat-Nasser and Hori (1993), Aboudi (1991), Chen et al. (1992), Dvorak (1992). Morphological descriptors of microstructures and their dispersion characteristics play a fundamental role in determining property values of composites, Pyrz (1994a), Pyrz and Bochenek (1998). This is particularly true for highly non-linear phenomena such as fracture, where the microfailure threshold is dominated by extreme fluctuations of the stress field, and these load hot spots are strongly influenced by a local disorder of the reinforcement, Axelsen and Pyrz (1995a), Pyrz and Bochenek (1994), Ostoja-Starzewski et al. (1994). It has been realized that the orientation dispersion of fibers in short-fiber composites is neither random nor aligned as assumed by some models. Orientation state of fibres influences overall properties, Curtis et al. (1982), Takao et al. (1982), Bozarth et al. (1987), and there exist several investigations to predict this morphological feature both during processing, Folgar et al. (1984), Jackson et al. (1986), Ausias et al. (1992), Becraft et al. (1992), and in a solid composite, Fischer and Eyerer (1988), Möginger and Eyerer (1991), Bay and Tucker (1992). Another important aspect of microstructure's morphology is shape and size of reinforcing constituents. In the case of short fibre composites the fibres are considered as being circularly cylindrical with a finite length. Although diameters of fibers as well as fibers' length are known to influence overall properties of composites, the aspect ratio of fibers remains the only parameter used in most models, Tandon and Weng (1984). Thus the details of diameters and length distribution are lost in this single parameter. The distribution of fibers' diameters may be easily inferred from planar sections of the material, where the intersections with fibers appear in a form of elliptical profiles. The minor axis of an ellipse represents the fiber diameter. A rather different situation arises when one wants to determine length distribution of short fibers embedded in the matrix material. The simplest way to extract fiber lengths is to burn off the matrix material and then perform length measurements in an image analyzer. However, the pyrolysis technique by itself may cause fibers' fragmentation, and its usefulness depends upon a particular combination of constituent phases. Also measurements of fiber lengths, made before processing, are doubtful as some processing techniques, like for example an injection molding, result in fiber breaking. The effect of inclusion shape and size is especially pronounced for non-elastic behavior such as viscoelastic behavior, yield and fracture, Lee and Mear (1991), Wang and Weng (1992), Derby (1995), Nakamura et al. (1998, 1999).

The probability of the direct interaction between the adjacent reinforcement's entities depends on their spacing and the decay length of disturbed stress field that is introduced by the very presence of the reinforcement embedded in a matrix material. The interaction length can be defined as the length at which the direct interaction between two inclusions provides a contribution to their own stress field disturbance. Therefore, inclusions' spatial locations and distributions are important for predicting overall behavior of a material at non-dilute concentrations of inclusions, Ju and Chen (1994), Ju and Tseng (1996), Axelsen and Pyrz (1995b), Ponte Castaneda and Willis (1995), Kwon and Liu (1998). When we examine the microstructure of a material we are looking at a very small sample of the structure. From this limited view we have tried to understand how the properties of the material relate to microstructure. The approach must necessarily be statistical in particular with respect to the measurement of the microstructure. A selection of representative volume elements where we seek to find microstructure-property relations must be based upon sound sampling techniques, Cruz-Orive and Weibel (1981), Gundersen and Jensen (1987), Gundersen et al. (1999). The size of representative volume element must be such that overall properties of a heterogeneous material are obtained by some volume or an ensemble-averaging procedure over the representative volume element, which length scale is much larger than the characteristic length scale of inclusions, but smaller than the characteristic length scale of a macroscopic sample. This description of the representative volume element must recognize the statistical nature of the microstructure of actual composites that is, to include a sampling of all possible microstructural configurations that may occur in the composite. Each microstructure configuration gives an estimate for the overall quantity studied, and the average of estimates should give accurate results with the scatter of the estimates as small as possible. This is a quite difficult task, and an alternative approach would be to require only that the average is accurate and the scatter left unconstrained, Drugan and Willis (1996). They showed that for a composite with elastic properties and spherical identical inclusions the minimum size of the representative volume element is surprisingly small. The effective elastic constants, calculated from ensemble averages of stress and strain field at finite lengths, converge quickly to those at an infinite length i.e. the overall ones. For a large class of constituents the accuracy of the effective elastic constants derived at a length equal to two sphere diameters were within a few percent as compared to overall elastic constants calculated at the infinite length. Gusev (1997) took another route studying periodic elastic composite with a disordered unit cell made up of a random dispersion of identical spheres. It was shown that already with a few dozen spheres in the unit cell, the scatter of the individual estimates obtained with the same number of spheres was small and the averages were practically stationary. Thus only a few realizations can reasonably predict the overall elastic constants of the periodic composite. Those results are valid for a particular family of composites and are aimed at the overall elastic properties. For other property considerations such as microcracking or fracture propagation, the size of the representative volume element must be related to the correlation length, which is the maximum length over which the fluctuations attributable to the individual defects are correlated, Stauffer and Aharoni (1992).

Spatial dispersion of fillers or inclusions is orchestrated during the manufacturing process and depends on the desired volume fraction of these entities, their shapes, sizes and orientations. These three factors are inherently related especially for the volume fractions approaching the maximum packing fraction achievable for a given shape, size and orientation. A general packing theory for particles with arbitrary shapes and general distributions of sizes does not yet exist. However, several specific investigations related to packing mechanisms of inclusions with clearly defined shapes may help to disclose range of geometrical configurations, which are not attainable due to geometrical constraints. Parkhouse and Kelly (1995) derived an approximate relationship between the limiting volume fraction and the aspect ratio of short fibers. Evans and Gibson (1986) performed a similar investigation. Toll (1998) estimated the maximum packing fraction of force free fibers based on a statistical analysis of the distribution of fiber-fiber contact points. The analysis allows also for a distribution in fiber length and orientation. Furthermore, other types of reinforcement such as woven fabrics were considered as well. The random packing of hard spheres has been dealt with both experimentally and theoretically, Tory et al. (1968), Finney (1970), Visscher and Bolsterli (1972), Powell (1980), Berryman (1983), Milne et al. (1993). It was found that the packing fraction of a random, close packing of equally sized, hard spheres in three dimensions is 0.637, and for loose, random packing approximately 0.60, where the definition of random, close packing means that the system's potential energy obtains a minimum. The packing fractions of hard discs in two dimensions are 0.89 for random, close packing. Another formulation of the packing problem is to derive the average excluded area, in two dimensions or volume in three dimensions, associated with two- and three-dimensional randomly oriented objects, Balberg et al. (1984). The excluded area (volume) of an object is defined as the area (volume) around an object into which the center of another similar object is not allowed to be located if overlapping of the two objects is not permitted. The usefulness of this kind of the analysis is related to the percolation properties of systems made of non-spherical objects such as short fibers that have random orientation in space. In general, the total excluded volume needed for the onset of percolation decreases with increasing degree of randomness.

Morphology is the study of form. The form of an object is the shape and structure of that object as distinguished from its material components. The structure may include dispersion and orientation distribution of features of the object. The shape or form of an object is not a well-defined concept. The notion of a geometrical structure, or texture, is not purely objective. The concept of an inclusion size is inherently connected to the notion of shape. Convex inclusions with geometrically regular shapes can, in principle, be assigned with linear sizes. For non-convex inclusions and inclusions with irregular shapes other measures of size such as surface area or volume must be used. The concept of size becomes critical if one wants to compare inclusions of unequal shape. In this situation statement defining a size for one type of shapes not necessarily holds true for another shape of inclusions. The boundary between computational geometry and morphology is therefore rather fuzzy. We use geometrical tools to extract and measure the shape and form of an object, and this might be referred to as a morphological characterization problem. A typical problem then would be to ask the following question: "What is the shape of a set of points and how can it be quantified". Another related problem is "Given two points sets how separated are they "or in other words "What is the metric and distance in a point space to which point sets belong". Any mathematical description of shape and form thus far conceived has always fallen short of what was hoped for. There is no one descriptor of form and shape that is best for all occasions. In the present context morphology of composite materials is related to the shape and structure of features on mesoscopic scale lengths. Thus, as a minimum the optical microscope is sufficient to reveal fine details in the sample at a range of magnifications from 10x to 4000x. Resolution of about 0.3 μm is possible, limited by the nature of the sample, the wavelength of light and the objective lens.

## 2 Measures of Spatial Distributions

Studies in spatial analysis rely on essential data that has been stored in a form, which emphasizes its spatial feature. In this framework objects from the material microstructure are stored as symbols, which represent their geometric characteristics. For example, fibers in unidirectional composites recorded on the transverse sections or spherical particles in particulate composite may be represented by a point process whereas cracks' traces, observed on the planar slices across a damaged material would be considered as a line process. The set of points in a region where each possible location is equally likely to be chosen and the location of each point is independent of the location of any other point, is defined as a completely random pattern. All statistical moments of this pattern are known analytically as its distribution function of inter-point distances is identical with the Poisson distribution, Hermann (1991). Very few point distribution models with known distribution function of distances have been proposed, and usually one has to turn to computer simulations. The Poisson set of points is physically unattainable due to a finite dimension of inclusions, but it may serve as a convenient dividing line between aggregated and more regular patterns.

### 2.1 First Order Quantities

There have been several attempts to quantify dispersion characteristics of point sets based upon distance methods, Bansal and Ardell (1972), Corti et al. (1974), Diggle et al. (1976), Davy and Guild (1988), Tewari et al. (1998), Karnezis et al. (1998). Different micro-morphological parameters derived from distance measures have been tested for their applicability to describe in an ambiguous way dispersion pattern of point sets, which exemplify distribution of fillers, Altan et al. (1990), Wimolkiatisak et al. (1990), Taya et al. (1991), Alberola et al. (1998).

In recent years a wide range of complex structures have been quantitatively characterized using the idea of fractal dimension: a dimension that corresponds in a unique fashion to the geometrical entity under study and often is not an integer. This concept provides a natural framework within which we may discuss a range of phenomena occurring in disordered systems. The fractal concept treats disorder as an intrinsic rather than a perturbative phenomenon. A system is weakly disordered if the disorder disappears when one samples the system at progressively smaller or larger scales. In strongly disordered systems the disorder persists at all scale lengths and may repeat itself at different length scales. This repetition is reflected in fractal dimension $D$, which characterizes space-filling ability of the system. Fundamentally, it is its geometric character, which gives fractals a great scope since fractal geometry forms the missing complement to Euclidean geometry and crystalline symmetry, Gouyet (1996). Recognizing that an object or phenomenon is fractal gives one a predictive tool over the range of scales, if the fractal dimension is known. The utility of being able to characterize quantitatively a random dispersion of fillers by the fractal dimension appears to be crucial in relating geometrical aspects of microstructure to physical properties. The fractal dimension may serve as an indicator of the stress variability giving a quantitative measure of the microstructure's load bearing capacity and its resistance to microcracking, Pyrz (1997a). Another important aspect of the fractal concept is its applicability to describe a complicated texture of materials' microstructure as recorded on captured images of the material, Super and Bovik (1991), Soille and Rivest (1996), Sandau and Kurz (1997).

A more informative way of the assessment of the point set arrangement is to map the points onto a set of contiguous polygons. If the point set represents the center points of inclusions that appear on the transverse section of two-phase material, then we can assign to each center all points of the planar area that are nearest to it. This is achieved by drawing the mid-perpendiculars to the segments joining any two neighboring centers. In this way a unique cell is constructed for each center, and defined as the smallest convex polygon surrounding it. Each polygon contains all points of the planar area that are closer to the corresponding center than to any other. This polygonal tessellation, the Dirichlet/Voronoi tessellation, covers the plane of observation without overlapping and assigns a unique area to each center point of the set, Getis and Boots (1978), Sibson (1980). In the case when the second phase inclusions differ significantly in size or shape then the tessellation must be extended to that of a "finite body" i.e. tessellations are constructed from the interfaces of each individual secondary phase body, Boselli et al. (1999). In any case, based on the tessellated cell structure, a number of measurements may be derived such as distribution of cell areas, perimeters, elongation, etc. The degree of regularity, clustering, randomness and anisotropy may be assessed via statistical analysis of tessellation, Spitzig et al. (1985), Billia et al. (1991), Brockenbrough et al. (1992), Everett and Chu (1993), Chiu (1993).

In geometrical terms, all random cellular structures generated by the same point set model are indistinguishable, even if they are not identical when superimposed. Therefore, the polygons cannot be defined and distinguished by their size or any other metric measure alone. Their distinctive feature is their shape, which is a topological property. Aiming to describe the topological randomness of the pattern, the topological shape of a polygon, i.e. the number of its sides, is considered as the random variable. Then the polygonal tessellation is described by the frequency $p_n$ of finding an n-sided polygon. The topological entropy, Rivier (1985), Weaire and Rivier (1984) Lemaitre et al. (1993), may characterize statistical status of the polygonal network. The topological entropy measures a degree of arbitrariness. The entropy value for a perfectly regular pattern is always equal to zero and grows with an increasing disorder. It is also a remarkable property of polygonal tessellation that the average number of polygon sides is 6, and this value is conserved regardless of the degree of disorder of the underlying point pattern. Thus the frustration of the random network of polygons, i.e. growing topological entropy, is caused by an appearance of polygons with a number of edges different from 6. Independent of the degree of disorder the mean area of n-sided polygons grows with the number. One could consider polygons with the number of sides different from six as topological defects which influence the value of topological entropy as well as enlarge the variability of the corresponding stress field depending on the degree of disorder, Pyrz (1996a), Pyrz and Bochenek (1998).

The idea of the Dirichlet tessellation has been used by Ghosh and co-workers (1997 a,b, 1998) to develop the Voronoi cell finite element model. By including the geometrical aspects of the microstructure in the hybrid finite element analysis, and developing element formulations which account for different constituent properties such as elastic, plastic and viscoplastic, the method provides very effective numerical tool for the determination of overall behavior of different classes of composite materials.

## 2.2 Second Order Quantities

There are two basic categories of classifying techniques to study point patterns, those dealing with absolute measurements of the dispersion of points about the area occupied by a point set, and those examining the association of points relative to other points in the set. The former category has been described in the previous sub-section, while the latter includes relative measures associated with point-to-point relationships.

Second-order characteristics of a point set describe variation and eventual correlation and association of points relative to other points of the set in an immediate neighborhood of the reference point. The second-order intensity function $K(r)$, for which the theoretical background has been established by Ripley (1977), is defined as the number of further points expected to lie within a radial distance $r$ of an arbitrary point and divided by the number of points per unit area. The calculation procedure amounts to the summation of a number of points in the circle with center at one point and radius $r$, and the successive repetition of this step for all points and radial distances limited to the radius range 0-0,3, considering the window of observation as a unit square, Diggle (1983), Pyrz (1994a). The second-order intensity function of the Poisson point set is equal to $\pi r^2$. For aggregated distributions, $K(r)$ function lies above the line corresponding to the Poisson set and diverging from it markedly as the clustering tendency becomes more pronounced. For point patterns tending towards a regular type distribution, the $K(r)$ function lies below the complete random and takes a stair-shaped form for perfectly regular distributions, where each stair indicates deterministic inter-point distances, Pyrz (1994b). The $K(r)$ function is relatively sensitive to local disturbances in the point pattern and can be utilized for characterization of inhomogeneous populations of points as well as mixture of several types of points, Lotwick and Silverman (1982), Diggle and Chetwynd (1991).

The pair correlation function *g(r)* is another important second-order quantity that might be used for exploratory statistical analysis of point sets, Everett (1996), Pyrz (1994b). The pair correlation function is related to the probability *g(r)dr* of finding a point whose center lies in an infinitesimal circular region with radius *dr* about the point *r*, provided that the coordinate system is located at the second point. Large values of *g(r)* indicate that point pairs of distance *r* appear frequently, whereas local minima correspond to the least frequent distances in the pattern. For statistically homogeneous pattern the pair correlation function tends to one for large distances. Local variations are smeared out in a continuum and the radius $r_o$ such that $g(r_o)=1$ determines the range of geometrical disorder in the point pattern. The functions *K(r)* and *g(r)* are related, but they provide quite different information.

The introduction of the mark correlation function *M(r)*, Pyrz (1996b), Stoyan (1984) may facilitate the characterization of more complex aspects of disorder. Attaching to each point of the set a typical feature expressible by numbers results, in multivariate characterization of a distribution. In addition to positions of points one introduces a further variable which may correspond to the size of inclusions, any field quantity such as stress, temperature, conductance etc. known at the inclusion spot, but also shape and type of inclusions may be used to "mark" the points. The mark correlation function corresponds to the conditional mean value of the mark product of a pair of inclusions positioned distance *r* apart. An interpretation of the mark correlation function is similar to the corresponding explanation of the pair correlation function. The local maxima indicate the attraction effect between inclusions' positions and corresponding marks, whereas local minima suggest repulsion between marks and distances. The mark correlation function couples physical and geometrical aspects of dispersions in composite materials disclosing a degree of physical interaction effects dependent upon a distribution disorder of interacting entities. For large distances $r_p$ the mark correlation function is equal to one, and at this distance a reference inclusion does not distinguish individual neighbors and "feels" the surroundings as a continuum. Thus a physical range of interaction $r_p$ may serve as an indicator for the selection of a proper size of the representative volume element for the physical problem under consideration.

Yet another second-order function is the covariance defined by a probability of finding two random points with specified distance both lying in the same phase. This function is suitable to describe structures where several population of objects coexist with rather distinct size distributions, multiphase materials and texture anisotropy, Serra (1982), Wiencek and Stoyan (1993).

Presented second-order quantities attempt to summarize the whole point set and are not able to localize sub-patterns within the point set, König et al. (1991), Mattfeldt et al. (1993), Karlsson and Liljeborg (1994). This is a serious drawback, since an immediate dispersion pattern of neighbors around reference inclusions determines local physical behavior of a material, which control highly non-linear and localized phenomena such as microcracking and fracture, Pyrz and Bochenek (1998). Furthermore, second-order quantities are important but not exhaustive since we may encounter different patterns, which have the same second-order properties. Then, either one turns to third-and higher-order characteristics, which enable refinements of the results or one uses various second-order quantities jointly with several first-order characteristics.

### 2.3 Higher Order Quantities

Third- and higher order characteristics of phase dispersion provide refinements of the results obtained by second-order characteristics. The $n$-point spatial correlation function is a measure of the probability of finding n points all lying in the region of space occupied by one constituent at a two-phase material. The first three correlation functions can be estimated from the micrographs taken on the cut through a material with a heterogeneous and random microstructure. Corson (1974) has made such measurements, and then the measurements can be automated using image-processing technique, Berryman (1985). Higher-order correlation functions contain much more information about the geometry of a microstructure, and this information reduces uncertainty concerning the effective property of the composite. A significant improvement of predicting bounds for effective properties of two-phase materials is already achieved by a consideration of the third-order correlation function. Introducing the concept of a cell material, where the cell is defined as being a mathematically closed surface containing a portion of the random medium of uniform property, Miller (1969), showed that the third-order correlation function has a simple geometrical significance on the improved bounds for effective electrical, thermal, magnetic and elastic properties of heterogeneous materials. Milton (1982) derived improved and simultaneously simplified bounds on the elastic and transport properties of two-phase composites. The usefulness of these bounds is dependent on the evaluation of certain constants, which are functions of the third-order correlation function. The same constants that characterize the micro-geometry of the composite can be implemented to calculate bounds on one effective material property from measurements of another. These include bounds correlating the effective electrical or thermal conductivity and the effective shear modulus with the effective bulk modulus. For porous materials, measurement of bulk modulus provides bounds on tortuosity and vice versa, Berryman and Milton (1988). Such cross-property relations are important to determine, which morphological features of a microstructure are common to physically different effective properties, Torquato (1994). In order to study interaction effect between different objects of a microstructure such as crack-to-crack, pore-to-pore or pore-to-crack interactions, it is necessary to introduce general n-point distribution functions, which correlate geometrical positions of those microstructural entities that are inherent to the phenomena under consideration. For instance, investigating interactions between pores it is more important to know statistics concerning distances between pores' surfaces rather than between pores' center points. General $n$-point distribution functions specialized for specific physical situations were investigated in Torquato (1986), and comprehensive reviews of advances in the characterization of heterogeneous random materials with implementation of n-point correlation functions are presented in Torquato (1991) and Quintanilla (1999).

In conjunction with $n$-point correlation functions an interesting inverse problem may be posed by generating a set of realizations of random heterogeneous microstructures with specified but limited, i.e. lower order, statistical correlation functions, Yeong and Torquato (1998 a,b), Cule and Torquato (1999). Reconstructing a microstructure in this way would provide answers to many problems. One would identify the class of microstructures that have precisely the same type of correlation functions and vice verse, one could determine the extent to which different correlation functions can reproduce the reference microstructure, thus shading light on the nature of the information contained in the correlation functions. Furthermore, it would be possible to construct microstructures that correspond to specified correlation functions and categorize random materials with respect to a class of correlation functions, used in reconstruction procedure.

### 2.4 Illustrative Examples

**Example 1.** Figure 1 illustrates five planar dispersions of points together with corresponding Dirichlet tessellations. Given distinct points in a two-dimensional area we can assign to each point a zone of influence consisting of that part of the area, which is closer to selected point than to any

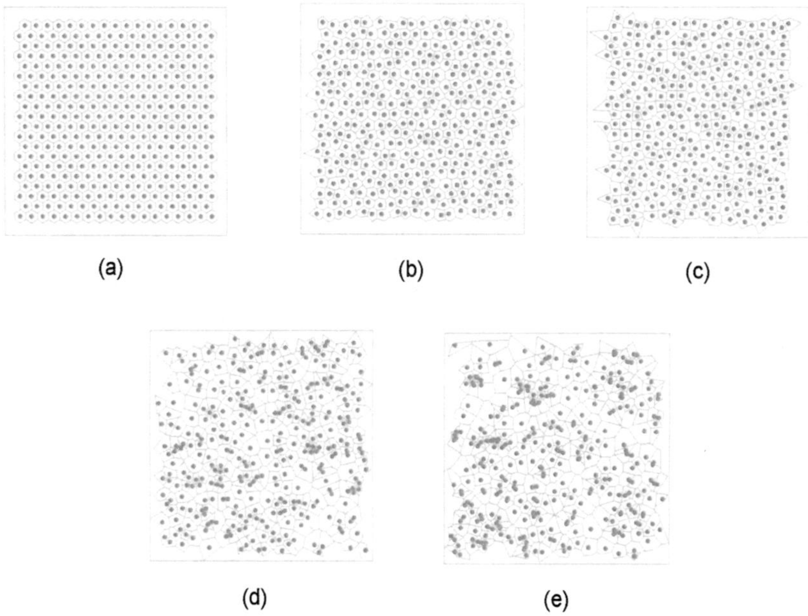


**Figure 1.** Point dispersions: regular hexagonal (a); disturbed hexagonal (b); homogeneous hard-core (c); non-homogeneous hard-core (d); Poisson (e).

other point. This tessellation covers the plane of observation without overlapping and assigns a unique area to each centre point of the set. Boundary polygons, which have a fragment of the window frame, are excluded from the analysis. This is because their topology and geometry is not solely defined by a point set, but is artificially constrained due to a finite size of the observation window. In geometrical terms, all random polygonal structures generated by the same point set model are indistinguishable, even if they are not identical when superimposed. Therefore, the polygons cannot be defined and distinguished by their size or any other metric measure alone. Their distinctive feature is their shape, which is a topological property. Aiming to describe the topological randomness of the pattern, the topological shape of a polygon, i.e. the number of its sides is considered as the random variable. Then the tessellation is described by the frequencies $p_n$ of finding $n$-sided polygon within an area of observation. Statistical status of the polygonal network may be characterized by the topological entropy $S$

$$S = -\sum_{n=1}^{k} p_n \ln p_n \tag{1}$$

The topological entropy measures a degree of arbitrariness. The entropy value for a perfectly regular pattern is always equal to zero and grows with an increasing disorder. Values of the topological entropy for point patterns illustrated in Figure 1 are correspondingly 0, 0.27, 1.24, 1.51 and 1.66.

**Example 2.** The pair correlation function *g(r)* is related to the expected number of inclusion centres to be found in a circular ring with radius *r* and thickness *dr*. Figure 1 shows the pair correlation function *g(r)* for the indicated point pattern. Some of its features can be interpreted in the following way. The pair correlation function is equal to zero for a distance smaller than minimal distance between a pair of fibres that is present within a pattern. It rises then to a peak at some distance that indicates a most frequent distance to near neighbours, which occur in the pattern. The existence of a sharp maximum indicates that the near neighbours have relatively uniform distance to the reference fibre. The more uniform distances are, the sharper is the first and the subsequent maximum. The pair correlation function for a perfectly regular pattern appears in a form of Dirac functions placed at points corresponding to regular distances of the pattern. The minima of the pair correlation function correspond to least frequent distances between relatively few fibres. For large radial distances the number of inclusions within the circle becomes more and more uncertain. The peaks broaden, merge with one another, and become lost in the continuum background where *g(r)* tends to unity. The pair correlation function for a completely random point pattern is identically equal to one. Local variations are smeared out in a continuum as $r$ grows, and the range of geometrical correlation $r_0$ can be defined such that $g(r) \approx 1$ for $r > r_0$.

The behaviour of pair correlation function is non-smooth which is inherently related to the way how this function is calculated. A new function may be introduced which preserves all characteristics of *g(r)* but represents smooth behaviour. The definition of the new function, an integral correlation function, is introduced as follows

$$X(r) = \int_{r}^{r_{max}} |1 - g(s)|\, ds \tag{2}$$

The comparison of *g(r)* and *X(r)* for a selected planar pattern is given in Figure 2 whereas the behaviour of *g(r)* and *X(r)* for spatial pattern is illustrated in Fig. 3.

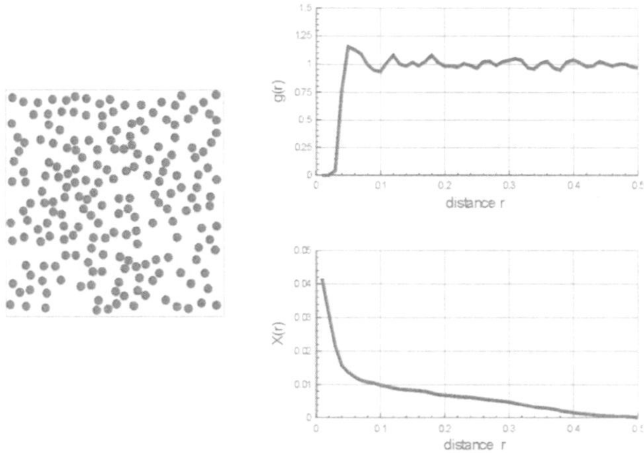


**Figure 2.** Planar dispersion and corresponding morphological functions.

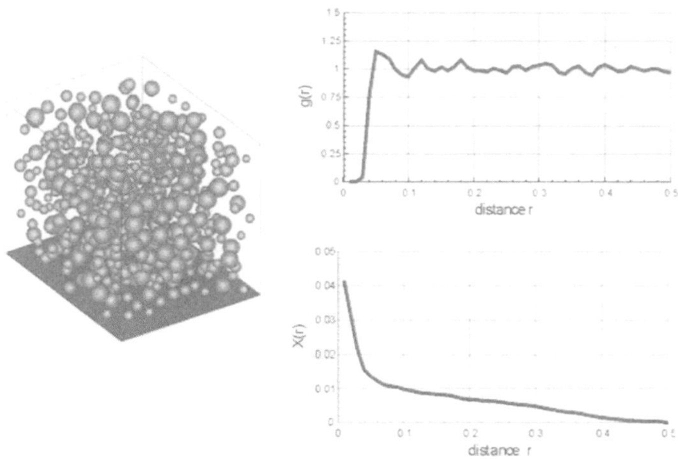


**Figure 3.** Spatial dispersion and corresponding morphological functions.

**Example 3.** The pair correlation function *g(r)* is frequently used to determine structural relaxation in amorphous solids, bulk viscosity and changes in local density, see e.g. Srolovitz et al., (1981), Okumura and Yonezawa, (2003). It indicates also configuration changes at molecular and atomic level caused by an external deformation. Figure 4 illustrates molecular dynamics simulation snap-shots of a model nano-composite composed of 16 polypropylene chains and a single C60 "bucky-ball" representing reinforcement. The initial cubic unit cell has dimensions38x38x38 Å and is extended by a unidirectional loading to the dimensions of 22x62x22 Å.

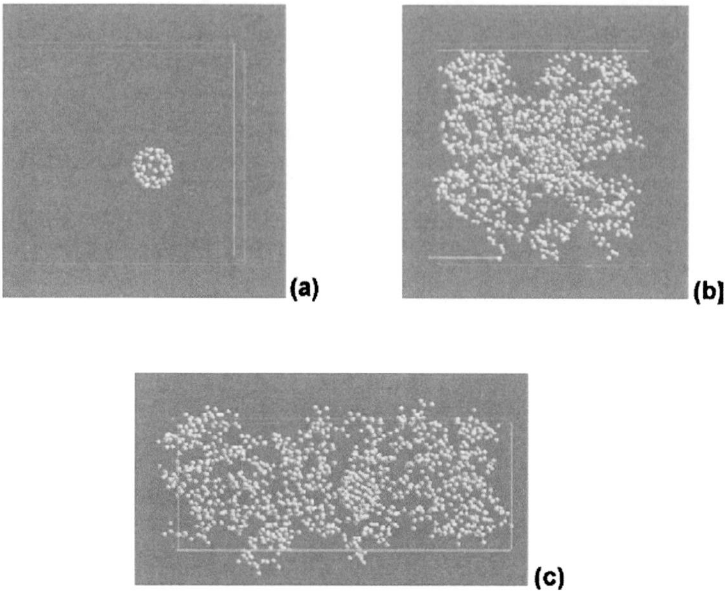


**Figure 4.** Initial unit cell with single C60 molecule (a); initial unit cell filled with C60 and polypropylene molecules (b); unit cell after deformation (c).

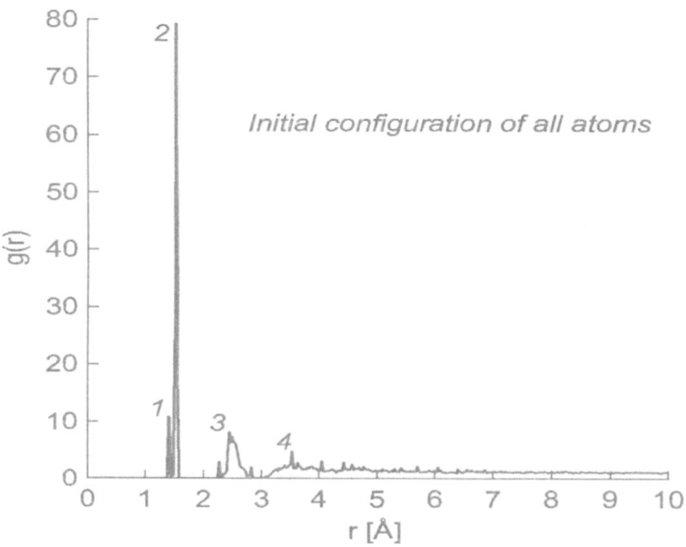


**Figure 5a.** Initial pair correlation function.

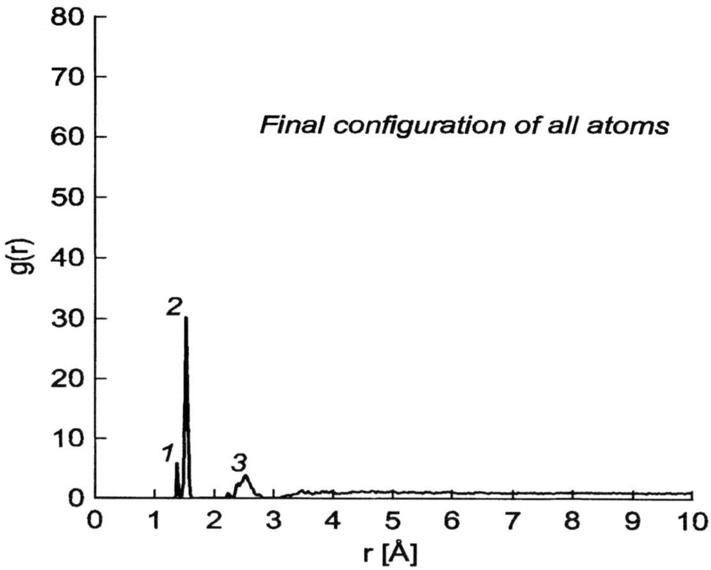


**Figure 5b.** Final pair correlation function.

The pair correlation function has been calculated for all carbon atoms contained within the unit cell. Figure 5(a,b) shows the pair distribution function at the initial and final interval of molecular dynamics simulation. The influence of the deformation is clearly indicated by lower values of first three peaks and disappearence of the fourth peak. The evolution of the pair correlation function during deformation can be used to determine local stresses at molecular level, see Pyrz (2003).

**Example 4.** The amount of constituent elements forming the material is in many cases known beforehand as for example the volume fraction of reinforcing phase in composite materials. However, the volume concentration of inclusions is merely a statement of the overall composition of the material rather than of its dispersion characteristics. The final architecture of a microstructure is controlled only to a limited extent, and on the micro scale the geometrical arrangement of second-phase inclusions is a result of complex and interacting processing micro mechanisms. If the processing conditions are stable and the variability of constituent properties are minimal then we may consider resulted microstructure as statistically homogeneous when sampled at different positions in a body of the material. In simulation experiments three model dispersions have been generated, Figure 6. A hard-core dispersion arises by the imposition of a minimum permissible distance between any two points, which are distributed randomly otherwise. The hard-core model is obtained by the imposition of the prohibitive distance between points to be significantly larger than the fibres diameter, whereas in the Poisson model a minimal inter-point distance is equal to the diameter of fibres. This restriction guarantees a large degree of randomness to be present in the Poisson pattern. The cluster model may be created in many different ways by distributing cluster centres randomly and placing offspring points randomly around parent points within a predetermined area. In this case, cluster area, number of parent points and number of daughter points within each cluster can be randomised. In the present case the cluster distribution possesses 30 parent points and 10 randomly distributed around offspring points. The hard-core and Poisson distribution models exemplify, respectively, well processed composite and a composite processed with insufficient consolidation pressure that results in decreased penetrability of fibres through the matrix material. The cluster model represents composite reinforced with randomly dispersed bundles of fibres.

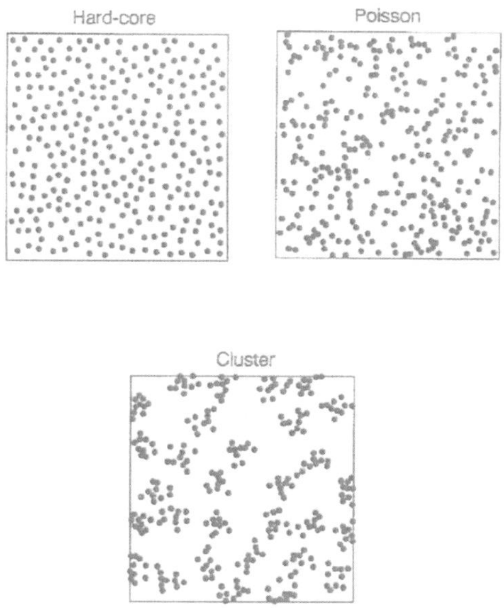


**Figure 6.** Dispersion patterns of fibres.

The local stress analysis of patterns is reduced to a two-dimensional plane strain problem as the fibres are assumed to be elastic parallelly aligned, and embedded in an elastic matrix with perfect bonding.

The stress field solution for a single inclusion embedded in a matrix may be obtained analytically from the Eshelby solution, Mura (1987). As for the Eshelby solution the eigenstrain analogy is also applied, but contrary to this solution where the Eshelby tensor is solved, an iterative procedure is derived. The heterogeneous solid is replaced by an equivalent homogeneous one where tractions are applied on the imaginary contour of the circular fibre and an unbalanced stress field is added inside this imaginary contour. The stress field inside the fibre is determined by an iterative procedure, and subsequently the stress field in the matrix can be calculated, see Axelsen and Pyrz (1995b). Three-dimensional case has been considered by Schjødt-Thomsen and Pyrz (2002).

In the solution for multiple fibres the stress field interaction between the fibres must be taken into account. The basic idea of accounting for this interaction is to determine the stress field inside a fibre, as it would be alone including, however, the interacting stress field from the neighbouring fibres. Then, the problem is divided into a number of sub problems corresponding to the number of fibres, Figure 7.

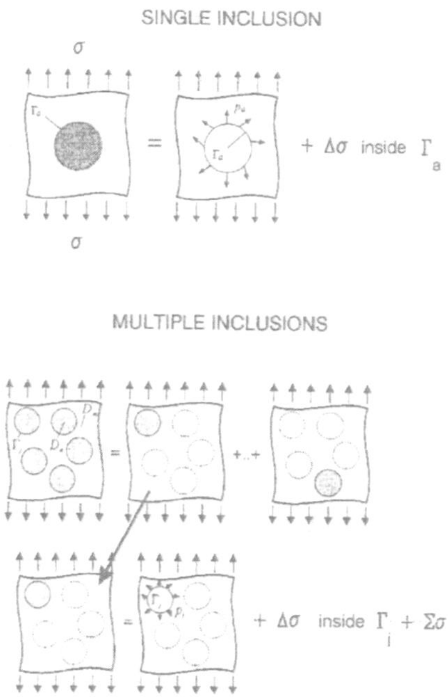


**Figure 7.** Superposition scheme for a solid containing multiple fibres.

Each sub problem is solved as for a single fibre and in addition the interacting stresses from the remaining fibres are included in the stress field calculation inside the area covered by the fibre. The plane solid containing multiple randomly dispersed fibres are subjected to unidirectional loading $\sigma_\infty$ at the remote boundaries. The solid contains $N$ fibres all of circular shape, and the boundary curved for the fibres are denoted $\ni_j$ where $j = 1,..., N$. All the fibres have the same elastic stiffness $\mathbf{D}_a$ and the matrix has the elastic stiffness $\mathbf{D}_m$. Radii of the fibres are all equal to $R = 5$ μm.

As a starting point in the iterative procedure the stress field in the whole solid is equal to remote loading at boundaries $\sigma = \sigma_\infty$. Using the theory of eigenstresses, Mura (1987), the unbalanced stress field for each fibre becomes

$$\Delta\sigma_j = (\mathbf{D}_m - \mathbf{D}_a)_a^{-1}\,\sigma_j \qquad j = 1,\ldots..,N \tag{3}$$

Tractions are applied at the imaginary contours of the fibres

$$\mathbf{p}_j = -\Delta\sigma_j \mathbf{n}_j \quad j = 1,....,N. \tag{4}$$

when $\mathbf{n}_j$ is an outward unit normal to the circular contour $\Gamma_j$. The boundary curve $\Gamma_j$, is subdivided into segments of length $ds$, and the tractions $\mathbf{p}_j$ are replaced by concentrated forces $\mathbf{p}_j ds$. By applying the solution for a concentrated force in homogeneous elastic solid the disturbance stress from $\mathbf{p}_j ds$ is denoted as $\mathbf{f}[\mathbf{p}_j]ds$ along the circular contours

$$\sigma_{ij} = \oint_{\Gamma_i} \mathbf{f}[\mathbf{p}_j]ds \quad j = 1,.....,N. \tag{5}$$

and the disturbance stress is calculated at an arbitrary point within the contour. The new stress field inside the fibres is obtained by

$$\sigma_j = \sigma_\infty + \sum_{i=1}^{N} \sigma_{ij} - \Delta\sigma_j \quad \text{inside } \Gamma_j. \tag{6}$$

From this expression the unbalanced stress is re-calculated according to Eq. (3), and the iterations are repeated until $\mathbf{p}_a$ dos not change significantly. The method converges quite rapidly and the stress field outside the fibre may be determined as follows:

$$\sigma_j = \sigma_\infty + \sum_{i=1}^{N} \sigma_{ij} \quad \text{outside } \Gamma_j. \tag{7}$$

The stress analysis is implemented in a numerical procedure and the stress field at any arbitrary point is obtained within reasonable computational time. It should be noted that every simulated pattern must be equipped with appropriate boundary conditions. Since the dispersion of fibres is observed within a finite window the edge effects may influence the stress field calculations as well as a calculation of geometrical and topological parameters of the dispersion. In order to avoid this problem periodic boundary conditions are applied to all calculations. It means that the window of observation has to be surrounded by a sequence f windows containing the same dispersion patterns as the original one. In other words, we enlarge the space of observation by assuming that the dispersion of fibres in a very large area may be reconstructed from a single window of observation by a sequential repetition of the observation window itself. This keeps the dispersion characteristics fixed and diminishes the edge effects.

Since the initiation of micro cracking usually occurs at the fibre-matrix interphase, maximal radial and tangential stresses in an annular layer around each fibre are of particular interest. Figures 8 and 9 illustrate variability of maximal radial and tangential stresses for all fibres of each dispersion types. The stresses have been calculated for both uniaxial as well as biaxial loading. The influence of the dispersion type on the maximal value of radial and tangential stresses is apparent. First of all, fluctuations of stresses are smallest for the hard-core distribution whereas the cluster distribution contains a number of fibres with significantly larger stress values than other fibres in the pattern. Secondly, maximal radial stresses for all fibres in considered dispersions are amplified with respect to remote unit load, whereas only a certain number of fibres in the Poisson pattern and a larger number in the cluster pattern are exposed to amplified tangential stresses. It is interesting to notice that the mean value of maximal radial and tangential stresses increases as the dispersion gets more disordered. The local variability of stress maxima is completely hidden in this quantity. Imposition of the biaxial loading to the patterns does not alter significantly maximal radial stresses, and if anything, the slight decrease of stress intensity is observed, however, the picture of maximal tangential stresses changes dramatically. The values of maximal tangential stresses are magnified almost twice as compared to the uniaxial loading. Thus, it is most likely that maximal radial stress hot spots alone generated under uniaxial loading are strong enough to initiate interfacial debonding, provided that the matrix strength is higher than the strength of the interphase. Furthermore, it is expected that increased intensity of both radial and tangential maximal stresses under biaxial loading would initiate radial matrix cracks

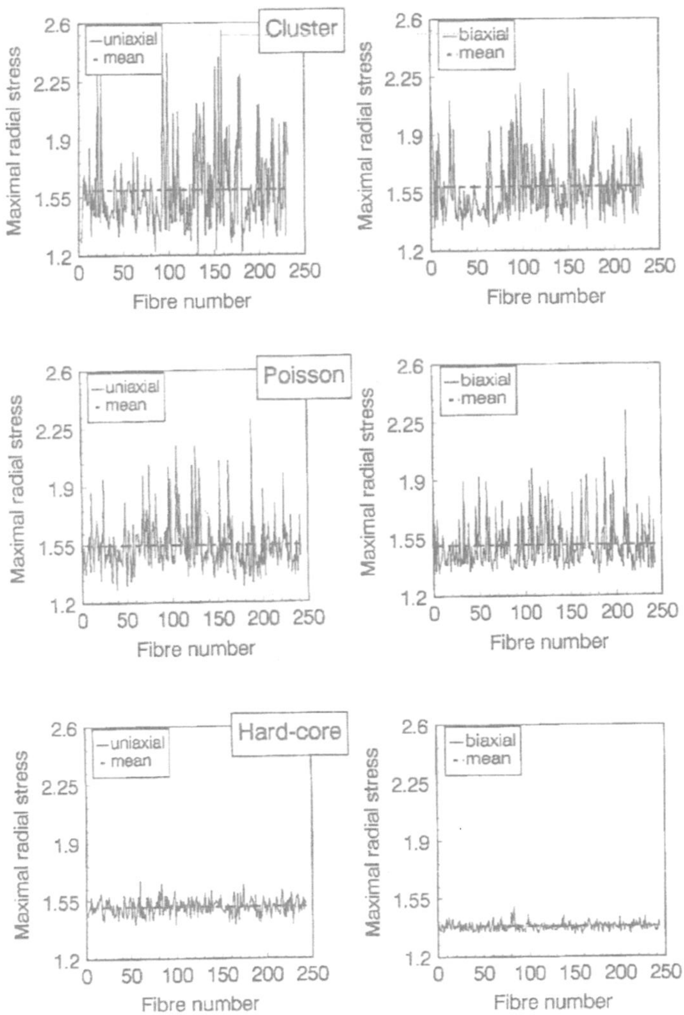


**Figure 8.** Variability of maximal radial stresses among fibres in uniaxial and biaxial loading.

combined with interfacial debonding. In the following analysis the attention will be paid exclusively to maximal radial stresses under uniaxial loading conditions. Amplification of local maximal radial stresses and their variability is related to nearest neighbour distances and nearest neighbour orientations of adjacent fibres. An average nearest neighbour distance between fibres in the hard-core model is larger than for two other models and this tendency is prevailed for all nearest neighbour orientations where the loading direction corresponds to $90^{\circ}$, Figure 10. The inter-fibre distances in the cluster model are significantly smaller and vary with orientation to a

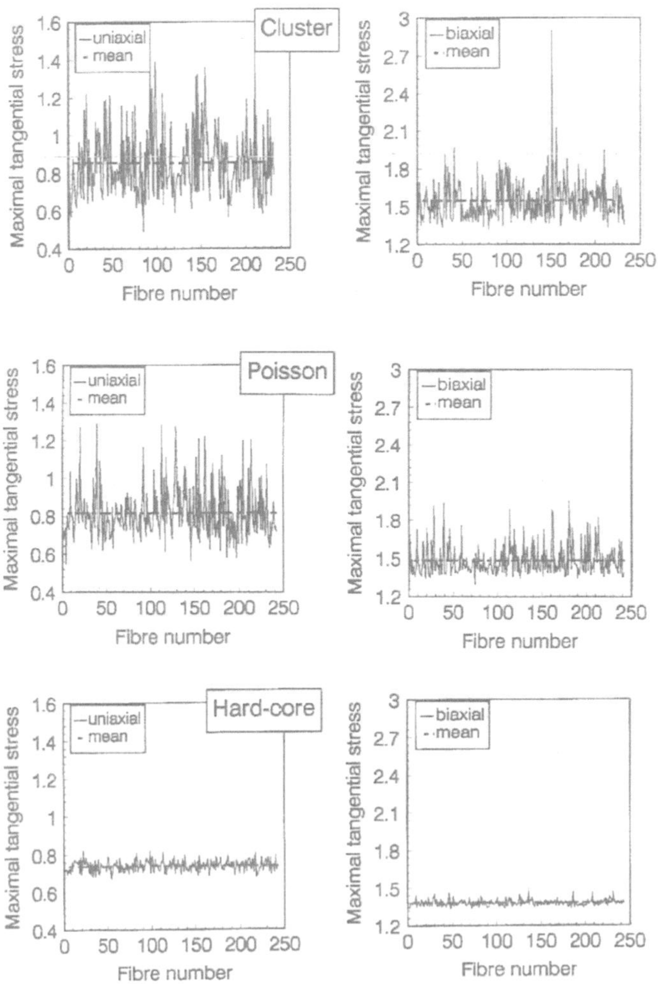


**Figure 9.** Variability of maximal tangential stresses among fibres in uniaxial and biaxial loading.

larger extent than in the hard-core distributions. Since the interaction effects are strongly influenced by the distances between neighbouring fibres, as will be shown in a sequel, the stress amplification is much more pronounced in the cluster model than in the other two models. Furthermore, the amplification of the maximal radial stresses is largest for fibres, which are aligned very close or along the loading direction. There are more fibres in the cluster model, which are mutually positioned along the loading direction at shorter distances than in other models. This has resulted in larger variability of maximal radial stresses among the fibres dispersed in the form of clusters.

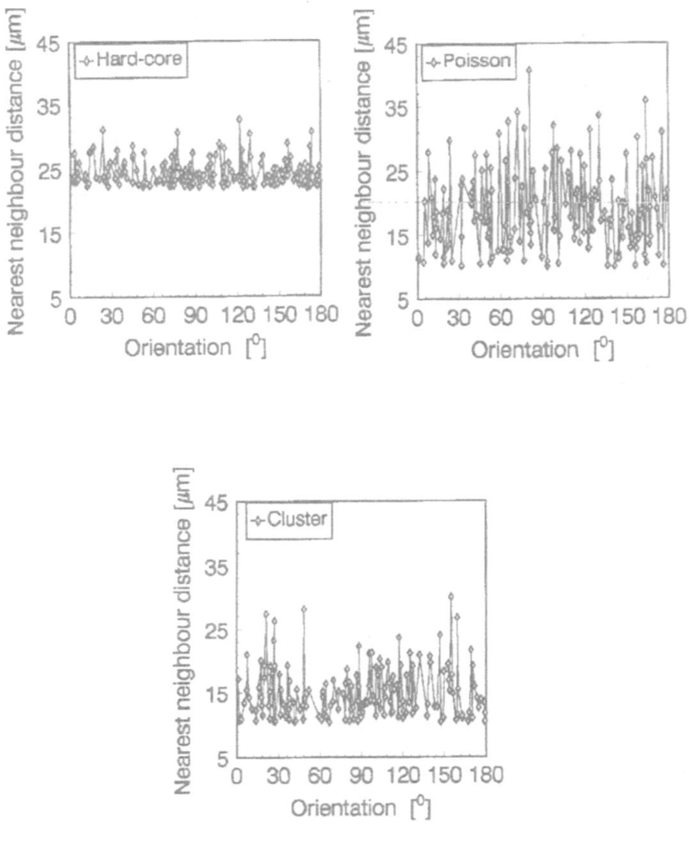


**Figure 10.** Variability of nearest neighbour distances vs orientation for different distributions.

Figure 11 illustrates transformation of the point sets onto the sets of contiguous polygons. The global correlation between stress field and topology expressed by the topological entropy and the coefficient of variation is shown in Figure 12. The coefficient of variation is defined as the ratio of standard deviation to mean value of maximal radial stresses. This figure illustrates the results of different simulations where several realizations of three models have been performed according to procedures described previously. Realizations shown in Figure 12 are marked by small triangles. Variability of the stress field increases with increasing topological disorder in agreement with results shown in Figure 8.

**Figure 11.** Tessellations for three distributions.

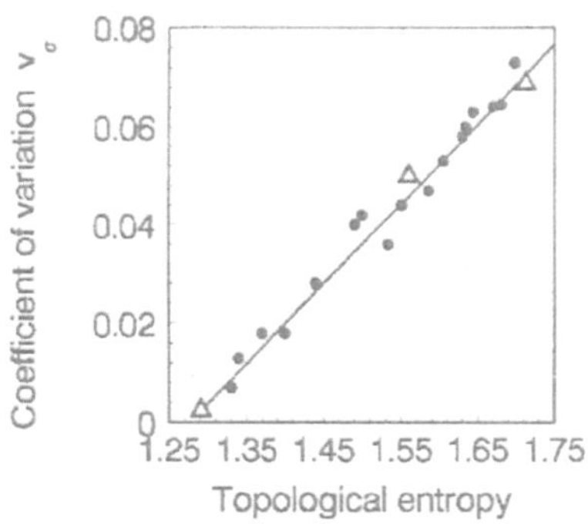


**Figure 12.** Correlation between topological entropy and maximal radial stresses.

The correlation between stresses and topology rely upon the coefficient of variation that is supposed to represent detailed variability of the stress field. This statement is, however, overestimated in the sense that the coefficient of variation includes only first two moments of the stress distribution whereas a more precise description of the stress distribution would necessitates an involvement of higher-order moments of the distribution or their combination. In any case, the global stress estimator should always have a clear physical interpretation which could be difficult to achieve with higher moments of the stress distribution, except perhaps first four statistical moments.

**Example 5.** Clustered distribution of inclusions has to be paid a special attention. Point pattern clustering is of importance in assessment of homogeneity of inclusions dispersion. Cluster analysis comprises the identification of groups of observations that are cohesive and separated from other groups. Figure 13 shows how significantly the behaviour of *X(r)* for clustered pattern differs from other types of dispersions, Bochenek and Pyrz (2003a).

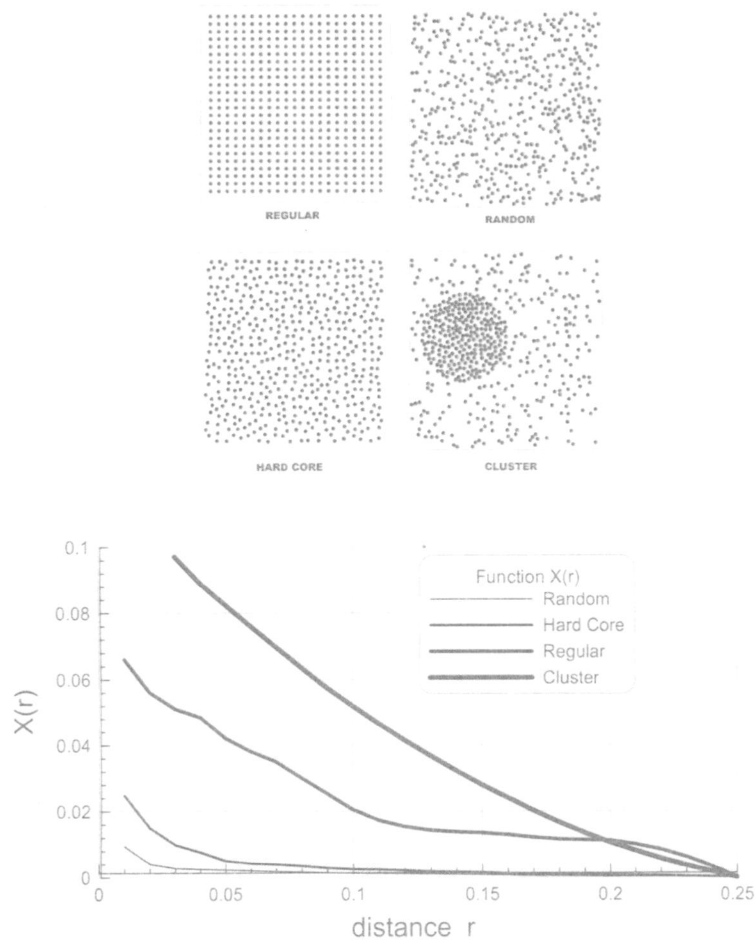


**Figure 13.** Integral correlation function for different dispersions.

The shape of function *X(r)* strongly depends on the number of points present in the pattern. Figures 14-17 illustrate this tendency. Increasing number of inclusions tends to lower the value of *X(r)* function except for clustered patterns where the tendency is opposite. Here the number of points indicated in Figure 17 is related to the number of points included in the cluster itself whereas a total number of points in the observation window is kept constant and equal to 500 i.e. 350, 200 and 50 points outside the cluster, respectively. That is to say more points in the cluster and less points outside it result in a larger value of the integral correlation function. However, if the number of points within the cluster is kept constant while changing the total number of points present in the pattern, the picture becomes similar as for other dispersions as illustrated in Figure 8. In this case the number of points in each cluster is equal to 250, whereas the number of points outside clusters are respectively 50, 250 and 500. The analysis indicates that identification of

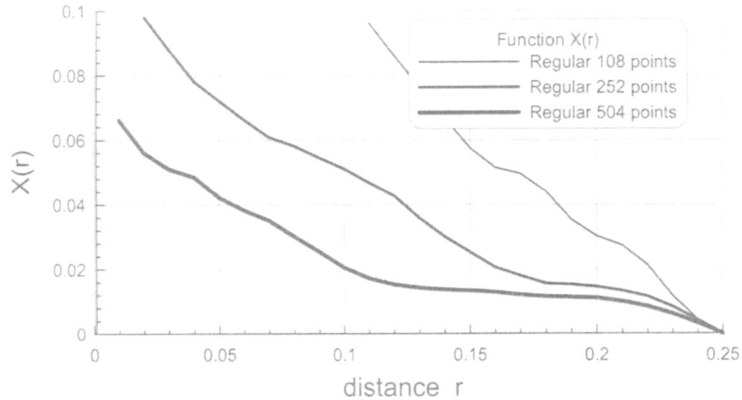


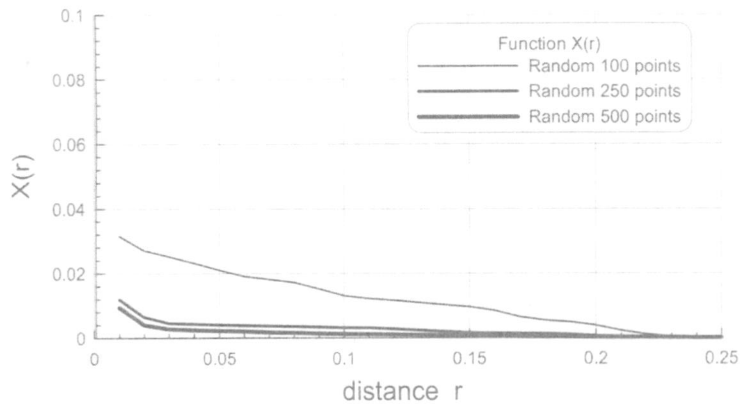


**Figure 15.** *X(r)* for random patterns.

clustered pattern and comparison with other types of dispersion must be carried out with caution as far as total number of points in the pattern and in the cluster is concerned. This is shown in Figure 19 where two cluster types have been generated with a total number of points equal 120. In cluster (A) there are 50 points within the cluster and 100 points within cluster (B). The increased density of cluster results in a larger value of the integral correlation function *X(r)*. The comparison with regular pattern having the same number of points is shown in this figure as well.

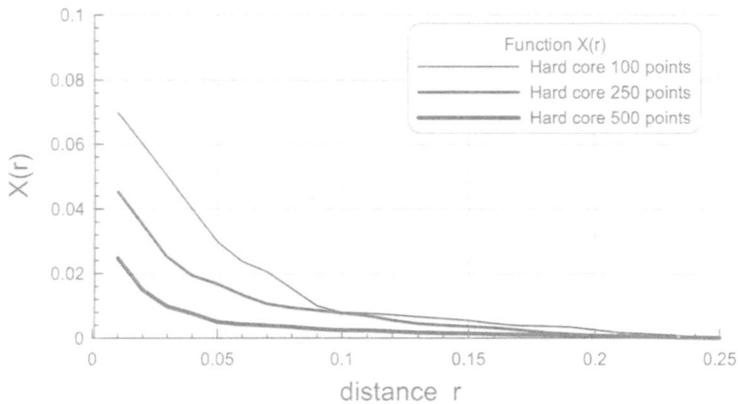


**Figure 16.** *X(r)* for hard-core patterns.

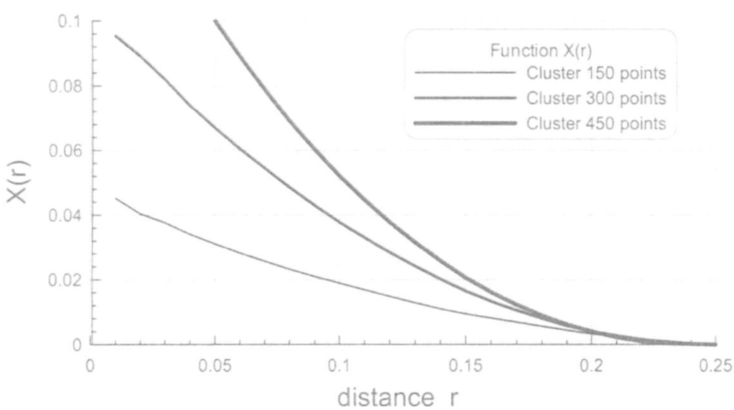


**Figure 17.** *X(r)* for clustered patterns.

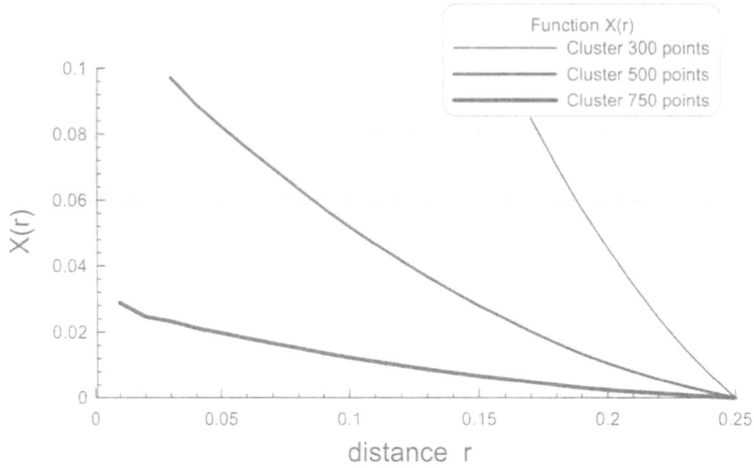


**Figure 18.** *X(r)* function for clustered patterns with the same number of points within the cluster.

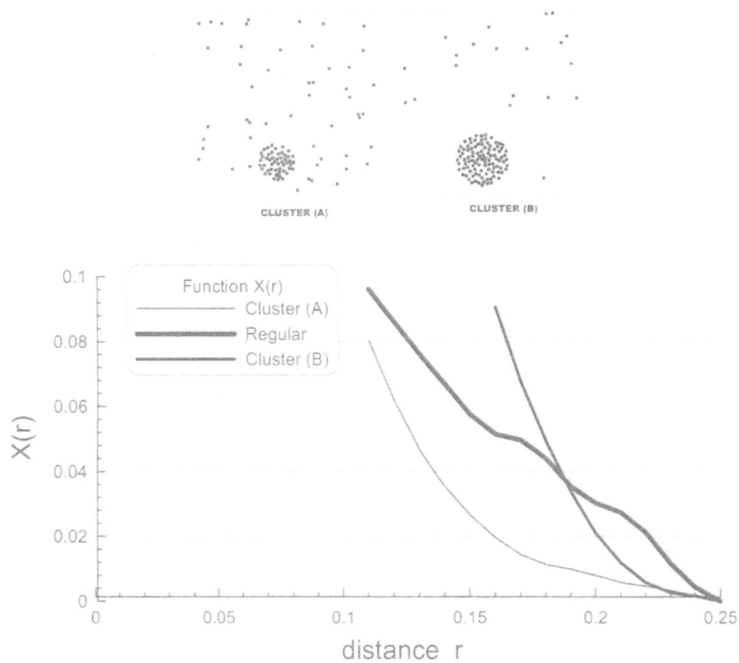


**Figure 19.** Clustered patterns and corresponding *X(r)* functions.

The analysis can be extended to the case of three-dimensional distribution of inclusions. Ten thousand points have been located in a unit cube according to the hard-core model, Figure 20. Then an inner cube 0.5x0.5x0.5 has been selected to calculate the integral correlation function thus avoiding problems with a necessity to correct for a boundary constraint.The same tendency prevails also for the 3D case as can be seen in Figure 21.The ability of *X(r)* function to distinguish between different spacial patterns is preserved.

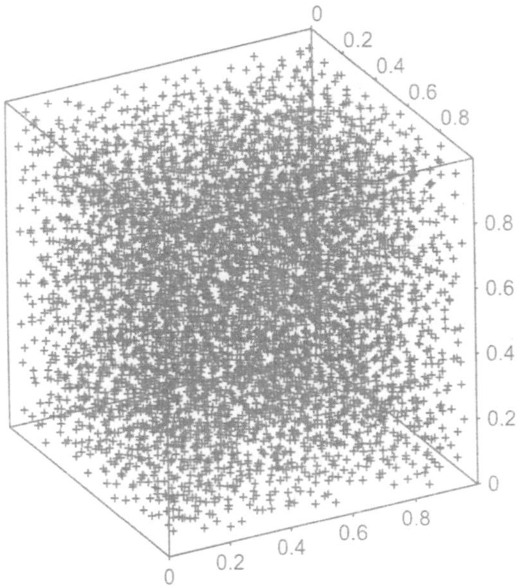


**Figure 20.** 3D hard-core dispersion of points.

Taking into account the above properties, the *X(r)* function has been chosen as the basis for description of clustered distributions. It is proposed then to introduce a new parameter defined as

$$X_C = \frac{1}{r_{\max}} \int_{r_{\min}}^{r_{\max}} X(r)\, dr \tag{8}$$

The larger value of the parameter the larger density of inclusions, and the more distinct is cluster. For random (Poisson) distribution the value of $X_C$ tends to 1.

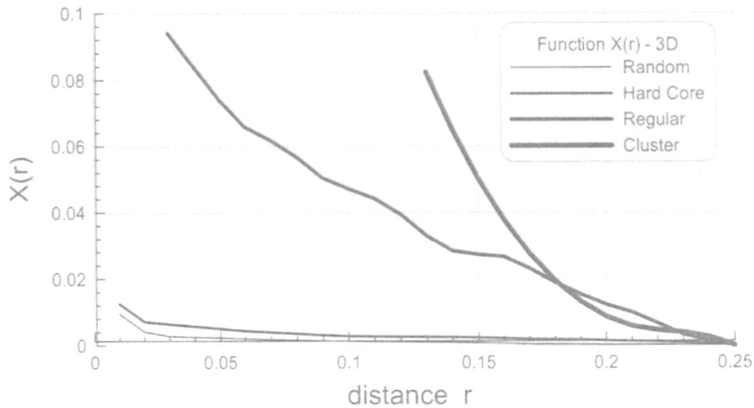


**Figure 21.** *X(r)* for different 3D dispersions.

The tessellation of the measure area is performed and the distribution of polygons associated with all points is found. It is observed that for points which form clusters areas of polygons are significantly smaller than for points situated outside clusters. Hence it is possible to define parameters that characterize density of points in specified region, see Figure 22, as the number of points divided by the total area of their polygons.

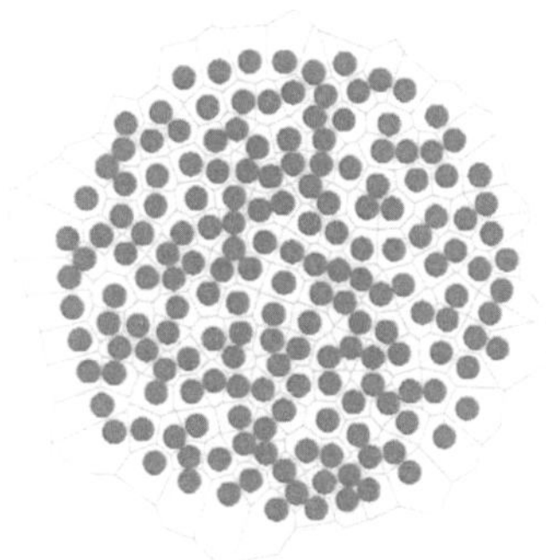

**Figure 22.** Cluster of points with corresponding tessellation.

Furthermore, the relative density can be introduced as the ratio of density calculated for a specified region to the density of points in the whole measure area. In what follows two relative density parameters are proposed, namely relative density within cluster

$$C_{IN} = \frac{N_{IN}}{A_{IN}} \frac{A_{ALL}}{N_{ALL}} \tag{9}$$

and relative density outside cluster

$$C_{OUT} = \frac{N_{OUT}}{A_{OUT}} \frac{A_{ALL}}{N_{ALL}} \tag{10}$$

where $A_{( )}$ and $N_{( )}$ correspond to respective areas and number of points. Based on the above quantities their ratio can be introduced as the third parameter

$$D^{IN}_{OUT} = \frac{C_{IN}}{C_{OUT}} \tag{11}$$

For the Poisson random distribution of pobints values of all parameters are close to 1. The more distinct cluster becomes the more significantly value of $D^{IN}_{OUT}$ retires from unity.

The values of the parameters are now compared for exemplary distributions of 600 points. First is the Poisson random distribution and the three others are clustered ones with one, two and three clusters respectively. The patterns are presented in Figure 23 and the values of parameters are given in the Table 1.

**Table 1**

| $N_{IN}$ | $N_{OUT}$ | $X_C$ | $C_{IN}$ | $C_{OUT}$ | $D^{IN}_{OUT}$ |
|---|---|---|---|---|---|
| 600 | | 0.917 | 1 | 1 | 1 |
| 160 | 440 | 15.25 | 2.05 | 0.843 | 2.43 |
| 320 | 280 | 21.94 | 1.98 | 0.639 | 3.10 |

**Table 2**

| $N_{IN}$ | $N_{OUT}$ | $X_C$ | $C_{IN}$ | $C_{OUT}$ | $D^{IN}_{OUT}$ |
|---|---|---|---|---|---|
| 450 | 250 | 26.19 | 1.74 | 0.565 | 3.09 |
| 2×225 | 250 | 21.41 | 1.72 | 0.569 | 3.03 |
| 3×150 | 250 | 14.96 | 1.72 | 0.570 | 3.02 |
| 4×112 | 252 | 13.35 | 1.76 | 0.565 | 3.11 |

What can be seen from the Table 1 is that the relative densities of points within clusters are close to each other whereas the ones calculated for points outside clusters vary. When the density $C_{OUT}$ gets smaller the clusters become more distinct what can be observed in the figures.

The problem of the influence of number of clusters on values of relative density parameters has to be investigated as well. For the fixed number of points within clusters and approximate values of relative density parameters the distributions with one, two and three clusters are analysed. Although the density parameters remain almost constant the „intensity of clustering" changes. It is expected that concentration of points is larger if they form smaller number of clusters. That effect can be recognized by calculating values of the $X_C$ parameter. It is shown in the Table 2 that the value of $X_C$ changes with the number of clusters. That indicates that number of clusters has to be taken into account when describing such patterns.

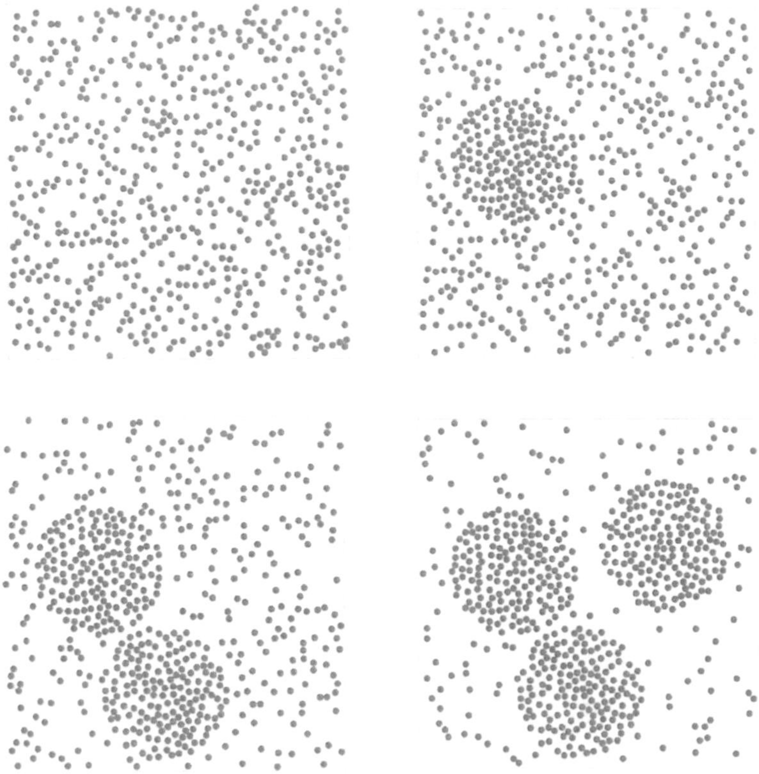

**Figure 23.** Point patterns for estimation of new parameters.

The distributions with one cluster have been analysed. Both relative density parameters and parameter $X_C$ are calculated. The obtained results are collected and presented as the graph that shows dependence of $D_{OUT}^{IN}$ with respect to $X_C$, Figure 24.

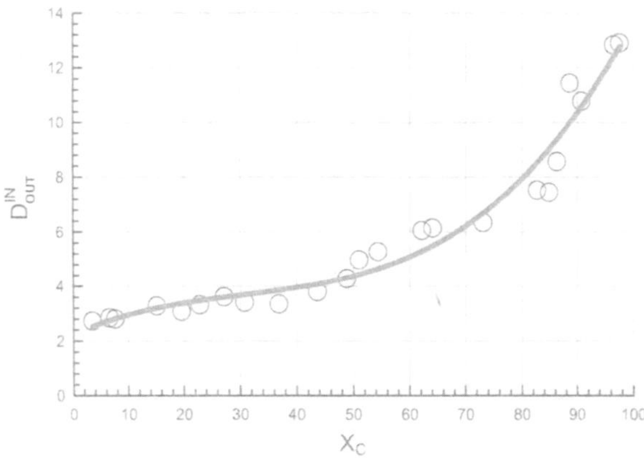


**Figure 24.** Correlation between cluster parameters.

There exists a clear correlation between cluster parameters and second-order correlation function *X(r)*. This correlation needs to be investigated further in order to disclose the relation between density and number of clusters in a pattern and the integral correlation function which will be used as a comparison function for reconstruction of microstructures.

# 3 Experimental Techniques

## 3.1 Confocal Raman Microspectroscopy

Micro-Raman imaging has emerged recently as a new technique for morphological characterization of microstructures. Although Raman spectroscopy has been known for more than seven decades its renaissance is due to latest developments in holographic filters and detection devices. Raman effect is based upon phenomenon of inelastic light scattering. In a basic Raman experiment, a sample is illuminated by a high energy monochromatic light source – typically from a laser. Some of the incident photons collide with molecules in the sample and are scattered with unchanged energy (frequency). When this happens, it is referred to as Rayleigh scattering and it occurs at the same energy (frequency) as the incident radiation. However, a second type of scattering can also occur. In Raman scattering, the molecule can either accept energy from the incident radiation being scattered – exiting the molecule into higher vibrational energy states; or give up energy to the incident photons – causing the molecule to return to its ground state. Thus, in the Raman experiment, the difference between the incident radiation and the Raman scattered radiation produces the vibrational spectrum of interest. The probing laser beam is focused to spot sizes of the order of 1 μm and by confocal point scanning one obtains Raman spectra at each scanned point. A schematic diagram of the Raman spectrometer layout is shown in Figure 25.

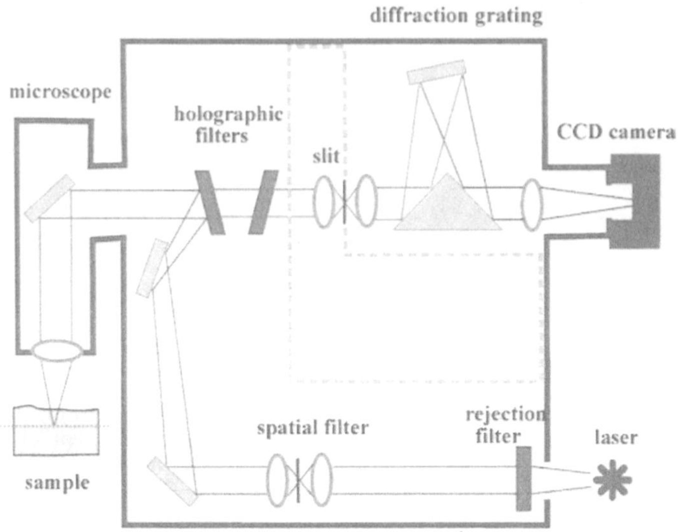


**Figure 25.** Schematic diagram of the Raman spectrometer.

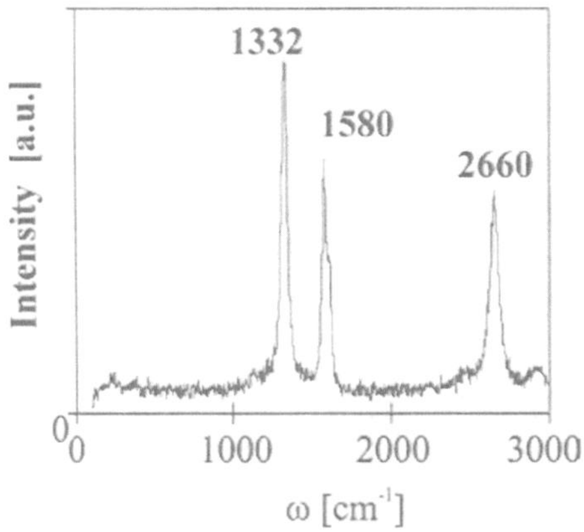


**Figure 26.** Raman spectrum of carbon fibre.

Raman spectroscopy detects changes in polarization, i.e. the "shape" of the electron distribution in the molecule as it vibrates. It provides key information on the structure of molecules. Molecular structure is studied by the position and intensity of features in the vibrational spectrum. Figure 26 illustrates typical Raman spectrum of carbon fibres recorded in air.

The unique feature of micro-Raman technique is its ability to measure in-situ stress (strain) transfer in composite materials. During deformation the energetic state of molecules is changed as compared to undeformed state. Scattering of light can be related to perturbations of the molecular vibrational state with strain. Measurement of strain in crystalline materials is based on the identification of a shift in the scattering frequency due to changes in the atomic spacing with strain. This provides a detailed insight into the relation between loading, material configuration and local strain state. Raman band shift of the carbon fiber loaded axially is shown in Figure 27 whereas a plot of residual strains in fragmented fiber is illustrated in Figure 28. The number of zero crossing corresponds to the number of fiber fragments.

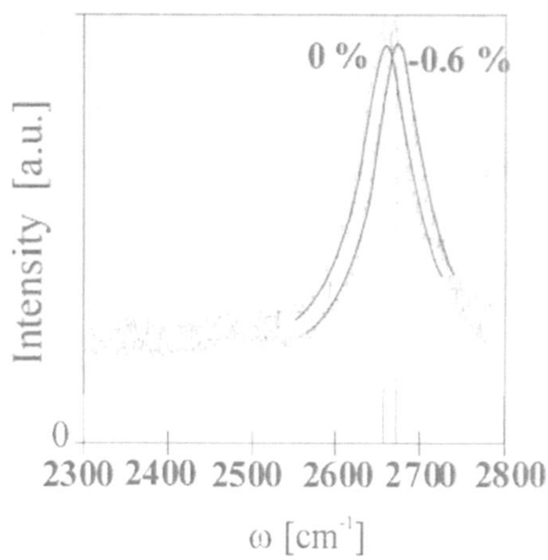


**Figure 27.** Shift of 2660 $cm^{-1}$ band with an applied compressive strain.

The mechanical properties of a composite depend critically on the load transfer efficiency at the interface between fibre and matrix. In composites with thermosetting polymers, such as carbon/epoxy, the interfacial properties can be enhanced by chemical treatment of the fibre either to enhance the reactivity of the fibre or to enhance the wetting of the fibre surface using sizings. For carbon fibre reinforced composites with non-reactive matrices, such as semicrystalline thermoplastic polymers, the adhesion is primarily a result of mechanical interaction. Composites based on such semicrystalline matrices are becoming increasingly important as a structural engineering materials because of the improved properties these polymers offer. For semicrystalline polymers like polypropylene, heterogeneous nucleation on the fibre surface may affect residual stresses. If the nucleation density is sufficiently high the spherulites impinge and can grow only in one direction, normal to the surface of the fibre, developing a transcrystalline interphase. The transcrystalline interphase is often large in diameter compared to the fibre and for a practical composite having a large volume fraction of fibres the bulk matrix may predominantly have a transcrystalline morphology. Due to the nucleation on the fibre surface and the anisotropy of this

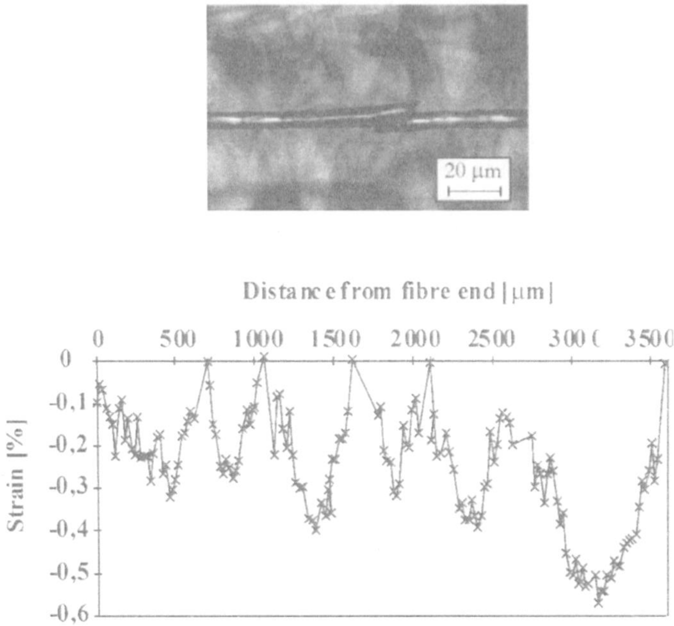


**Figure 28.** Strain profile along fragmented fiber.

phase, the transcrystalline interphase may be expected to affect the properties of the composite by influencing thermal residual stresses and the mechanical properties of the surrounding matrix. Thus the fundamental tasks in the prediction of thermal stresses are to determine parameters describing crystallization kinetics, thermorheological behaviour of the matrix and properties of the interphase being formed between the fibre and the solidifying matrix, see Nielsen and Pyrz (1997, 1999) and Nielsen, Batchelder, Pyrz (2001).

Model composites with a transcrystalline interlayer can be manufactured by embedding single carbon fibres (unsized PAN based high modulus) between isotactic-polypropylene films. The assembly is heated to melt (205 °C) for 5 min under a light pressure to exclude air and subsequently cooled at 30 °C/min to a crystallisation temperature of 130 °C and held for 30 min. After this isothermal crystallisation period the sample is finally quenched to room temperature. This process results in a well defined transcrystalline interlayer, Figure 29A. The sample is carefully polished to obtain three perpendicular free surfaces on which Raman spectra of the six different polarisations could be acquired as shown in Figure 29B.

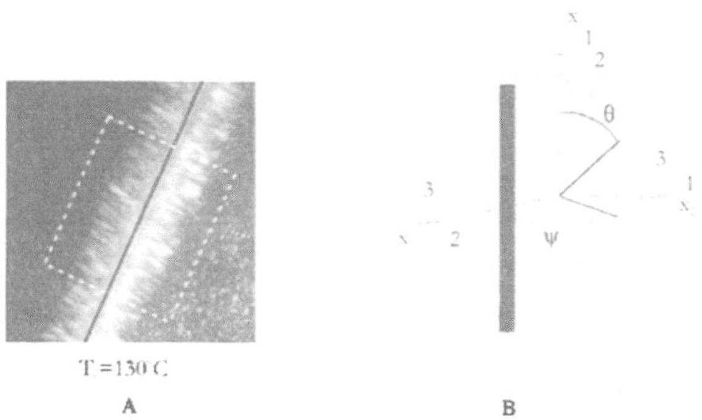


**Figure 29.** Morphology of isothermally crystallised samples (A); Three orthogonal planes used to measure crystalline orientation (B).

The crystalline polymer is considered as a homogeneous assembly of partially oriented crystalline chain segments. The orientation of chain segments is described in terms of the Euler angles $\theta$ and $\psi$ shown in Figure 29B. Associated with each chain segment is the second-rank polarisability (or Raman) tensor $p_{ij}$ which determines the intensity of Raman scattering as

$$I_{ij} = I_0 \int_0^{2\pi} \int_0^{\pi} (e_i e_j p_{ij})^2 N(\theta,\psi) \sin\theta \, d\theta \, d\psi \tag{12}$$

where $I_o$ is a constant depending on instrument, $N(\theta,\psi)$ is the orientation distribution function and $e_i$ and $e_j$ are unit normals.

The chain orientation distribution function can be expressed in terms of spherical harmonics

$$N(\theta,\psi,\phi) = \sum_{l=0,2,4} \sum_{m=-l} \sum_{n=-l} W^{lmn} Z^{lmn} (\cos\theta) e^{-im\psi} e^{-in\phi} \tag{13}$$

where $Z^{lmn}$ are generalized Legendre functions and $W^{lmn}$ are the expansion coefficients. Crystals of isotactic polypropylene have monoclinic symmetry. Their elastic properties can be theoretically determined and thermal expansion for the monoclinic unit cell has been measured by Isasi et al. (1997) using X-ray diffraction experiments. Experimental results show that characteristic Raman band 808 $cm^{-1}$ is assigned to the carbon-carbon backbone stretching of helical chains in the crystal unit cell and is sensitive to the crystal c-axis orientation. Thus the principal direction of the Raman tensor is aligned with the crystal c-axis and is rotation symmetric with respect to this axis i.e. the orientation distribution function (11) does not depend on the angle $\phi$. It follows from the orthogonality of the Legendre polynomials that the expansion coefficients $W^{lmn}$ are the moments of the orientation distribution. In order to determine the anisotropy of the fourth-rank elasticity tensor it is necessary to determine up to the fourth-order moments ($l=4$). Inserting equation (13) into equation (12) one obtains six equations from which six moments of the orientation distribution can be calculated knowing scattering intensity on three orthogonal planes. Figure 30 shows crystals orientation distribution function in the transcrystalline layer as determined with the help of Raman microspectroscopy. It can be seen that there are preferred orientation directions of the c-axis which results in a significant anisotropy of the transcrystalline layer. Knowing the orientation distribution of crystals in the transcrystalline layer it is possible to calculate elastic stiffness of the interphase.

**Figure 30.** Orientation distribution of crystal c-axis.

Spatial orientation averaging can be performed using following formula

$$<C_{ijkl}>=\int_0^{2\pi}\int_0^{2\pi}\int_0^{\pi} N(\theta,\psi,\phi)\, a_{ip}\, a_{jq}\, a_{kr}\, a_{ls}\, C_{pqrs} \sin\theta\, d\theta\, d\psi\, d\phi$$

where $a$ elements represent unit normals and $N$ is the orientation distribution function of Euler angles for a crystal having stiffness $C_{pqrs}$. A single crystal of polypropylene shown in Figure 31

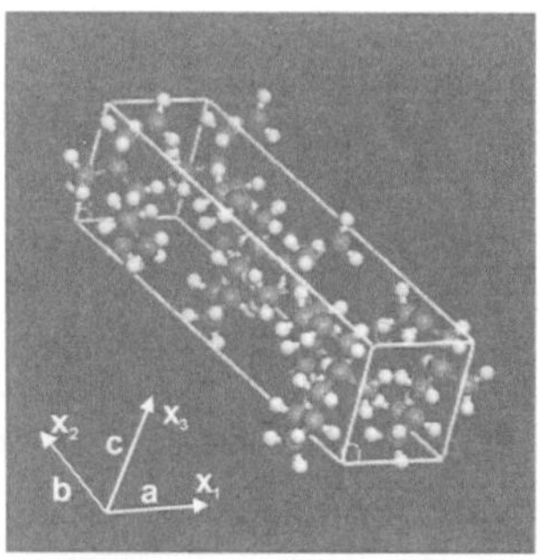


**Figure 31.** Single crystal of polypropylene.

has α-monoclinic symmetry and a full population of stiffness elements can be calculated using molecular dynamics simulations. The principal stiffness components along crystal axes read 12, 8 and 42 GPa respectively. Figure 32 shows elastic moduli of the transcrystalline interphase. It is clearly seen that the interphase is anisotropic and non-homogeneous and this can significantly influence the overall stiffness of the composite. Polarization Raman spectroscopy can be also used to measure orientation distribution of molecular chains in an amorphous polymer which is directly related to measurement of the strain profile in the matrix material. In amorphous materials, strain alters the molecular orientation distribution of molecules which affects the interaction with light by changing the state of polarization.

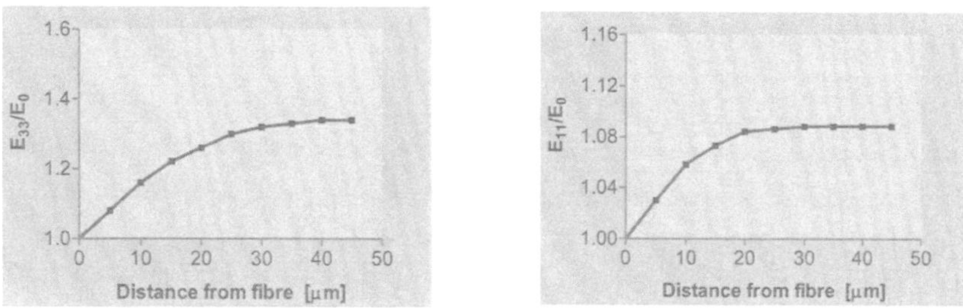


**Figure 32.** Elastic moduli of the interphase.

Figure 33A shows the thermal strain distribution around the embedded fibre, see Nielsen and Pyrz (2002). The strain level indicates a zone of shear yielding for a shear yield limit of 36 MPa. The length of the yield zone is approximately 40-50 μm, see Figure 33B, which is in quantitative agreement with fibre strain measurements.

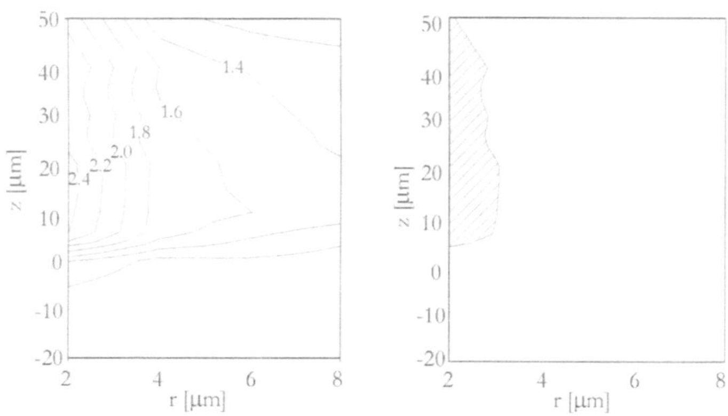


**A** **B**

**Figure 33.** Isoplot of matrix shear strain (A); indication of the zone of plastic yielding (B).

The Raman experiment is also available for another mode of operation allowing for the Raman imaging. A schematic diagram for the imaging of sample is shown in Figure 34.

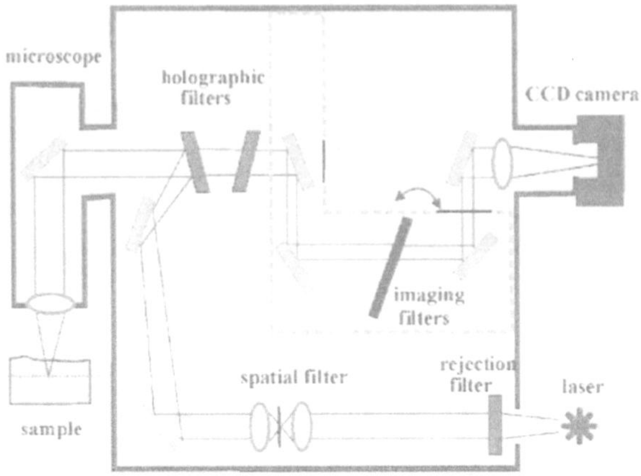


**Figure 34.** Schematic diagram of the Raman spectrometer for imaging mode.

In the imaging mode the Raman light is diverted through imaging filters, set at a wavelength which corresponds to a band characteristic to the material component to be imaged. In this mode the CCD camera is used as a two-dimensional detector with *x* and *y* coordinates representing the special dimensions of the sample under the microscope.

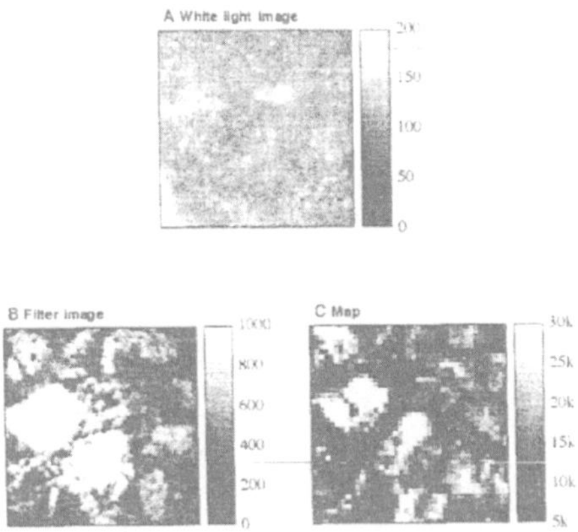


**Figure 35.** White light image of the surface (A); Raman image of the band 1335 cm$^{-1}$ (B); point by point Raman mapping (C).

Figure 35 illustrates the result of the Raman imaging mode for a diamond film place on the silicin substrate. The film thickness is 225 μm and the measurements are taken confocally at the bottom of the film. Adjusting imaging filters to the band of diamond, see Figure 36, one obtains the Raman image of diamond phase distribution in the film which is impossible to detect with white light microscopy as shown in Figure 35A. In this mode of operation the laser spot is enlarged and may cover required area. The image in Figure 35 has dimensions 40x40 μm. A more detailed Raman imaging can be achieved by making point-by-point mapping with very small laser spot size, see Figure 35C.

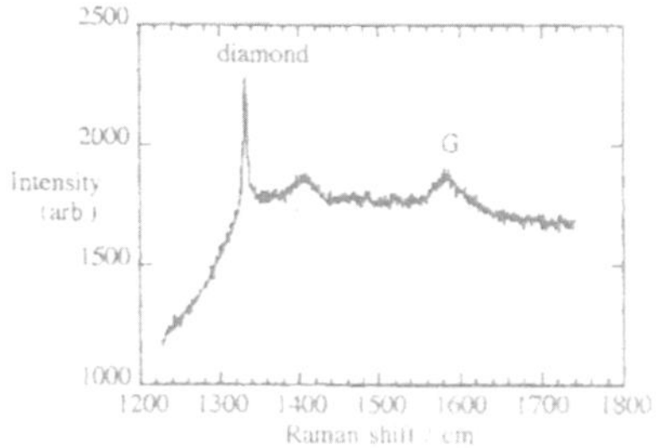


**Figure 36.** Raman spectrum of diamond film.

### 3.2 X-Ray Microtomography

Development of X-ray microscopy and computer tomography provides a new tool for the assessment of materials morphology. X-ray microscopy is a relatively new technique that has not been applied to any significant extent in materials science, see Pyrz (1999, 2001). Most X-ray microscopy development has so far been made using large synchrotron sources. This has limited X-ray microscopy to a research tool available only at the major synchrotron facilities. The use of X-ray tubes with a very small focus and an energy in the order of 20-100 keV together with very sensitive recording devices enable the design of a bench-top X-ray microscope with a spatial resolution less than 8 μm, Sasov and Van Dyck (1998). A common problem in morphological analysis of non-homogeneous materials is that three-dimensional information of microstructure is required, but its images are two-dimensional. Monitoring materials′ microstructure using X-ray micro-tomography allows us to begin to bridge this gap since a three-dimensional image of the specimen can be reconstructed from non-destructive, serial sections and can be processed to show and measure three-dimensional features.

In X-ray microtomography, the object is rotated so as to obtain radiographic projections from different viewing angles. The projections are the measured values of the overall attenuation that X-rays undergo when they travel through the object, Figure 37. During radiography the rotation of the object is carefully controlled with a minimal rotation step of 0.45°. The object can be moved up and down depending on the axial position of the volume element to be irradiated. An

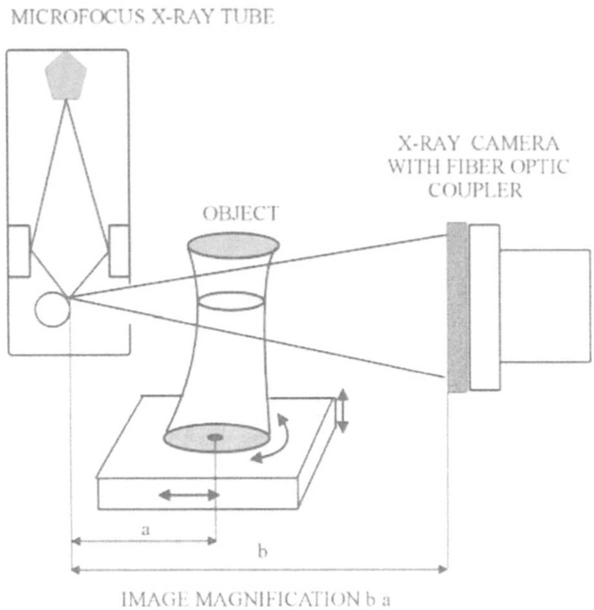


**Figure 37.** Schematic diagram of X-ray scanner.

image magnification is achieved by moving the object horizontally between the source-detector set. The attenuation of each ray in a beam result from interactions between the radiation used for imaging and the substance of which the object is composed. X-rays passing through material are absorbed according to a linear attenuation coefficient that has some spatial variation depending on the local composition and the density. These differences in the linear attenuation coefficient provide the contrast necessary to form an image. In an idealized case of parallel X-ray illumination a profile of the attenuation through the sample is obtained, Figure 38. Practically, the attenuation measurement is averaged over a finite-sized volume element in the sample since the detector has finite spatial resolution to discriminate between closely spaced ray paths. By rotating the sample in discrete angular increments through $180^{o}$, sufficient data are obtained from the projection images to reconstruct slices of the three-dimensional object. A reconstruction procedure is based on the backprojection principle. The attenuation recorded in each projection is due to the structure of the object along the individual lines. It is not possible to know from one projection where along the line the attenuation occurs. However, it is possible to distribute the measured attenuation evenly along the line. If it is done along projections from several views, the super-position of the attenuation values should correspond to the features present in the structure. Figure 39 illustrates this result for the slice through a cylindrical object. Data from several views overlap to delineate the cylinder's slice in the reconstruction image.

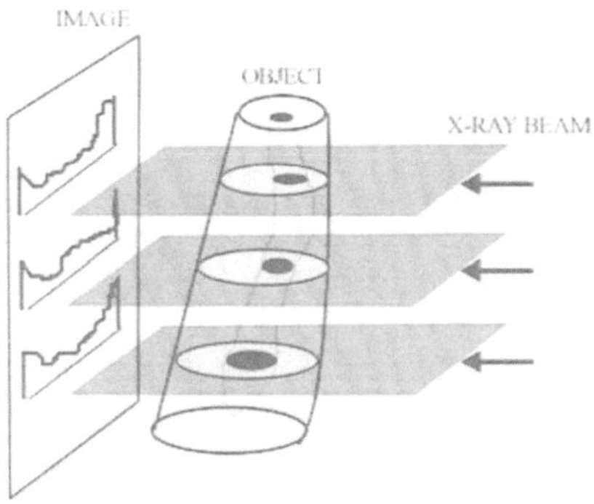


**Figure 38.** A set of projection lines through an object.

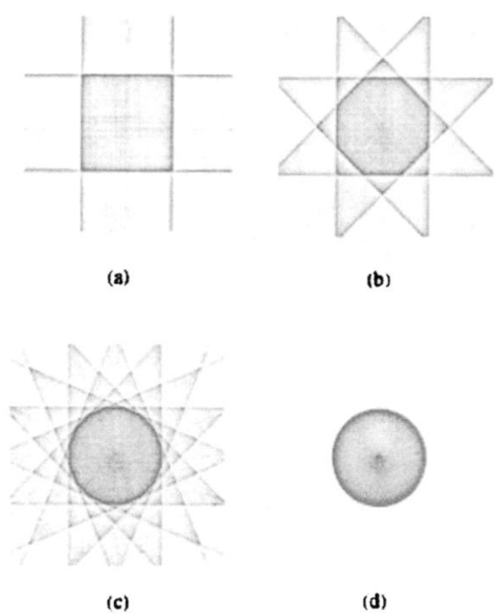


**Figure 39**. Backprojection procedure: 2 views (a); 4 views (b); 8 views (c); 200 views (d).

A problem of reconstruction of spherical particle sizes has been frequently treated in the literature since mid twenties, see Weibel (1980). The concept of size becomes critical if one type of shape not necessarily holds true for another shape of inclusions. In any case, the reconstruction technique may be used to overcome these difficulties. The radiographic view and consecutive reconstructed sections through the compound of glass particles are shown in Figure 40. The glass spheres were embedded in an epoxy matrix. The mean diameter of the spheres was 75 μm, and the distance between slices is equal to 25 μm. With an appropriate 3D algorithm which created spatial visualization from reconstructed microtomographic images it is possible to measure such microstructural features as distances, particles' volume and particle numbers per volume.

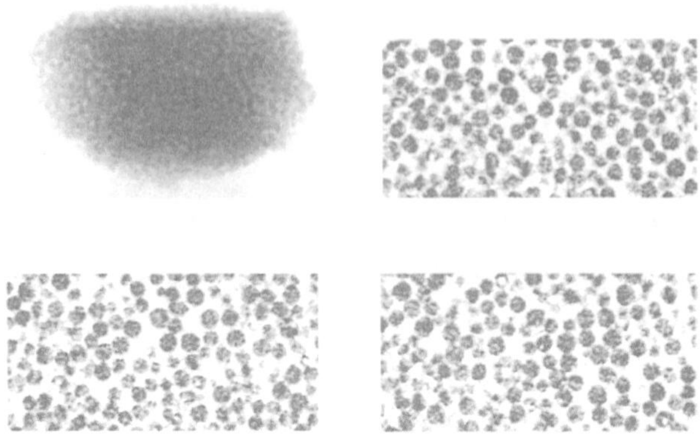

**Figure 40.** Glass spheres embedded in an epoxy matrix.

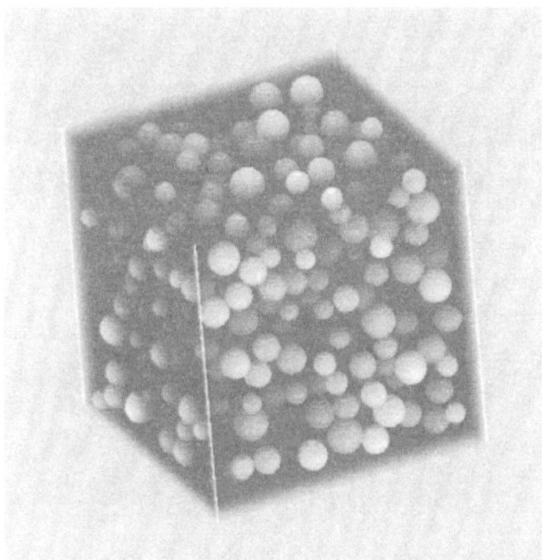

**Figure 41.** 3D configuration of spherical inclusions.

The radiographic image of unidirectional fibre glass epoxy composite together with two microtomographic slices reconstructed at 102 μm apart are shown in Figure 42. These non-destructive images allow determination of the centre position for all fibres in order to characterize a dispersion pattern in quantitative terms, simultaneously giving a visual impression about the homogeneity of fibres' dispersion. After having the microtomographic layers digitized the coordinates of fibre centres could be determined. Although the microtomographic slices apparently look the same, there exist subtle differences of fibres' relative position on consecutive slices. This effect is detailed in Figure 43, where marked fibres change position disclosing fibre waviness, a

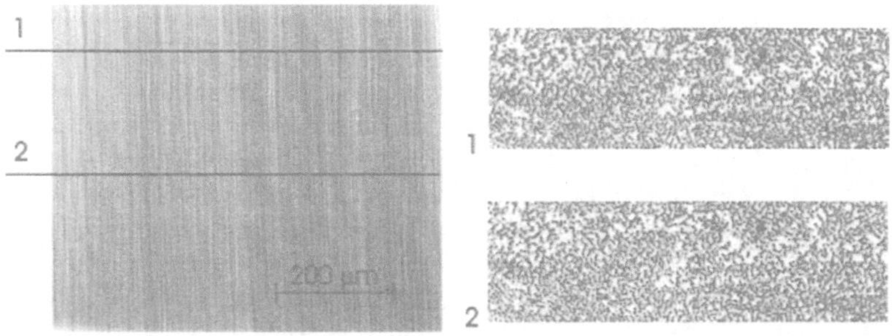


**Figure 42.** X-ray projection image of unidirectional composite and two reconstructed sections.

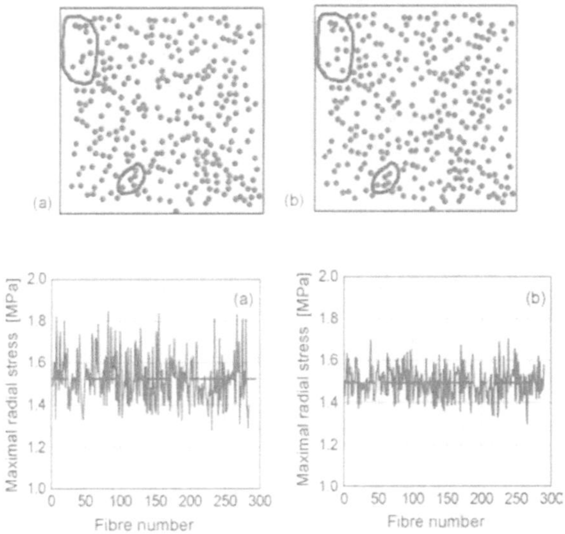


**Figure 43.** Selected areas from reconstructed cross sections and corresponding diagrams of maximal radial stress variability.

geometrical factor that controls compressive properties of the unidirectional composite. The same figure presents variability of maximal interfacial radial stresses for all fibres under shear loading. The transverse remote loading is equal to 1 MPa and the stresses have been calculated using the procedure outlined previously. It should be observed that all maximal interfacial radial stresses are magnified as compared with the remote stress. Secondly, the fluctuation of stresses and their mean values are smaller for fibres' pattern in section (b) than in section (a). This indicates that section (b) can sustain higher transverse shear loads before the interfacial radial stresses reach the failure threshold. The arrangement of fibres in unidirectional composites has a pronounced effect on transverse properties of these materials. The distribution of radial and circumferential stresses at the interface between the fibres and the matrix is influenced by the pattern of fibres' dispersion. This strong correlation has a significant impact on failure modelling where the local geometrical disorder cannot be neglected.

This technique allows also determine an orientation and length distribution of short fibres embedded in a polymeric matrix. The distribution of fibre diameters can easily be inferred from planar sections of the composite, where the intersections with fibres appear in a form of elliptical profiles. The minor axis of an ellipse represents the fibre diameter. A rather different situation arises when one wants to determine length distribution of short fibres. The simplest way to extract fibre lengths is to burn off the matrix material and then perform length measurements in an image analyzer. However, the pyrolysis technique by itself may cause fragmentation of fibres and its usefulness depends upon particular combination of constituent phases. Also measurements of fibre lengths made before processing are doubtful as some processing techniques, such as injection moulding, result in breakage of fibres. Thus a non-destructive and three-dimensional reconstruction of short fibres morphology in composite materials becomes a crucial issue. Figure 44 shows radiographic image of glass fibre polyester composite together with four reconstructed cross sections 4 μm apart and the reconstructed 3D image of short fibres. Importance of the fibre orientation distribution has been investigated by Schjødt-Thomsen and Pyrz, (2000, 2001).

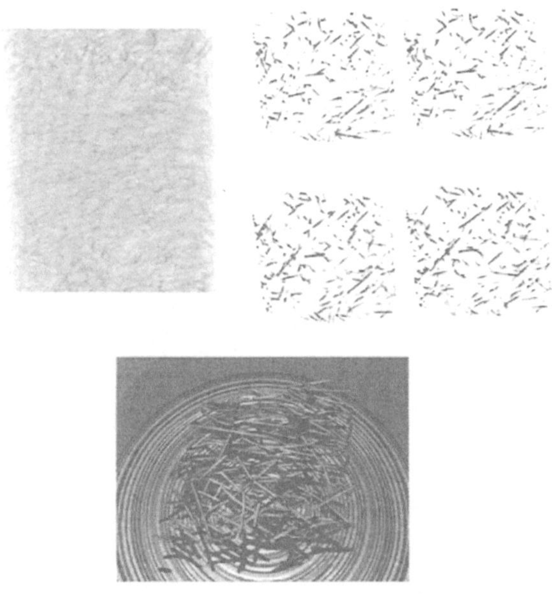

**Figure 44.** Three-dimensional reconstruction of short fibers.

# 4 Computer Assisted Morphological Characterization

Mesoscopic modelling of microstructures can be performed for different systems. We may draw a major difference between those systems for which an averaging is sufficient, and those for which one has to construct many realizations of the microstructure. Averaging is often done by an effective medium approach, and is suitable for predicting properties such as stiffness or conductivity. However, fracture or electrical breakdown will depend on specific details in the microstructure and usually averaging is not acceptable. In such cases, one must check the specific observable property for many realizations of the microstructure. Furthermore, only a tiny fraction of the original material specimen can actually be analysed and therefore, it is mandatory to reconstruct the dispersion pattern that would have similar geometrical characteristics as the dispersion of inclusions acquired from the limited experimental observations. From this limited view we have tried to understand how the properties of the material relate to microstructure. The approach must necessarily be statistical in particular with respect to the measurements of the microstructure. A virtual experiment that allows to reconstruct the microstructure and subsequently to find an ensemble average of a given property provides means for this statistical assessment. The principle of this experiment is that we want to produce a statistical sample of the microstructure, Bochenek and Pyrz (2003b).

The problem of reconstruction of dispersion pattern can be formulated as an optimization task. A sum of squared differences of nodal points of pair correlation function $g(r)$ for a reference and optimized pattern is then the objective function and coordinates of inclusions center points are treated as the design variables for that problem. The objective function is minimized subject two geometrical constraints. The first constraint limits a difference between values of topological entropies for the reference and reconstructed patterns to be less than 5%. The second constraint can be taken either as a pre-selected difference between a mean and standard deviation of distances of neighbouring inclusions or as a fulfillment of statistical T-tests and F-test for mean and standard deviation of distances, respectively. This allows to reach configurations, which are geometrically close to the reference dispersion. The reference image has been taken from X-ray microtomographic image of composite material, as illustrated in Figure 45.

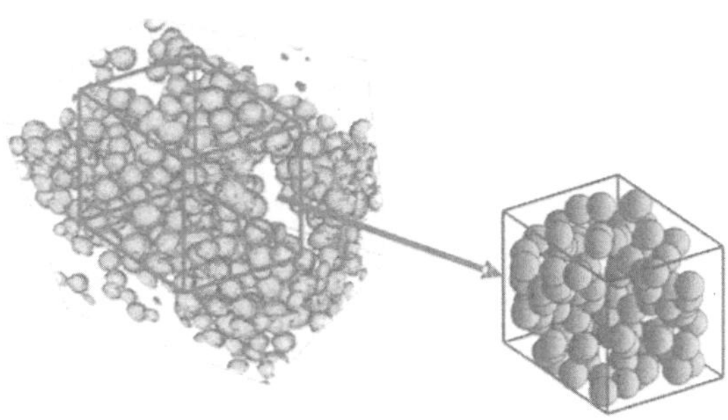

**Figure 45.** Reference image taken from X-ray image of a real material.

Figure 46 shows reference and reconstructed patterns for spherical inclusions having Gauss radii distribution. The reference pair correlation function used as the objective function and its counterpart for reconstructed distribution is shown in Figure 47 together with resulted integral correlation functions and a good agreement is observed. Values of topological entropy, mean neighbour distance and its standard deviation for the reference pattern differ from corresponding values of the reconstructed pattern by 3.6%, 2.6% and 0.6%, respectively.

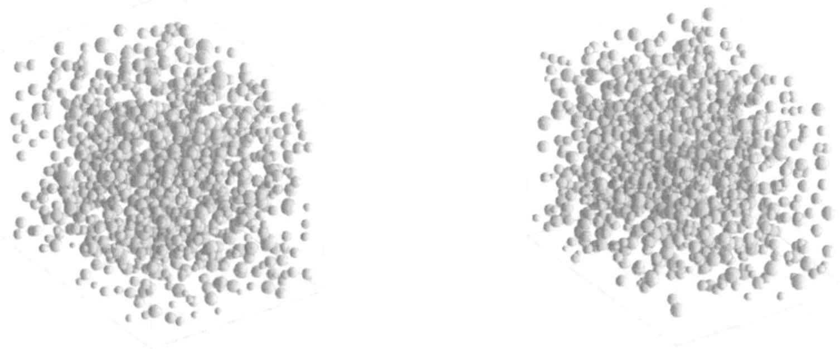

**Figure 46.** 3D dispersion with Gauss radii distribution – reference (left); reconstructed (right).

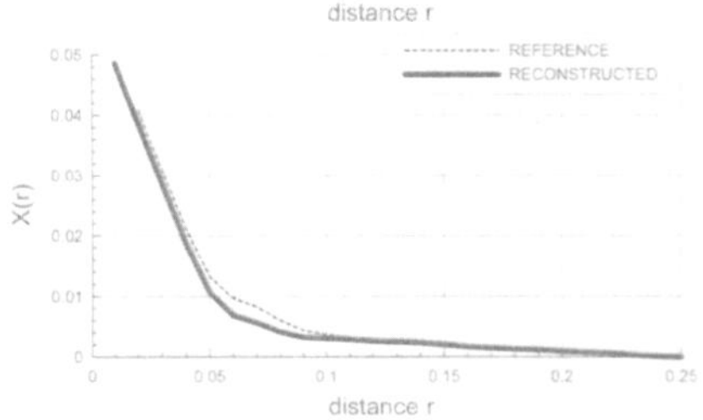


**Figure 47.** Correlation functions for special patterns with Gauss radii distribution.

A proximity of reconstructed pattern to its reference counterpart can be validated by comparing the density distribution function of distances to the near neighbours as seen in Figure 48. Since the stress interaction between neighbouring inclusions is strongly dependent on inter-inclusion distances it may be expected that the reconstructed pattern will generate a stress field, which is representative for material microstructure.

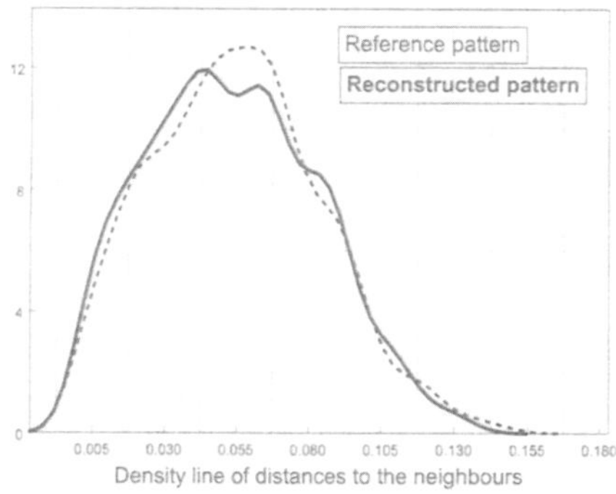


**Figure 48.** Density distances to the near neighbours for patterns with Gauss radii distribution.

Similar analysis has been performed for the representative volume element containing 512 spherical inclusions with lognormal radii distribution. The results are presented in Figures 49, 50 and 51. This example serves again to illustrate that the procedure for reconstruction of spatial dispersions of spherical inclusions with an arbitrary distribution of sphere radii is capable of generating structures that match reference configurations with respect to selected criteria.

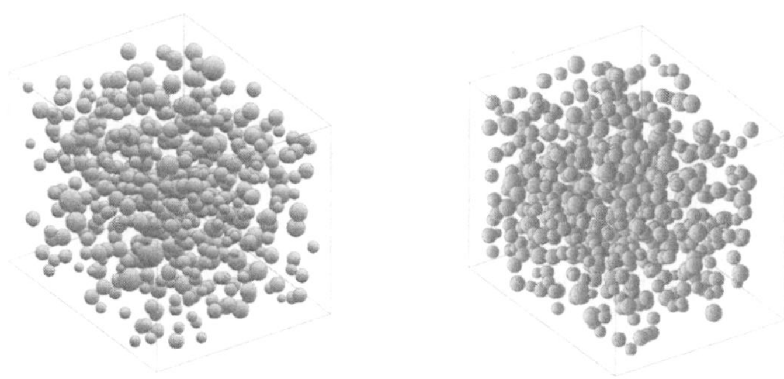

**Figure 49.** 3D dispersion with lognormal radii distribution-reference (left), reconstructed (right).

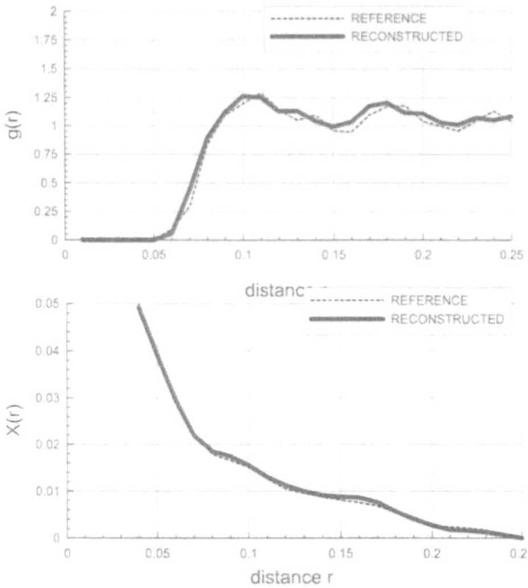


**Figure 50.** Correlation functions for spatial patterns with lognormal radii distribution.

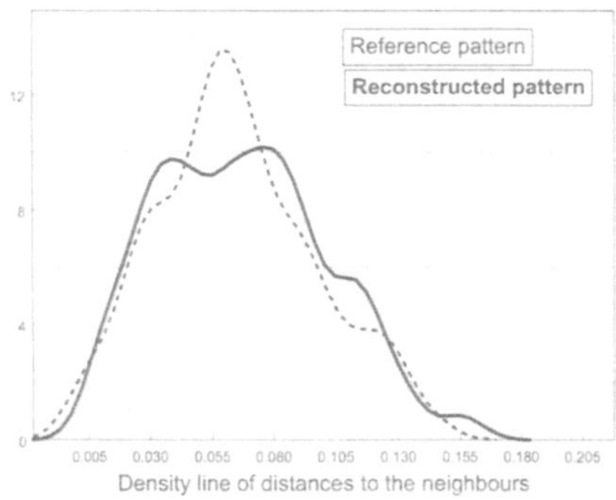


**Figure 51.** Density distances to the near neighbours for patterns with lognormal radii distribution.

The procedure for reconstruction of spatial dispersions is slightly different as compared with similar procedure for planar dispersions. For illustrative purposes, the integral correlation function *X(r)* is selected as an objective function for the optimization procedure. The objective function is minimized subject to two constraints. First constraint is imposed on values of topological entropy *S*, which represents first order correlation of the pattern. A physical interaction between embedded inclusions is controlled by a second constraint imposed on stress interaction parameter *C*, which represents maximal radial stresses at inclusions interfaces (see Pyrz and Bochenek, 1998). The reference image has been taken from X-ray microtomographic image of unidirectional glass-epoxy composite material, as illustrated in Figure 52. After several image transformations, the centre points of fibres are recovered, which constitutes reference dispersion of points.

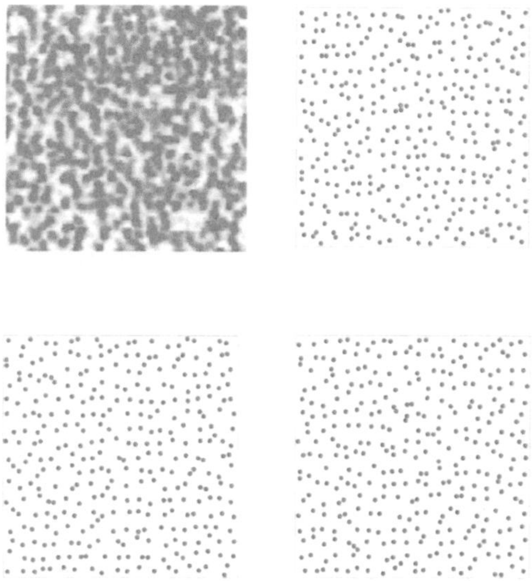

**Figure 52.** Original reference image and three reconstructed patterns (a, b, c).

As can be seen from Figure 53, the optimization procedure leaves integral correlation functions of the reference and reconstructed pattern very close to each other. The constraints for reference pattern were found to be $S$=1.37 and $C$=13.6. The same parameters found for reconstructed patterns a, b and c are respectively 1.31, 1.38 and 1.32 for the topological entropy $S$ and 12.2, 13.0 and 12.4 for the stress interaction parameter $C$. Lower values of $C$ for reconstructed patterns indicate that we may expect slightly higher maximal radial interfacial stresses present in the reconstructed dispersion as compared with reference dispersion. This conclusion is supported in Figure 54, which shows longer tail values of the distribution functions for reconstructed patterns than for the reference pattern.

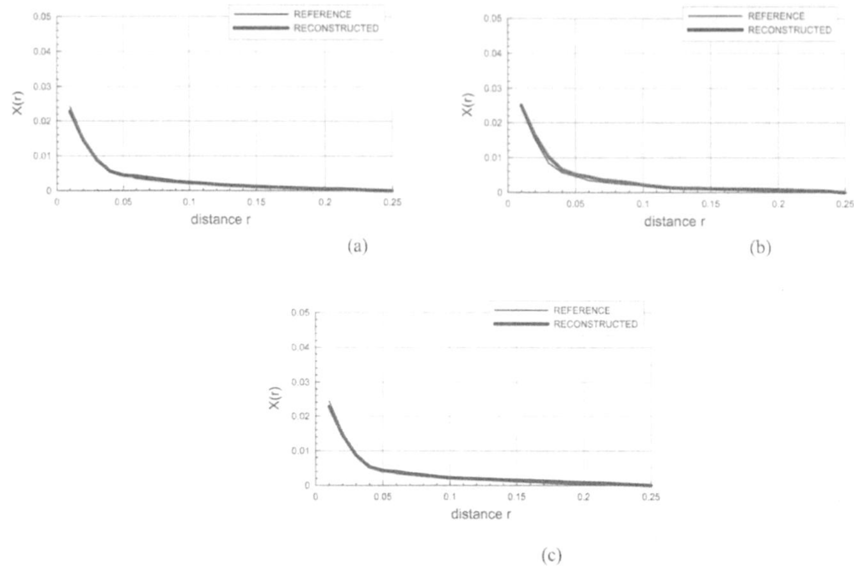


**Figure 53.** Comparison of $X(r)$ functions for reference and reconstructed patterns.

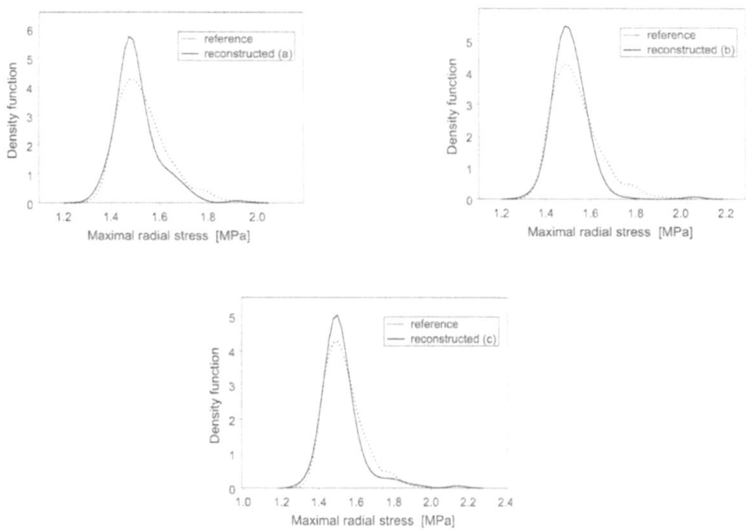


**Figure 54.** Stress density distribution for reconstructed patterns.

If generated patterns are subject to the only constraint of non-overlapping, see Figure 55, the corresponding density functions for the maximal radial stresses at fibre interfaces are shown in Figure 56 for three simulated patterns. There is a significant difference between stress density functions for this case and corresponding functions for reconstructed patterns. Tails of simulated dispersions are wider as compared with the reference pattern. It means that there are more fibres in the simulated dispersions than in the reference pattern, which are subject to interfacial stresses having values from the tails of the density distribution.

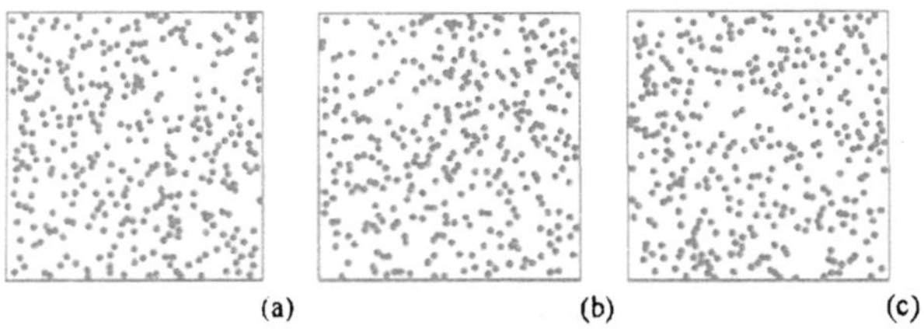


**Figure 55.** Three simulated hard-core patterns.

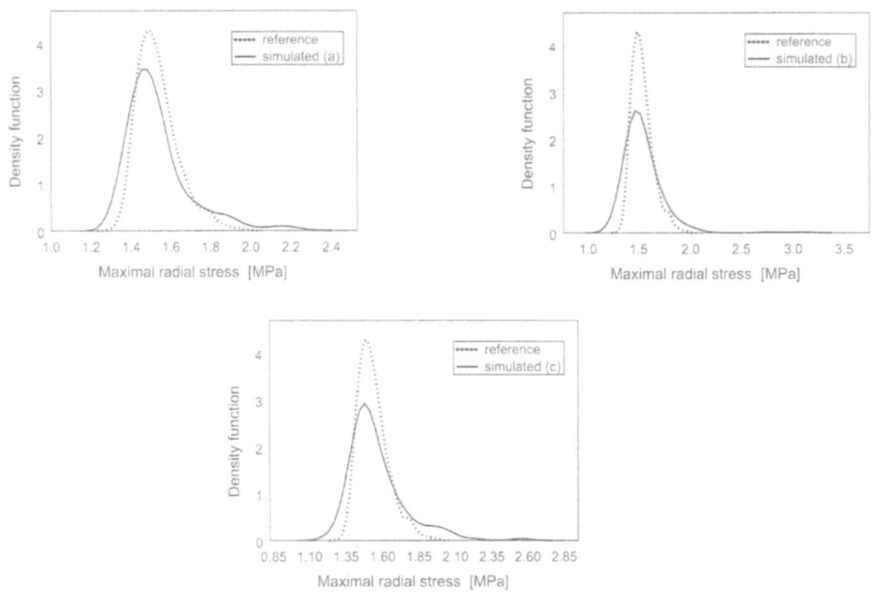


**Figure 56.** Stress density distribution for simulated patterns.

Further discrepancies between reconstructed and simulated dispersions can be observed on the diagrams representing nearest neighbour distance distribution, Figure 57. While reconstructed dispersions provide reasonable distribution of nearest neighbour distances, the simulated patterns yield far too short nearest neighbour distances. The consequence of this fact is that corresponding stress density distribution exhibits longer and higher tails for larger stress ranges as observed in Figure 56.

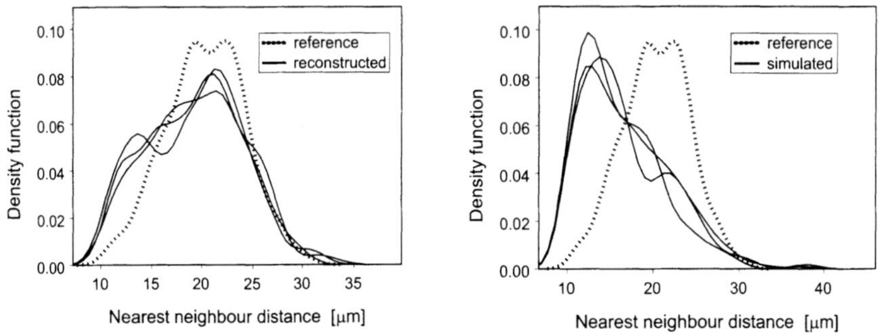


**Figure 57.** Nearest neighbour distance distribution for reconstructed and simulated patterns.

We have analysed reconstruction algorithm for planar and spatial dispersions of inclusions. It can be used both as a method of data analysis or as a structural modelling tool to make predictions about the system, which are based on ensemble averaging methodology. The obtained solutions are not unique and a true test of their usefulness has to be performed on locally dependent physical phenomena such as nucleation and evolution of damages and initiation of failure.

The reconstruction algorithm can be further extended to take into account other lower or higher order statistical descriptors that characterize the microstructure. Identification and reconstruction of microstructures with several descriptors would determine the extent to which a variety of statistical descriptors can reproduce the reference microstructure, thus shedding light on the nature of information contained in the descriptors. Furthermore, one can construct microstructures that correspond to a given set of descriptors and employ them to investigate and simulate physical phenomena where spatial patterns are of primary importance.

Presented reconstruction procedure necessitates at least one three-dimensional structure to be known from non-destructive evaluation or serial sectioning. This is not always possible either due to high cost of the experiment or due to time limitations and a labor with serial sectioning. In such circumstances the only physical information available is a plain cross section throughout material e.g. circular traces on a plane section traversing composite material reinforced with spherical inclusions. This information might be enough to estimate three-dimensional radii distribution of spheres provided we have created a simulation model that allows find a proper correlation.

The basic idea is to simulate spatial microstructures with known inclusion size distributions and to make plane sections on which 2D inclusion size distributions are measured. For each of the sphere radii distributions used to simulate the spatial structures, a continuous function describing the plane profile radii distribution can be constructed. Sphere radii can be simulated with different distribution functions, e.g. constant, uniform, Poisson, Gaussian, lognormal, etc. in order to create a library of 2D-3D correlations for distributions most frequently encountered in real materials.

Each distribution is characterized by a set of parameters describing its shape and position relative to the axis, as e.g. mean value $\mu$ and standard deviation $\sigma$ for Gaussian distributions. Similar parameters exist for all distributions used to generate the sphere radii and those describing its counterpart on plane sections. When parameters from the spatial distribution function are correlated with parameters for the corresponding plane function, it is possible to construct correlation maps that combine 3D and 2D radii distributions. Before such correlation maps can be constructed with a reliable accuracy, it is necessary to make a large number of simulations for each of the five sphere radii distributions with varying distribution parameters.

A Poisson distribution is discontinuous as it is only defined for integer values. Inclusion sizes in real material have of course all possible sizes and not only integer values. If inclusion sizes are divided into intervals with an interval midpoint for each integer value the distribution can be characterized by integers alone, and then described by a Poisson distribution. The probability density function for Poisson distributed sphere radii is given in equation 14. $\lambda_p$, the mean sphere radii, is defined for all values and not only integers.

$$f_3(R)=\frac{\lambda_P^R}{R!}\exp(-\lambda_p) \qquad R\in \text{ integer} \tag{14}$$

An example of Poisson distributed sphere radii is illustrated in Figure 58. The micrograph in Figure 58(left) is from a section through a structure with volume containing 16,000 inclusions and a mean sphere radii $\lambda_p$ = 15.0. Figure 58 (right) is the corresponding profile radii distribution estimated from profiles measured on $m$ = 120 sections, which are the number of sections necessary for achieving the predetermined accuracy.

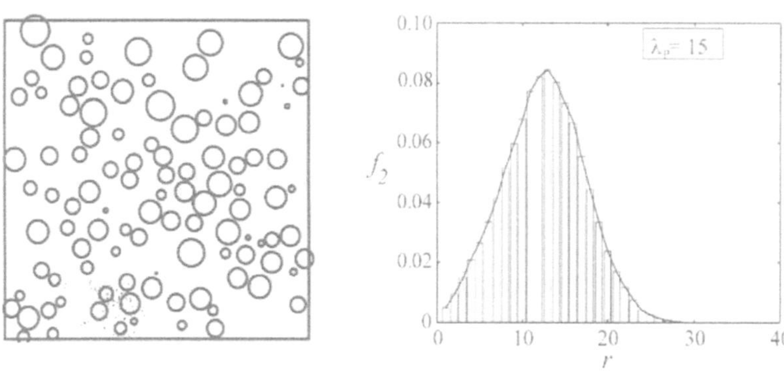


**Figure 58.** Profiles on a random section through a structure with Poisson distributed sphere radii (a); the profile radii probability density curve and histogram (b).

Profile distributions as shown in Figure 58 are more frequent in real materials than those for constant and uniform radii distribution. Sections through spheres with integer sizes give not only integer profile sizes. Therefore, the expression used to describe the profile probability must be defined for all values. It is convenient, if the profile radii distributions (2D) are identical to the sphere radii distributions (3D), but as illustrated for a number of simulations in Figure 59 this is not the case.

For different $\lambda_p$ the sphere radii distributions start from different radii due to the definition of the Poisson distribution. However, profile radii distributions all starts from zero value. Therefore assumptions of a direct comparison between the 3D and 2D radii distribution can give biased results. However, a correlation between the spatial and plane radii distributions is possible if the profile radii distribution is described accurately by a continuous function. The distribution parameters from this function can then be related to the spatial radii distribution parameter $\lambda_p$.

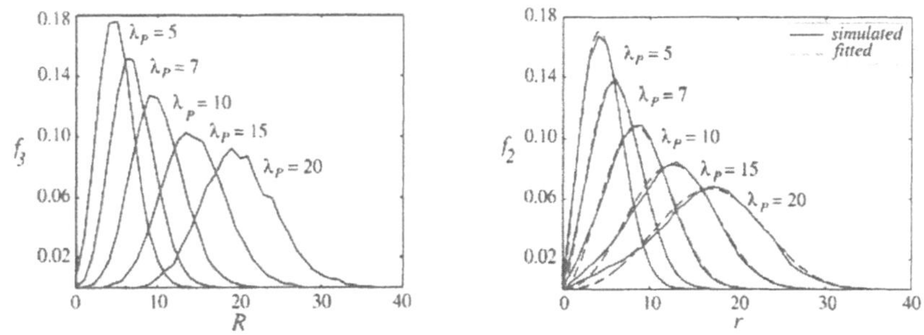


**Figure 59.** Simulated sphere radii distribution and profile radii distributions measured on plane sections.

It is very likely that one function is able to characterize all profile radii distributions estimated from sections through structures with all possible Poisson distributed sphere radii. If the sphere radii distribution $f_3(R)$ is unchanged and only the parameters vary, the resulting profile radii distribution $f_2(r)$ is also unchanged. The function used to describe the profile radii distribution for the Poisson distributed radii must contain the exponential function. Among several functions a probability density function in the Weibull family is found to characterize the profile radii probability density with good accuracy

$$f_2(r) = A_p C_p r^{(C_p - 1)} \exp(-A_p r^{C_p}) \tag{15}$$

When equation 15 is fitted to the simulated profile radii distribution in Figure 59, good agreement is found between the simulated radii distribution, illustrated by solid lines, and the fitted distributions, illustrated by dashed lines. For larger $\lambda_p$-values the fitted and simulated distribution curves differ slightly.

Both large and small sphere radii distributions, equation 15, describe the corresponding profile radii distribution with good accuracy. For the statistical mean value, the largest difference between the fitted and the simulated distributions is about 3%, and for the statistical variance the difference is maximum 8%.

The idea is to correlate $A_p$ and $C_p$ from equation 15, fitted to the profile radii distribution, with $\lambda_p$ that characterizing the spatial sphere radii distribution. If this is possible, the 3D sphere radii distribution can be determined from the 2D profile radii distribution. There is a relatively simple correlation between $A_p$ and $\lambda_p$, which is best illustrated in a semi log plot because $A_p$ varies over a large range of values, $10^{-5}$ – $10^{-2}$, and $\lambda_p$ only varies from 5 to 20, Figure 60. $A_p$ and $C_p$ are not independent on each other. This is expected because only one distribution parameter describes the spatial radii distribution, viz. $\lambda_p$. This gives a simple linear correlation between $C_p$ and $\lambda_p$ in a double log-plot, Figure 60, estimated by least square method.

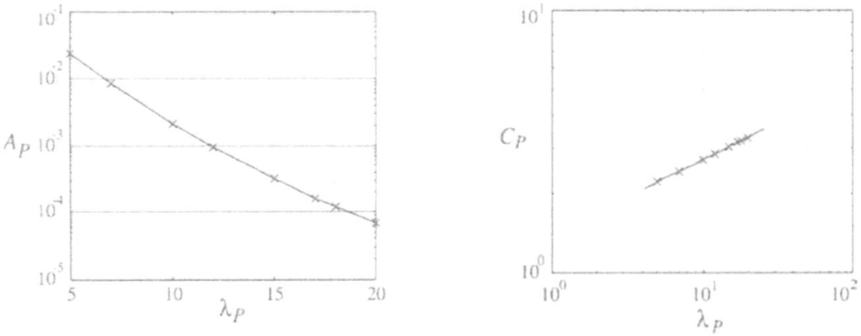


**Figure 60.** Correlation between $\lambda p$ and $Ap$ and correlation between $\lambda p$ and $Cp$.

If equation 15 is fitted to a profile radii distribution with good accuracy the corresponding sphere radii distribution is most probably similar to a Poisson distribution. From the fitted values $A_p$ and $C_p$ it is possible to estimate $\lambda_p$ and the probability density function for the spatial radii distribution given in equation 15. When the reconstruction approach is used, only 15 size classes for the profile radii are possible and as the result a discontinuous distribution is obtained. Using the described method, these limitations are eliminated, e.g. 30 size classes are used in Figure 58.

In the next particular example the sphere radii were distributed according to the Gauss distribution function

$$f_3(R) = \frac{1}{\sigma_G \sqrt{2\pi}} \exp\left[ -\frac{(R-\mu_G)^2}{2\sigma_G^2} \right] \tag{16}$$

where the mean value $\mu_G$ and standard deviation $\sigma_G$ are distribution parameters. The resulting plane radii distribution function $f_2(r)$ is not symmetrical and follows very well the extreme value function

$$f_2(r) = \frac{1}{B_G} \left\{ \exp\left( \frac{r - A_G}{B_G} \right) \exp\left[ -\exp\left( \frac{r - A_G}{B_G} \right) \right] \right\} \tag{17}$$

where $A_G$ is a location parameter and $B_G$ is scale parameter. Equation 17 has been plotted in Figure 61 against calculated probability density functions from many 3D simulations with different $\mu_G$ and $\sigma_G$. For a given mean sphere radii $\mu_G$, the location of the profile radii $A_G$ is independent of

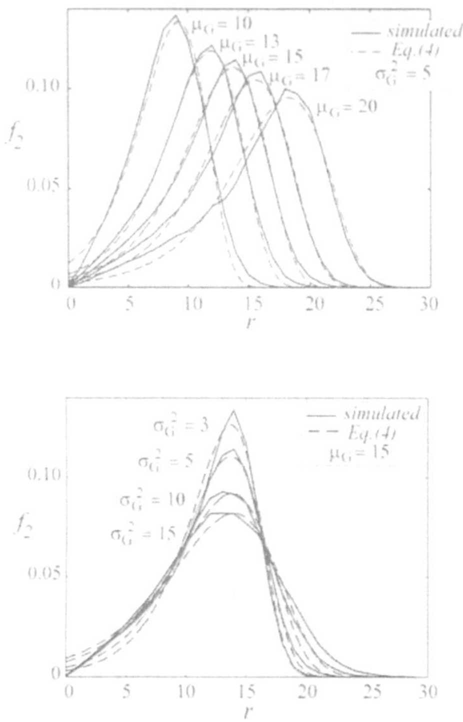


**Figure 61.** Simulated and analytical probability density function $f_2(r)$.

$\sigma_G$, and the relation between $A_G$ and $\mu_G$ is linear. Similarly, the relation between $B_G$ and $\sigma_G^2$ is linear; however, it shifts parallel towards larger $B_G$ values as $\mu_G$ gets larger. Correlating the spatial and plane distribution parameters from many simulations with varying $\mu_G$ and $\sigma_G$ one obtains the correlation map as illustrated in Figure 62. Similar correlation maps can be constructed for other distribution types such as uniform, Poisson, lognormal and constant distributions of sphere radii.

The 3D sphere radii distribution of glass spheres embedded in an epoxy matrix is estimated from plane profiles and the correlation maps. The radii of spheres had been collected before embedding them into the matrix material and then, after preparing a set of specimens, the plane sections were taken to measure the density function of sphere profiles, Figure 63. It appears that the 2D-function parent to the 3D Gauss distributed sphere radii describes in a best way the measured function of profiles *f(r)*, see Figure 64. Thus from Eq. 17 the parameters $A_G$ and $B_G$ were estimated and related to the three-dimensional distribution parameters $\mu_G$ and $\sigma_G$ using the

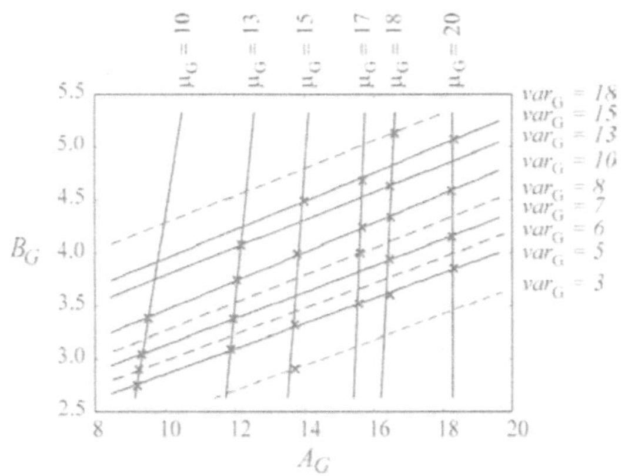


Figure 62. Correlation map between 2D and 3D parameters for Gauss distribution.

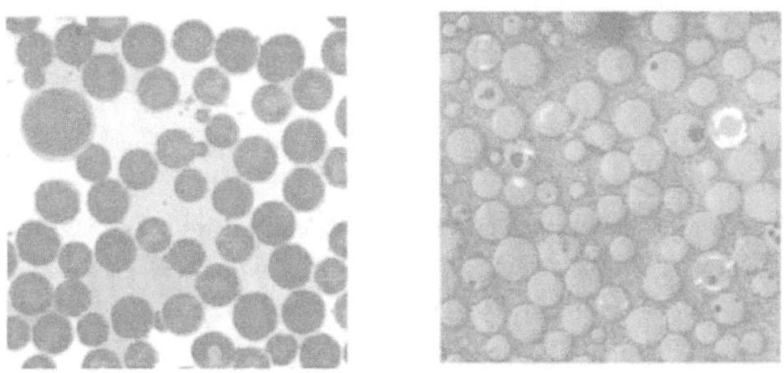

(a) (b)

**Figure 63.** 3D spheres (a) and sphere profiles (b) visible on a plane section.

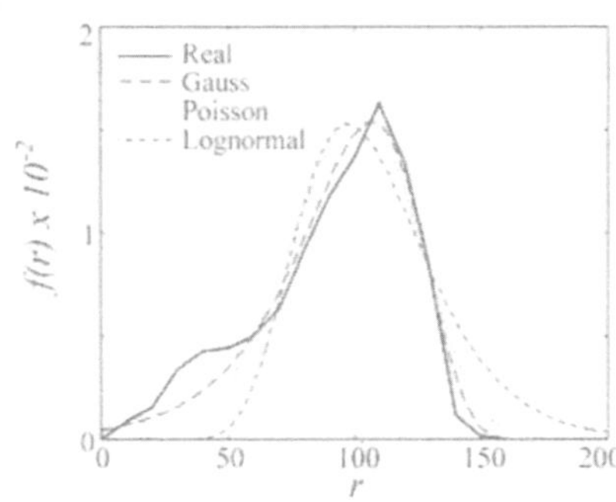


**Figure 64.** 2D probability functions parent to listed 3D distributions superposed on measured probability density function of profiles.

the correlation map. The reconstructed 3D probability density function *f(R)* compares very well with real density function of glass spheres, see Figure 65. The reconstruction procedure based on the correlation maps seems to be reliable and proves to be valid for other distributions, as well.

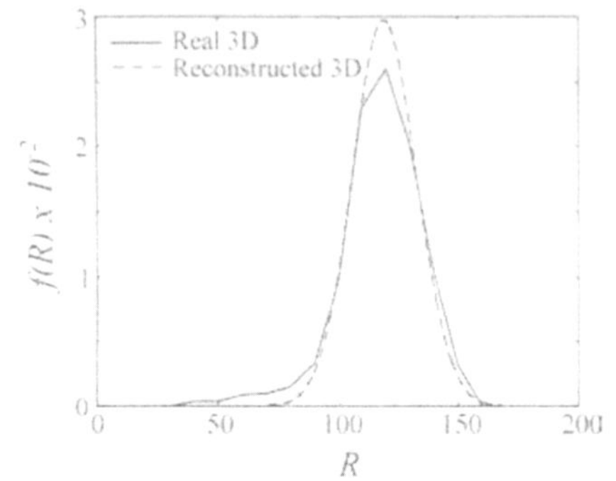


**Figure 65.** The real and reconstructed 3D density functions for sphere radii.

The knowledge of probability density function of spherical particles is necessary for an explanation of different microstructural mechanisms. For instance, the toughening mechanism in epoxy-rubber particulate composites is related to the size distribution of crack bridging particles, Kunz-Douglass et al. (1980), Saito et al. (1996). Particles' and voids' size distribution affects fracture behaviour of particulate composite materials, Steele (1989), Manoharan et al. (1989) and it may also affect sintering rates of the ceramic compacts, Chappel et al. (1986). The case with short fibre composites has been evaluated in From and Pyrz (1999).

# References

J. Aboudi. *Mechanics of Composite Materials*. Elsevier, Amsterdam, 1991.

N.D. Alberola, G. Merle, and K. Bendarti. Unidirectional fibre-reinforced polymers: Analytical morphology approach and mechanical modelling based on the percolation concept. *Polymer*, 40:315–328, 1999.

B. Altan, Z. Misirli, and H. Yorucu. The determination of the homogeneity in multiphase mixtures. *Z. Metall.*, 81:221–228, 1990.

G. Ausias, J.F. Agassant, M. Vincent, P.G. Lafleur, P.A. Lavoie, and Carreau P.J. Rheology of short glass fiber reinforced polypropylene. *J. Rheol.*, 36:525–542, 1992.

M. Axelsen and R. Pyrz. Correlation between fracture toughness and the microstructure morphology in transversely loaded unidirectional composites. In R. Pyrz, editor, *Microstructure–Property Interactions in Composite Materials*, pages 15–26. Kluwer Academic Publishers, Dordrecht, 1995a.

M. Axelsen and R. Pyrz. Effect of microcrack formation in transversely loaded unidirectional composites. In N. Sottos, editor, *Durability and Damage of Composite Materials*, MD-Vol. 69–1, pages 239–248. ASME, New York, 1995b.

I. Balberg, C.H. Anderson, S. Alexander, and N. Wagner. Excluded volume and its relation to the onset of percolation. *Phys. Rev. B*, 30:3933–3943, 1984.

P.P. Bansal and A.J. Ardell. Average nearest-neighbor distances between uniformly distributed finite particles. *Metallography*, 5:97–111, 1972.

R.S. Bay and C.L. Tucker. Fibre orientation in simple injection mouldings. Part II: Experimental results. *Polym. Engng. Sci.*, 32:240–253, 1992.

M.L. Becraft and A.B. Metzner. The rheology, fiber orientation, and processing behavior of fiber-filled fluids. *Rheol,*, 36:143–174, 1992.

J.G. Berryman. Random close packing of hard spheres and disks. *Phys. Rev. A*, 27:1053–1061, 1983.

J.G. Berryman. Measurement of spatial correlation functions using image processing techniques. *J. Appl. Phys.*, 57:2374–2384, 1985.

J.G. Berryman and G.W. Milton. Microgeometry of random composites and porous media. *J. Phys. D*, 21:87–94, 1988.

B. Billia, H. Jamgotchian, and H. Nguyen Thi. Statistical analysis of the disorder of two-dimensional cellular arrays in directional solidification. *Metall. Trans. A*, 22:3041–3050, 1991.

B. Bochenek and R. Pyrz. Identification of clustered distributions. Submitted to *Comp. Sci. Techn.,* 2003a.

B. Bochenek and R. Pyrz. (2003). Reconstruction of random microstructures – a stochastic optimization problem. Submitted to *Comput. Mat. Sci.*, 2003b.

J. Boselli, P.D. Pitcher, P.J. Gregson, and I. Sinclair. Random close packing of hard spheres and disks. *J. Microsc.,* 195:104–112, 1999.

N.J. Bozarth, J.W. Gillespie, and R.L. McCullough. Fiber orientation and its effect upon thermoelastic of short carbon fiber reinforced poly(etheretherketone). *Polym. Compos.*, 8:74–81, 1987.

J.R. Brockenbrough, W.H. Hunt, and O. Richmond. A reinforced material model using actual microstructural geometry. *Scripta Metall. Mater.*, 27:385–390, 1992.

J.S. Chappel, T.A. Ring, and J.D. Birchall. Particle-size distribution effects on sintering rates. *J. Appl. Phys.*, 60:383–391, 1986.

T. Chen, G.J. Dvorak, and Y. Benveniste. Mori-Tanaka estimates of the overall elastic moduli of certain composite materials. *J. Appl. Mech.*, 59:539–546, 1992.

S.N. Chiu. Aboav-Weaire's and Lewis' laws – A review. *Mat. Char.*, 34:149–165, 1995.

D.B. Corson. Correlation functions for predicting properties of heterogeneous materials. *J. Appl. Phys.*, 45:3159–3170, 1974.

C.W. Corti, P. Cotterill, and G.A. Fitzpatrick. The evolution of the interparticle spacing in dispersion alloys. *Int. Metall. Rev.*, 19:77–88, 1974.

L.M. Cruz-Orive and E.R. Weibel. Sampling designs for stereology. *J. Microsc.*, 122:235–257, 1981.

D. Cule and S. Torquato. Generating random media from limited microstructural information via stochastic optimization. *J. Appl. Phys.*, 86:3428–3437, 1999.

A.C. Curtis, P.S. Hope, and I.M. Ward. Modulus development in oriented short-glass-fiber-reinforced polymer composites. *Polymer Composites*, 3:138–145, 1982.

P.J. Davy and F.J. Guild. The distribution of interparticle distance and its application in finite-element modelling of composite materials. *Proc. R. Soc. Lond. A*, 418:95–112, 1988.

B. Derby. Microstructure and fracture behavior of particle-reinforced metal-matrix composites. *J. Microsc.*, 177:357–368, 1995.

P.J. Diggle, J. Besag, and J.T. Gleaves. Statistical analysis of spatial point patterns by means of distance methods. *Biometrics*, 32:659–667, 1976.

P.J. Diggle. *Statistical Analysis of Spatial Point Pattern*, Academic Press, London, 1983; p. 70.

P.J. Diggle and A.G. Chetwynd. Second-order analysis of spatial clustering for inhomogeneous populations. *Biometrics*, 47:1155–1163, 1991.

W.J. Drugan and J.R. Willis. A micromechanics-based nonlocal constitutive equation and estimates of representative volume element size for elastic composites. *J. Mech. Phys. Solids*, 44:497–524, 1996.

G.J. Dvorak. Transformation field analysis of inelastic composite materials. *Proc. R. Soc. Lond. A*, 437:311–327, 1992.

K.E. Evans and A.G. Gibson. Prediction of the maximum packing fraction achievable in randomly oriented short-fibre composites. *Compos. Sci. Techn.*, 25:149–162, 1986.

R.K. Everett and J.H. Chu. Modeling of non-uniform composite microstructures. *J. Comp. Mat.*, 27:1128–1144, 1993.

R.K. Everett. Quantification of random fiber arrangements using a radial distribution function approach. *J. Comp. Mat.*, 30:748–758, 1996.

J.L. Finney. Random packings and the structure of simple liquids. *Proc. R. Soc. Lond. A*, 319:479–493, 1970.

G. Fischer and P. Eyerer. Measuring spatial orientation of short fibre reinforced thermoplastics by image analysis. *Polym. Comp.*, 9:297–304, 1988.

F. Folgar and C.L. Tucker. Orientation behavior of fibers in concentrated suspensions. *J. Reinf. Plast. Comp.*, 3:98–119, 1984.

P.S. From and R. Pyrz. Computer assisted stereology of short fibre composites: (I) determination of fibre length distribution. *Sci. Engng. Comp. Mat.*, 8:143–157, 1999.

A. Getis and B. Boots. *Models of Spatial Processes.* Cambridge University Press, Cambridge, 1978.

S. Ghosh, Z. Nowak, and K. Lee. Quantitative characterization and modeling of composite microstructures by Voronoi cells. *Acta Mat.*, 45:2215–2234, 1997a.

S. Ghosh, Z. Nowak, and K. Lee. Tesselation-based computational methods for the characterization and analysis of heterogeneous microstructures. *Comp. Sci. Techn.*, 57:1187–1210, 1997b.

S. Ghosh and S. Moorthy. Particle fracture simulation in non-uniform microstructures of metal-matrix composites. *Acta Mat.*, 46:965–982, 1998.

J.-F. Gouyet. *Physics and Fractal Structures*, Springer, New York, 1996.

H.J.G. Gundersen and E.B. Jensen. The efficiency of systematic sampling in stereology and its prediction. *J. Microsc.*, 147:229–263, 1987.

H.J.G. Gundersen, E.B. Jensen, K. Kiêu, and J. Nielsen. The efficiency of systematic sampling in stereology reconsidered. *J. Microsc.*,193:199–211, 1999.

A.A. Gusev. Representative volume element size for elastic composites: A numerical study. *J. Mech. Phys. Solids*, 45:1449–1459, 1997.

H. Hermann. *Stochastic Models of Heterogeneous Materials*, Trans Tech Publications, Zurich, 1991.

J.R. Isasi, R.G. Alamo and L.A. Mandelkern. A study of the dilation of the unit cell of metallocene isotactic poly(propylenes): The monoclinic form. *J. Polym. Sci. B*, 35:2511–2522, 1997.

W.C. Jackson, S.G. Advani, and C.L. Tucker. Predicting the orientation of short fibers in thin compression mouldings. *J. Comp. Mat.*, 20:539–557, 1986.

J.W. Ju and T.M. Chen. Effective elastic moduli of two-phase composites containing randomly dispersed spherical inhomogeneities. *Acta Mech.*, 103:123–144, 1994.

J.W. Ju and K.H. Tseng. Effective elastoplastic behavior of two-phase ductile matrix composites: A micromechanical framework. *Int. J. Solids Struct.*, 33:4267–4291, 1996.

L.M. Karlsson and A. Liljeberg. Second-order stereology for pores in translucent alumina studied by confocal scanning laser microscopy. *J. Microsc.*, 175:186–194, 1994.

P.A. Karnezis, G. Durrant, and B. Cantor. Characterization of reinforcement distribution in cast Al-alloy $SiC_p$ composites. *Mat. Char.*, 40:97–109, 1998.

D. König, S. Carvajal-Gonzales, A.M. Downs, J. Vassy, and J.P. Rigaut. Modelling and analysis of 3-D arrangements of particles by point processes with examples of application to biological data obtained by confocal scanning light microscopy. *J. Microsc.*, 161:405–433, 1991.

S. Kunz-Douglass, P.W.R. Beaumont, and M.F. Ashby. A model for the toughness of epoxy-rubber particulate composites. *J. Mater. Sci.*, 15:1109–1123, 1980.

Y.W. Kwon and C.T. Liu. Effects o non-uniform particle distributions on damage evolution in pre-cracked particulate composite specimens. *Polym. & Polym. Comp.*, 6:387–397, 1998.

B.J. Lee and M.E. Mear. Effect of inclusion shape on the stiffness of non-linear two phase composites. *J. Mech. Phys. Solids*, 39:627–649, 1991.

J. Lemaitre, A. Gervois, J.P. Troadec, N. Rivier, M. Ammi, L. Oger, and D. Bideau. Arrangement of cells in Voronoi tessellations of monosize packing of disks. *Phil. Mag. B*, 67:347–362, 1993.

H.W. Lotwick and B.W. Silverman. Methods for analyzing spatial processes of several types of points. *J. Roy. Statist. Soc. B*, 44:406–413, 1982.

M. Manoharan, C. Liu, and L.L. Lewandowski. Microstructure and particle size effects on fracture in aluminum metal matrix composites. In K. Salama et al., editors, *Advances in Fracture Research*, pages 2977–2985. Pergamon Press, Oxford, 1989.

T. Mattfeldt, H. Frey, and C. Rose. Second order stereology of benign and malignant alterations of the human mammary gland. *J. Microsc.*, 171:143–151, 1993.

M.N. Miller. Bounds for effective electrical, thermal, and magnetic properties of heterogeneous materials. *J. Math. Phys.*, 10:1988–2004, 1969.

S.J. Milne, M. Patel, and E. Dickinson. Experimental studies of particle packing and sintering behavior of monosize spherical silica powders. *J. European Ceram. Soc.*, 11:1–7, 1993.

G.W. Milton. Bounds on the elastic and transport properties of two-component composites. *J. Mech. Phys. Solids*, 30:177–191, 1982.

B. Möginger and P. Eyerer. Determination of the weighting function $g(\beta_i,r,v_f)$ for fibre orientation analysis of short fibre-reinforced composites. *Composites*, 22:394–399, 1991.

T. Mura. *Micromechanics of Defects in Solids*. Kluwer, Dordrecht, 1987.

Y. Nakamura S. Okabe, N. Yoshimoto, and T. Iida. Effect of particle shape on the mechanical properties of particle-filled PVC. *Polym. & Polym. Comp.*, 6:407–414, 1998.

Y. Nakamura, S. Okabe, and T. Iida, T. Effects of particle shape, size and interfacial adhesion on the fracture strength of silica-filled epoxy resin. *Polym. & Polym. Comp.*, 7:177–186, 1999.

S. Nemat-Nasser and M. Hori. *Micromechanics: Overall Properties of Heterogeneous Materials*, North-Holland, Amsterdam, 1993.

A.S. Nielsen and R. Pyrz. In-situ observation of thermal residual strains in carbon/thermoplastic micro-composites using Raman spectroscopy. *Polym. & Polym. Comp.*, 5:245–257, 1997.

A.S. Nielsen and R. Pyrz. Study of the influence of thermal history on the load transfer efficiency and fiber failure in carbon/polypropylene microcomposites using Raman spectroscopy. *Composite Interfaces*, 6:467–482, 1999.

A.S. Nielsen and R. Pyrz. A novel approach to measure local strains in polymer matrix systems using polarized Raman microscopy. *Comp. Sci. Techn.*, 62:2219–2227, 2002.

A.S. Nielsen, D.N. Batchelder, and R. Pyrz. Estimation of crystallinity of isotactic polypropylene using Raman spectroscopy. *J. Polymer*, 43:2671–2676, 2001.

H. Okumura and F. Yonezawa. New expression of the bulk viscosity. *Physica A*, 321:207–219, 2003.

M. Ostoja-Starzewski, P.Y. Sheng, and I. Jasiuk. Influence of random geometry on effective properties and damage formation in composite materials. *J. Engng. Mat. Tech.*, 116:384–391, 1994.

P.G. Parkhouse and A. Kelly. The random packing of fibres in three dimensions. *Proc. R. Soc. London A*, 451:737–746, 1995.

P. Ponte Castañeda and J.R. Willis. The effect of spatial distribution on the effective behavior of composite materials and cracked media. *J. Mech. Phys. Solids*, 43:1919–1951, 1995.

M.J. Powell. Computer-simulated random packing of spheres. *Powder Techn.*, 25:45–52, 1980.

R. Pyrz. Quantitative description of the microstructure of composites. Part I: Morphology of unidirectional composite systems. *Comp. Sci. Techn.*, 50:197–208, 1994a.

R. Pyrz. Correlation of microstructure variability and local stress field in two-phase materials. *Mater. Sci. Engng. A*, 177:253–259, 1994b.

R. Pyrz and B. Bochenek. Statistical model of fracture in materials with disordered microstructure. *Sci. Engng. Comp. Mat.*, 3:95–109, 1994.

R. Pyrz. Pattern analysis and topological effects in transversely loaded unidirectional composites. In W. Tzuchizing and T.W. Chou, editors, *Proceedings of the International Conference Progress in Advanced Materials and Mechanics,* pages 581–586. Peking University Press, Beijing, 1996a.

R. Pyrz. Disorder and fracture model for transversely loaded composite materials. In H. Nisitani et al., editors, *Computer Aided Assessment and Control of Localized Damage*, pages 385–392. Computational Mechanics Publications, Southhampton, 1996b.

R. Pyrz. Fractal characterization of second phase dispersion in composite materials. *Sci. Engng. Comp. Mat.*, 6:141–150, 1997a.

R. Pyrz, Recent trends in morphological characterization of microstructures in polymer composites. In S/I/ Andersen et al., editors, *Proceedings of the 18th Risø International Symposium on Materials Science*, page 81. Risø National Laboratory, Roskilde, 1997b.

R. Pyrz and B. Bochenek. Topological disorder of microstructure and its relation to the stress field. *Int. J. Solids Struct.*, 35:2413–2427, 1998.

R. Pyrz. Application of X-ray microtomography to the study of polymer composites. In *Proceedings of the 12th International Conference on Composite Materials*, paper no. 909. Paris, 1999.

R. Pyrz. X-ray microtomography of composite materials. In S. Bandyopahyay et al., editors, *Proc. ACUN-3 International Conference on Technology Convergence in Composite Applications*, pages 185–190. Sydney, 2001.

R. Pyrz. Calculation of local stresses using molecular dynamic simulations. Submitted to *Comp. Sci. Techn.*, 2003.

J. Quintanilla. Microstructure and properties of random heterogeneous materials: A review of theoretical results. *Polym. Engng. Sci.*, 39:559–585, 1999.

N. Rivier. Statistical crystallography. Structure of random cellular networks. *Phil. Mag. B*, 52:795–819, 1985.

B.D. Ripley. Modelling spatial patterns. *J. Roy. Statist. Soc. B*, 39:172–212, 1977.

K. Saito, S. Araki, and T. Nakamura. Stereological modeling and micromechanical analysis of rubber-particle reinforced epoxy composite materials. *Mech. Comp. Mater.*, 32:317–329, 1996.

K. Sandau and H. Kurz. Measuring fractal dimension and complexity – An alternative approach with an application. *J. Microsc.*, 1997, 186, 164–176, 1997.

A. Sasov and D. van Dyck. Desktop X-ray microscopy and microtomography. *J. Microsc.*, 191, 151–158, 1998.

J. Schjødt-Thomsen and R. Pyrz. Overall creep modelling of short fibre reinforced composites with weakened interfaces and complex fiber orientation distributions. *Mech. Mater.*, 32:349–359, 2000.

J. Schjødt-Thomsen and R. Pyrz. The influence of stochastic interfacial parameter distribution on the stiffness of composites with complex fiber orientation. *Comp. Sci. Techn.*, 61:697–704, 2001.

J. Schjødt-Thomsen and R. Pyrz. Stress fields in heterogeneous solids with spatially dispersed inclusions. In R. Pyrz et al., editors, *New Challenges in Mesomechanics,* pages 75–82. Aalborg University, 2002.

J. Serra. *Image Analysis and Mathematical Morphology*, Academic Press, London, 1982; p. 271.

R. Sibson. The Dirichlet tessellation as an aid in data analysis. *Scand. J. Statist.*, 7:14–20, 1980.

P. Soille and J.F. Rivest. On the validity of fractal dimension measurements in image analysis. *J. Visual Comm. Image Represent.*, 7:217–229, 1996.

W.A. Spitzig, J.F. Kelly, and O. Richmond. Quantitative characterization of second-phase populations. *Metallography*, 18:235–261, 1985.

D. Srolovitz, T. Egami, and V. Vitek. Radial distribution function and structural relaxation in amorphous solids. *Phys. Rev. B,* 24:6936–6944, 1981.

D. Stauffer and A. Aharoni. *Introduction to Percolation Theory.* Taylor & Francis, Washington, 1992.

J.H. Steele, A stereological analysis of ductile fracture by microvoid coalescence. In K. Salama et al., editors, *Advances in Fracture Research*, pages 3449–3456. Pergamon Press, Oxford, 1989.

D. Stoyan. On correlations of marked point processes. *Math. Nachr.*, 116:197–204, 1984.

B.J. Super and A.C. Bovik. Localized measurement of image fractal dimension using Gabor filters. *J. Visual Comm. Image Represent.*, 1991, **2**, 114–128, 1991.

Y. Takao, T.W. Chou, and M. Taya. Effective longitudinal Young's modulus of misoriented short fiber composites. *J. Appl. Mech.*, 49:536–540, 1982.

G.P. Tandon and G.J. Weng. The effect of aspect ratio of inclusions on the elastic properties of unidirectionally aligned composites. *Polymer Composites*, 5:327–333, 1984.

M. Taya, K. Muramatsu, D.J. Lloyd, and R. Watanabe. Determination of distribution patterns of fillers in composites by micromorphological parameters. *JSME Int. J. I*, 34:198–206, 1991.

A. Tewari, M. Dighe, and A.M. Gokhale. Quantitative characterization of spatial arrangement of micropores in cast microstructures. *Mat. Char.*, 40:119–132, 1998.

S. Toll. (1998). Packing mechanics of fibre reinforcements. *Polym. Engng. Sci.*, 38:1337–1350.

S. Torquato. Microstructure characterization and bulk properties of disordered two-phase media. *J. Stat. Phys.*, 45:843–873, 1986.

S. Torquato. Random heterogeneous media: Microstructure and improved bounds on effective properties. *Appl. Mech. Rev.*, 44:37–76, 1991.

S. Torquato. Unified methodology to quantify the morphology and properties of inhomogeneous media. *Physica A*, 207:79–91, 1994.

E.M. Tory, N.A. Cochrane, and S.R. Wadell. Anisotropy in simulated random packing of equal spheres. *Nature*, 220:1023–1024, 1968.

W.M. Visscher and M. Bolsterli. Random packing of equal and unequal spheres in two and three dimensions. *Nature*, 239:504–507, 1972.

Y.M. Wang and G.J. Weng. The influence of inclusion shape on the overall viscoelastic behavior of composites. J. Appl. Mech., 59:510–518, 1992.

D. Weaire and N. Rivier. Soap, cells and statistics – Random patterns in two dimensions. *Contemp. Phys.*, 25:59–99, 1984.

E.R. Weibel. *Stereological Methods*. Academic Press, London, 1980.

K. Wiencek and D. Stoyan. Spatial correlations in metal structures and their analysis. *Mat. Char.*, 31:47–53, 1993.

A.S. Wimolkiatisak, J.P. Bell, D.A. Scola, and J. Chang. Assessment of fiber arrangement and contiguity in composite materials by image analysis. *Polym. Comp.*, 11:274–279, 1990.

C.L. Yeong and S. Torqato. Reconstructing random media. *Phys.Rev. E*, 57:495–506, 1998a.

C.L. Yeong and S. Torqato. Reconstructing random media. II. Three-dimensional media from two-dimensional cuts. *Phys.Rev. E*, 58:224–233, 1998b.

# Processing-Microstructure-Property Relationships in a Short Fiber Reinforced Carbon-Carbon Composite System

T. Siegmund, R. Cipra, J. Liakus, B. Wang
School of Mechanical Engineering, Purdue University, West Lafayette, Indiana, U.S.A.

M. LaForest
Honeywell Aircraft Landing Systems, South Bend, Indiana, U.S.A.

A. Fatz
National Composite Center, Kettering, Ohio, U.S.A.

**Abstract.** Manufacturing of composite preforms by use of a programmed and controlled reinforcement spray deposition process presents itself as an attractive approach to produce short fiber reinforced composite structures. To predict properties of the final composites structure, simulations of the reinforcement deposition process are conducted to obtain the reinforcement orientation distribution. A micromechanics analysis incorporating the Mori-Tanaka method and texture tensors, as well as a composite strength model are used to predict the properties of the final consolidated composite parts. This processing-microstructure-property prediction scheme is applied to the analysis of composite structures in the carbon-carbon system. The effects of variations in reinforcement length in the spray deposited preform, and boundary effects as occurring in a near net shape composite disk are discussed.

# 1 Introduction

Composite parts made from short fibers have emerged as alternatives to long fiber reinforced composites due to the attractive manufacturing processes and cost advantages associated with shorter fiber length. Among the manufacturing processes considered for short fiber reinforced composite are injection molding, squeeze casting, freeform molding and tape casting. In addition – as of focus in the present study - manufacturing routes are of interest where chopped segments of initially continuous fiber or tow products are cut and deposited by a spray process. Such an approach is common in low cost composites application. It is also of interest to high-end structures in automotive and aerospace products, Cordell et al. (2000), Corum et al. (2001). Depending on the densification route, this technique has been applied to the production of polymer matrix composites (Cordell et al., 2000, Corum et al., 2001) and recently has been extended to produce preforms for carbon-carbon composites, (Siegmund et al., 2002).

The spray deposition process becomes attractive for the production of composite structures requiring high levels of durability if a programmable robotic system is used to move the chopper head in a controlled fashion along predefined paths along the deposition target. A manufacturing method that can accomplish this is the so-called Programmable Powder Preform Process (P4) (Cordell et al., 2000, Siegmund et al., 2002). The P4 technology offers the potential to dramatically reduce the cost of composite components and can be coupled with several different resin infusion methods. This process was originally intended for automotive applications with glass fibers preform. Recent developments lead to the "P4 for Aerospace (P4A)" process and its application to aerospace composite material structures (Cordell et al., 2000). In all these applications the P4 process was intended for thin walled structures, and typically preforms are made by deposition of a single layer of fibers. Recently, the P4 process has extended to produce thick composite parts for carbon-carbon composite brakes (Siegmund et al., 2002). For brake manufacturing applications the P4 process provides not only the opportunity to lower cost but also to introduce new concepts in the design of the composite microstructure.

A typical installation of a P4 system for the production of thin composite parts consists of the following components. The robotic fiber chopping system attached to a robotic arm is used to cut the continuous fiber tow, and spray it onto a screen through which a vacuum is drawn. At the same time, a dry, thermoset powdered binder is also applied. The next step is the consolidation of the preform by the passage of heated air through the preform. This cures the binder and holds the fibers in position in the compacted preform. The preform is then cooled by passing room temperature air through the preform. Once the preform is cooled, it is removed from the tool and is capable of holding its shape for further handling and processing by infiltration with a resin. Typical fiber volume fractions used in preforms produced in the automotive and aerospace P4 process are between 33 to 57 % volume fraction of fibers and with 3-5 % binder weight. With appropriate automated control of the fiber deposition the P4 process becomes a point-by-point method of fiber placement, and also allows for the creation of complex preform shapes with e.g. integrated stiffeners. Controllable process parameters include the control of the fiber lengths, the reinforcement amount and potentially its orientation. For polymer matrix composite the subsequent densification can be accomplished by the use of the SRIM process. In past applications the P4 process has been applied to thin wall structures where a single layer of chopped fibers is deposited. The typical wall thickness values are of the order of 1.5 to 3.2 mm.

In the development of the P4 process to aircraft brake fabrication significant extensions of the original P4 process are undertaken. Preforms for aircraft brakes are thick (up to 60 mm). Epoxy binders must be substituted by phenolic resin binders to increase carbon yield, as needed in the further densification C-C processes. The volume fractions of both binder and fiber to be deposited are considerable higher than in conventional P4 processes. Brake preforms ideally should possess equal approximately fractions of fiber and binder. The fiber microstructures obtained by the P4 process, and the simulation of the microstructures is discussed in the next section of the paper. For carbon-carbon composite a process route including steps of resin char, resin transfer molding and chemical vapor deposition has recently been proposed.

For the properties of the composite structure produced by the approach described above, reinforcement volume fraction, and reinforcement length can be regarded as primary input parameters to the spray deposition process. On the other hand, the orientation of the reinforcements in the composite part depends on a number of parameters and process conditions, including the deposition rate, the tow segment length, the path of the robot guiding the chopper head, the distance of the chopper head to the deposition area, and the interaction of the spray deposition with boundaries. It is well recognized that the orientation of the reinforcement is of great importance in determining the properties of the final composite part (Dunn et al.,

1996). Image analysis approaches to as-manufactured parts can be applied to determine the orientation of reinforcements (Bay and Tucker, 1992). Figure 1 depicts two typical examples of tow segments deposited by the spray system along a circular path.

Case 1

Case 2

**Figure 1.** Typical fiber spray patterns obtained with different manufacturing conditions.

In Figure 1, case 1 and 2 were produced with slight modifications in the orientation of the vector of the spray processes relative to the deposition plane. Tow segments are outlined in red, and their orientation relative to the radial direction was measured. Figure 2 depicts the histogram of the frequency of the radial distribution of fibers. Case 1 contains a significantly larger amount of tangentially oriented tows than case 2.

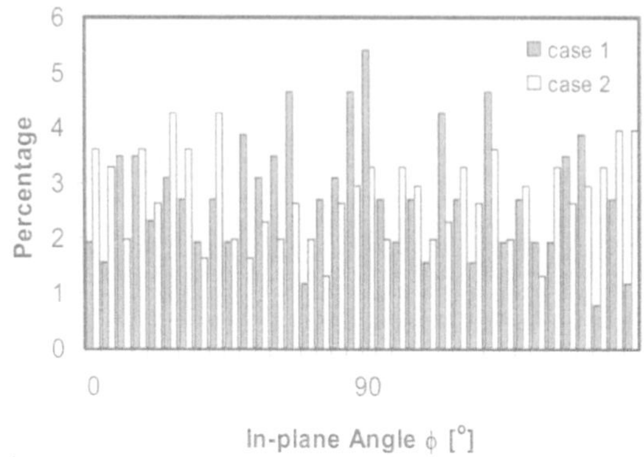


**Figure 2.** Orientation distributions for cases 1 and 2 in Figure 1.

In view of a modeling and simulation based engineering approach – the need for quantitative capabilities for the prediction of the reinforcement orientation distribution arises. The present paper focuses on the development of a link between the reinforcement deposition process and the final material properties in the composite part. In the first step a model of the reinforcement deposition process is described. This model is based on the geometric considerations of the robot motion and the kinematics of the reinforcement deposition process. The reinforcement orientation distribution information obtained from the deposition simulation is subsequently used in the predictions of the thermomechanical, transport, and strength properties of the final composite part. Properties are calculated by the use of a multilevel level approaches which account for the inhomogeneous distribution of the fibers in the final composite.

The processing-microstructure-property prediction scheme developed here is applied to two scenarios. First, the effects of variations in reinforcement length are investigated. For this part, the focus is on the analysis of the out-of-plane orientation effects due to the entanglement of the short fiber tows during the spray deposition process. The second example investigates a preform for a near-net shape part. Here, the interaction between a mold wall and the reinforcements during the deposition process become important, and the effects of the resulting reinforcement orientation distribution on the in-plane properties of the composite part are discussed.

# 2 Process Simulation

A simulation code for the prediction of distribution of the chopped fiber tow was developed (Liakus et al., 2003). The code allows for a prediction of the position and orientation of the as-deposited reinforcements in the preform, and under the assumption that the densification does not change the orientation distribution, also in the final composite part. The code developed in MATLAB includes four main components:

(1) Simulation of the chopper head path relative to the deposition plane,
(2) Reinforcement deposition within a given spray radius,

(3) Interaction of the tows with a preform mold wall, and
(4) Interactions between the tows, including the sliding of tows relative to each other.

Thus the simulation will provide data on: the percent coverage for each layer; the distribution of in-plane angles; and the distribution of out-of-plane angles. The path of the chopper head is described on the basis of geometric arguments only. Here, motion of the chopper head along concentric circles with predefined radii, $r_p$, is considered. Reinforcements in the form of fiber tow cut to predefined length, $L$, are deposited within a circle of given spray radius, $r_s$, with the center of the circle moving along the circle with radius $r_p$. The geometry of the problem is depicted in Fig. 3(a). Tow segments are assumed to possess a circular cross section with diameter $d_t$. The spray radius depends on the choices made in the distance of the chopper head from the deposition plane, the chopper nozzle geometry, and the rate at which tow is fed through the chopper. A random number generator is used to position the center of gravity of a tow segment at position $r_f$ in the currently active deposition circle. Then a random number is assigned to the angle between the tow axis and the XY coordinate axes spanning the deposition plane. Once a fiber tow is deposited, the chopper increments along the spray path. The process repeats until the entire path has been traversed, resulting in the deposition of a single layer.

As preforms for near-net shape parts are being produced, the reinforcement deposition process has to account for the interaction between tow segments and a mold wall.

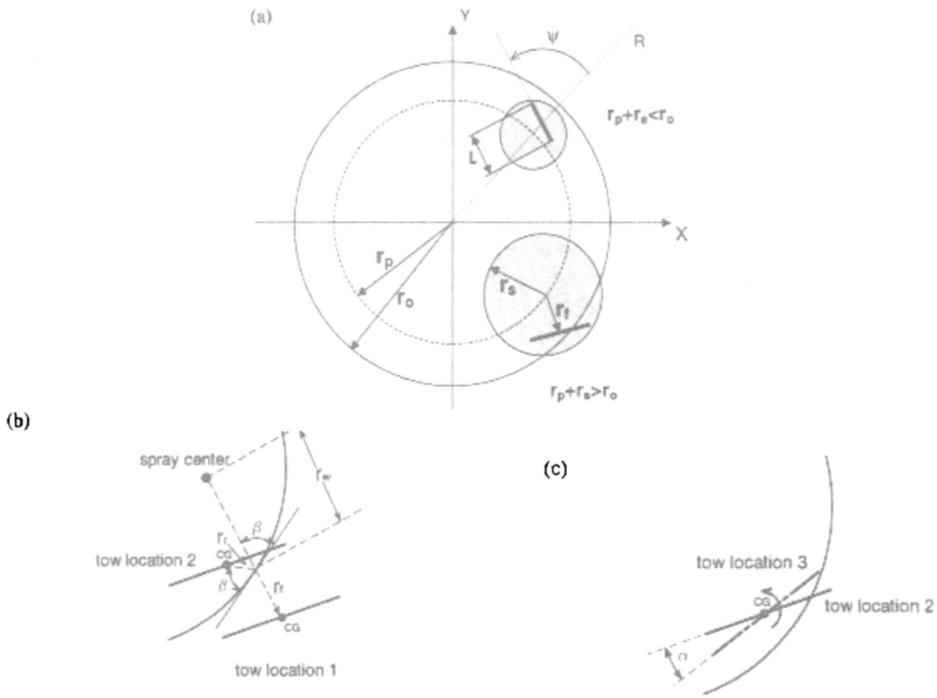


**Figure 3.** In-plane geometry: (a) Definition of target area and robot path; (b) Interaction of tow and wall; (c) Rotation of tow due to interaction with wall.

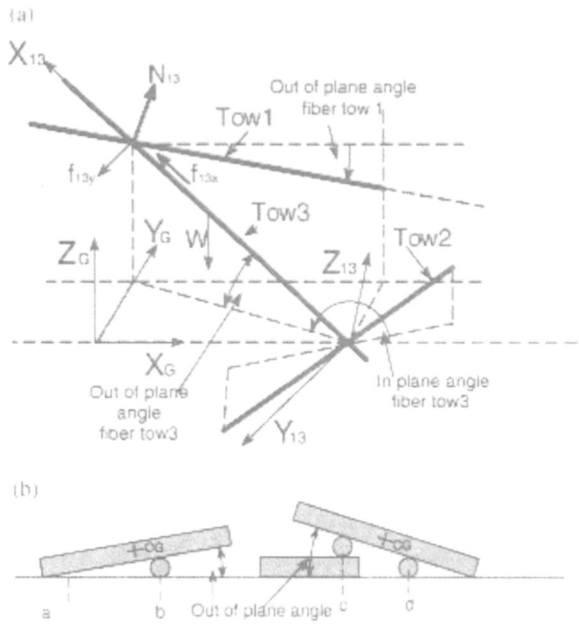


**Figure 4.** Out-of-plane geometry: (a) Definition of angles and forces in the analysis of tow sliding; (b) Definition of the out-of-plane angle.

A composite disk (and composite ring) are considered and a circular (concentric) mold with the radius $r_o$ is assumed. The algorithm used to account for the interaction between tow and wall within the simulation is based on reflecting the fiber tow inside the wall if its center of mass (CG) is located outside the wall. As shown in Figure 3(b), the reflected distance $r_r$ is a percent of the distance that the tow was outside the boundary, and the reflected angle ($\beta$) is as shown. Next, if either end of the fiber tow is still outside the wall, the tow is rotated about the CG through an angle ($\alpha$) until the fiber tow is inside the wall as shown in Figure 3(c).

The simulation accounts for the interaction between fiber tows by determining if a fiber tow will slide on the fibers it lands on and after it comes to rest, calculates the angle of the tow relative to the deposition plane, thus the height distribution of the deposited layer is accounted for. As a first step, after a tow segment is deposited, it is determined if any fibers are underneath it and how it is initially supported on top of the already existing tows (see Figure 4(b)). Subsequently, sliding between the tow segments can occur due to the weight of the tow segments and the coefficient of friction. The free body diagram accounting for the sliding interaction between tow segments is depicted in Fig. 4, as one tow segment (tow 3 in Fig. 4) lands on two other tows already deposited (tows 1 and 2 in Fig. 4). For the contact between tow 3 and tows 1 and 2, the following relationships are developed in order to determine if tow 3 will slide and if so, in what direction, by solving for the friction forces required to keep tow 3 in equilibrium. For contact between tow 3 and tow 1, the normal at the contact point is perpendicular to both the axis of fiber tow 3 and the axis of fiber tow 1 and is denoted by $N_{13}$ in the $Z_{13}$ direction while $X_{13}$ is defined along fiber tow 3. The friction forces at this contact point are $f_{13x}$, along the fiber tow, and $f_{13y}$, tangent to the fiber tow. The same occurs at the contact between tow 3 and tow 2, with friction forces $f_{23x}$ along tow 3 and $f_{23y}$ tangent to tow 3. Thus three conditions are addressed: tow 3 slides along its own axis, tow 3 slides tangent across tow 1, tow 3 slides tangent across tow 2, or a combination of these motions occurs. Next, the friction forces required to keep tow 3 in equilibrium are solved for. In these considerations $\varphi_1$, $\varphi_2$, and $\varphi_3$ are the out-of-plane angles of tow segment 1, 2 and 3, respectively, and $W$ is the weight acting at the center of gravity. Summing moments about $Y_{13}$ and $Z_{13}$ yields the normal force and the friction force required to prevent sliding:

$$N_{13} = W\frac{L_{2c}}{L_{21}}S_{13}\cos\varphi_1\cos\varphi_3\sin\psi_{d31} \tag{1a}$$

$$f_{13y} = W\frac{L_{2c}}{L_{21}}S_{13}\left(\sin\varphi_1\cos^2\varphi_3 - \cos\varphi_1\sin\varphi_3\cos\varphi_3\cos\psi_{d31}\right) \tag{1b}$$

where $L_{21}$ is the length between tow 2 and tow 1 along tow 3, and $L_{2c}$ is the length between tow 2 and the center of tow 3 along tow 3. The angle $\psi_{d31}$ is the difference between the in-plane angle of tow 3 and the in-plane angle of tow 1. We define $S_{13} = -1$ if $\psi_{d31} < 0$, else $S_{13} = 1$.
Summing moments about $Y_{23}$ and $Z_{23}$ yields:

$$N_{23} = W\frac{L_{1c}}{L_{21}}S_{23}\cos\varphi_2\cos\varphi_3\sin\psi_{d32} \tag{2a}$$

$$f_{23y} = W\frac{L_{1c}}{L_{21}}S_{23}\left(\sin\varphi_2\cos^2\varphi_3 - \cos\varphi_2\sin\varphi_3\cos\varphi_3\cos\psi_{d32}\right) \tag{2b}$$

where $L_{1c}$ is the length between tow 1 and the center of tow 3 along tow 3. The angle $\psi_{d32}$ is the difference between the in-plane angle of tow 3 and the in-plane angle of tow 2. We define $S_{23} = -1$ if $\psi_{d32} < 0$, else $S_{23} = 1$.

In order to prevent sliding of tow 3 tangentially along tow 1, the required coefficient of friction is:

$$\mu_{13y} = \frac{f_{13y}}{N_{13}} = \frac{\tan\varphi_1\cos\varphi_3 - \sin\varphi_3\cos\psi_{d31}}{\sin\psi_{d31}} \tag{3a}$$

To prevent sliding of tow 3 tangentially along tow 2, the required coefficient of friction is:

$$\mu_{23y} = \frac{f_{23y}}{N_{23}} = \frac{\tan\varphi_2\cos\varphi_3 - \sin\varphi_3\cos\psi_{d32}}{\sin\psi_{d32}} \tag{3b}$$

Finally, summing forces along $X_{13}$ yields:

$$f_{13x} + f_{23x} = W\left|\sin\varphi_3\right|. \tag{4a}$$

Thus, the required coefficient of friction to prevent sliding of tow 3 along its axis is:

$$\mu_{3x} = \frac{W\left|\sin\varphi_3\right|}{N_{13} + N_{23}} \tag{4b}$$

which after substituting equations 1a and 2a into 4b can also be expressed as:

$$\mu_{3x} = \frac{|L_{21}\tan\varphi_3|}{S_{12}L_{2c}\cos\varphi_1\sin\psi_{d21} + S_{23}L_{1c}\cos\varphi_2\sin\psi_{d32}} \qquad (4c)$$

If any of the above values for the coefficient of friction are greater than the actual coefficient of friction, the fiber tow is allowed to slide a short distance and the friction coefficients are again calculated. This process continues until the tow reaches equilibrium or encounters another fiber tow, ground, or the wall (boundary).

Input parameters for the simulation code are used to define the chopper head path, the spray radius, the geometry of the mold wall, the reflected distance percent, the actual coefficient of friction, the mass per unit length of the tow, tow segment length, and the amount of reinforcements deposited per layer. Predictions are displayed in form of the actual simulated reinforcement structure, as well as histograms of the angle distribution.

In this section the influence of the reinforcement length on the properties of the composite layer is investigated. Stacking such individual layers can subsequently allow the construction of composite structures with a through thickness gradient in reinforcement length. The resulting property gradients can be beneficial in applications where indentation type loads are applied to the composite structure (Jørgensen et al., 1998). Also, in cases where different parts of a composite structure possess distinctly different tow segment lengths, a gradient can alleviate the stress concentrations associated with the property change (Miyamoto, 1999).

Six deposition processes were considered with the tow segment lengths of the layers $L$=0.5 $L^*$, 0.75 $L^*$, $L^*$, 1.5 $L^*$, 2 $L^*$, 3 $L^*$ ($L^*$=25.4 mm=1 in.). With the tow diameter $d_t$=0.7 mm, the aspect ratio of the reinforcements are 18.1, 27.9, 36.3, 54.4, 72.4 and 108.8. For each layer 47 g of fiber tow, corresponding to a total tow length of 85 m, was deposited by a process with spray radius $r_s$=2 $L^*$ along five circular paths of $r_p$= 1.5 $L^*$, 3 $L^*$, 4.5 $L^*$, 6 $L^*$, 7.5 $L^*$. No mold walls were considered. Each layer was deposited individually onto a flat surface. Furthermore, to access the statistical effects of the random number generation scheme on final predicted composite properties the simulation of the deposition process on the flat surface for $L$=1.5 $L$* was repeated several times.

Figure 5 depicts two examples of the as deposited tow segments, $L$=0.5 $L$* and 2 $L^*$. The figures demonstrate that in-plane orientation of the reinforcement is random. The coverage of the deposition plane with reinforcement is nearly uniform for most of the deposition area with some larger amount of reinforcements accumulated in areas of overlaying spray. However, near the perimeter of the deposition area of radius $r_p + r_s = 9.5L^*$ some reduction of the amount of reinforcement occurs due to the lack of a specified mold.

The distributions of the computed out-of-plane angle, $\varphi$, for the six different tow segment lengths and a friction coefficient of $\mu = 0.5$ are given in Fig. 5(c). The simulations predict that the fraction of tow segments with their axis in the deposition plane decreases with $L$. For the tow segment lengths $L$=0.5 $L^*$, 0.75 $L^*$, $L^*$, 1.5 $L^*$, 2 $L^*$, 3$L^*$ 90% of all tows possess out-of-plane angles less than 11°, 11°, 11°, 12°, 14°, 28°, respectively. The average predicted out-of-plane angles are 3.9°, 4.1°, 4.1°, 5.1°, 6.1°, 12.5° for these tow segment length values. For $\mu = 0.1$ sliding of fibers relative to each other becomes easier. As a consequence the predicted out-of-plane angles are reduced. For $L$=0.5 $L^*$, 0.75 $L^*$, $L^*$, 1.5 $L^*$, 2 $L^*$, 3$L^*$ 90% of all tows possess out-of-plane angles less than 7°, 8°, 8°, 9°, 9°, 12°, respectively. The average predicted out-of-plane angles are 2.2°, 2.7°, 3.0°, 3.6°, 3.6°, 5.2°, respectively.

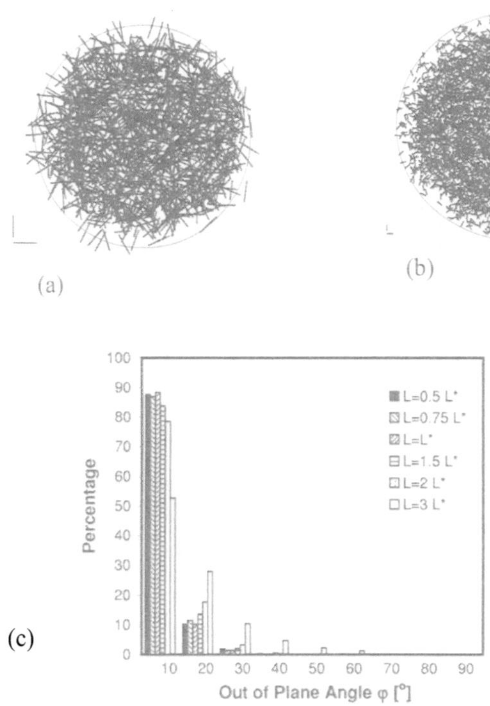


**Figure 5.** (a) Results of deposition simulation for $L$=2$L$*, (b) Results of deposition simulation for $L$=0.5 $L^*$, (c) Computed values of the out-of-plane orientation in dependence of the tow segment length.

To determine the variation in the prediction between individual runs of the fiber deposition simulation, the computations for the case of $L$=2.0 $L^*$ and $\mu = 0.5$ were performed at total of five times. These computations predict the average out-of plane angle to be 6.54° with a standard deviation of 0.58°. On the average, 90% of all fibers possessed an out-of-plane angle of less than 14.8° with a standard deviation of 1.3°.

In advanced manufacturing it is desirable to produce complex parts with near net shape geometry, in order to avoid scrap, machining cost and long process cycles. As can be seen from Figs. 5, a deposition process without an appropriate boundary does not lead to a preform with a well-defined preform shape. In order to simulate a near-net shape composite part, a cylindrical wall, or mold, is now considered, and the interaction of the spray process with the wall is accounted for.

Figure 6 depicts a typical microstructure obtained in the simulation for a total tow amount of 1071 $L^*$ deposited. Due to the interaction with the walls, tow segments close to the perimeter The structure considered is a disk produced from a preform for which a single layer of tow segments with length $L$=2 $L^*$ was deposited. The deposition process used the same parameters for $r_s$ and $r_p$ as described in the previous section, with the addition of the mold wall at radius

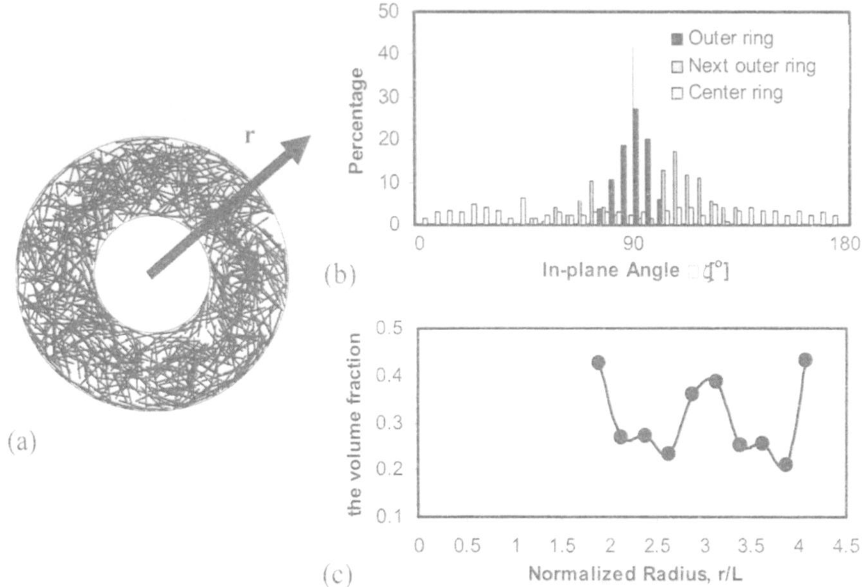


**Figure 6.** (a) Results for deposition simulation of a ring; (b) Orientation distribution; (c) Reinforcement volume distribution.

$r_o$=8.5 $L^*$. Again, it is assumed that the preform is densified such that the reinforcement volume fraction $V_r$=0.3 is obtained.

In the analysis of the near net shape disk preform, the out-of plane orientation of the tow segments was set to $\varphi = 0^\circ$ .are oriented dominantly in the tangential direction, and a boundary layer at the mold walls is evident. Further quantitative analysis of the wall effects was conducted for a simulation with a total reinforcement amount deposited of 6430 $L^*$.The orientation distribution of the angle, $\psi$ , i.e. the angle between the radial direction and the reinforcement axis, was determined within a series of concentric rings.

For the outermost ring, $r_o - 0.25L < r < r_o$ , the majority of reinforcements is oriented a tangential direction, $\psi = 90^\circ$ , with essentially all reinforcements in the range between $45^\circ < \psi < 135^\circ$ . For the ring $r_o - 0.5L < r < r_o - 0.25L$ the orientation distribution possesses two peaks, at $45^\circ$ and at $135^\circ$ . For the third ring located at the center a random distribution of the angle $\psi$ is predicted by the simulation.

# 3 Property Prediction

## 3.1 Thermomechanical Properties

**Theory.** The final microstructure of a composite with reinforcement volume fraction $V_r$, is comprised of regions of high fiber concentration corresponding to the original tows impregnated by the matrix. These regions of high reinforcement concentration are then embedded in matrix only regions. A three-level modeling approach is applied to calculate the elastic constants, the coefficient of thermal expansion and the thermal conductivity of the final composite. Figure 7 depicts a schematic drawing of the modeling approach. In level I, individual fibers are located in a unidirectional orientation within the tow region. Fibers are assumed packed with a packing factor, *P,* in each individual tow. The volume fraction of inclusions and matrix on level I, $V_I^I, V_M^I$, depend on packing factor of the fibers in the tow, only.

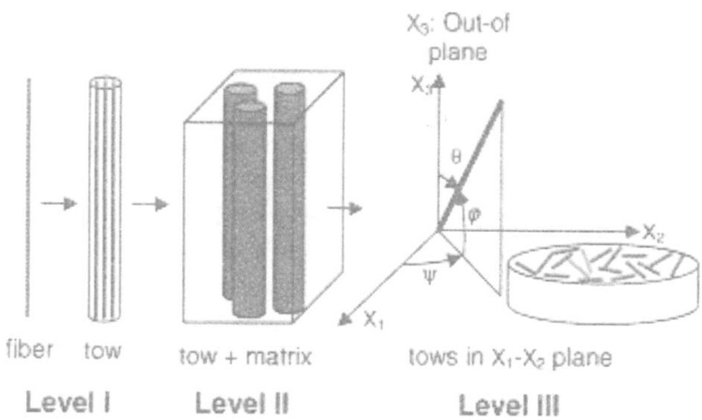


**Figure 7.** Modeling approach.

Throughout the paper a dense packing with $P$=0.7 is assumed such that the volume fractions of inclusions and matrix for the level I analysis become, $V_I^I = P = 0.7$, $V_M^I = 1 - P = 0.3$. On level I, fibers are approximated as spheroids with large aspect ratio.

In level II, the impregnated tows are then placed unidirectionally aligned within the matrix. The volume fraction of inclusions, i.e. the tow regions, and matrix of the level II composites, $V_I^{II}, V_M^{II}$, depend on the volume fraction of reinforcement, $V_r$, and the packing factor of the fibers in the tow at level I via $V_I^{II} = V_r / P$, $V_M^{II} = 1 - V_I^{II}$. The shape of the inclusions on level II was assumed to be spheroidal with the aspect ratio $L/d_t$.

For levels I and II the computations of properties are based on the Mori - Tanaka method (Nadeau and Ferrari, 1999). The stiffness tensor of a composite material, $\bar{\boldsymbol{C}}^a$, in which all inclusions are uniaxially aligned is given by (Beneviste, 1987):

$$\begin{aligned} \bar{C}^a &= C_M + V_I(C_I - C_M)A_1 \\ A_1 &= A^{dil}\left[(1-V_I)I + V_I A^{dil}\right]^{-1} \\ A^{dil} &= \left[I + SC_M^{-1}(C_I - C_M)\right]^{-1} \end{aligned} \tag{5}$$

where $S$ is the Eshelby tensor, $I$ is the forth-rank identity tensor, $C_M$ and $C_I$ are the stiffness of matrix and inclusion, respectively. Similarly, the coefficient of thermal expansion of a composite with unidirectional aligned inclusions, $\bar{\alpha}^a$, are given by (Rosen and Hashin, 1970):

$$\bar{\alpha}^a = \alpha_M + (\alpha_I - \alpha_M):(D_I - D_M)^{-1}:(\bar{D}^a - D_M) \tag{6}$$

where $\alpha_I$ and $\alpha_M$ denote the coefficients of thermal expansion of inclusion and matrix. $D_I$, $D_M$, $\bar{D}^a$ denote the compliance of inclusion, matrix and composite, respectively. Finally, the thermal conductivity, $\bar{k}^a$, of a composite with unidirectionally aligned inclusions is given by (Hatta and Taya, 1986):

$$\begin{aligned} \bar{k}^a &= k_M + V_I(k_I - k_M)a_1 \\ a_1 &= a^{dil}\left[(1-V_I)i + V_I a^{dil}\right]^{-1} \\ a^{dil} &= \left[i + sk_M^{-1}(k_I - k_M)\right]^{-1} \end{aligned} \tag{7}$$

where $k_I$ and $k_M$ are the conductivities of inclusion and matrix, respectively. $s$ is the second-rank Eshelby tensor, and $i$ is the second-rank identity tensor.

On level III the properties of composite parts under consideration of the actual reinforcement orientation distribution are then calculated. While full three dimensional orientation distributions can be accounted for with the method used in the present investigation, the focus of the discussion is on two special cases of orientation distributions. First, for composite preforms where boundary effects during the spray deposition can be neglected, the orientation of the reinforcement is axisymmetric with respect to the normal of the deposition plane, and only the angle of the tows with respect to the deposition plane (the out-of-plane angle) is accounted for. Second, in case of near-net shape preforms, the reinforcement orientation due to boundary effects dominates over the out-of-plane orientation. In this case only the in-plane reinforcement orientation is accounted for, and all reinforcements are assumed to have their axis in the deposition plane. For the case of the disk shaped preform discussed here the reinforcement orientation is orthotropic in the axisymmetric coordinate system of the disk.

Starting from the properties predicted in level II for aligned reinforcements, the properties of composites with specific reinforcement orientation distributions, $\bar{C}, \bar{\alpha}, \bar{k}$, are obtained as orientational averages. The averaging process is performed by the use of the forth and second order texture matrices, $T^4$ and $T^2$, respectively (Dunn et al., 1996):

$$\bar{C} = T^4\bar{C}^a \quad , \quad \bar{\alpha} = T^2\bar{\alpha}^a \quad , \quad \bar{k} = T^2\bar{k}^a \tag{8}$$

The components of the texture matrices depend on the orientation distribution function only. Following Fig. 3, for reinforcements with circular cross section, two Euler angles, $\theta$ and $\psi$, are being used to describe the orientation of the reinforcement. Following a framework established in texture analysis (Roe, 1965), and applied to composite materials (Dunn et al., 1996), the orientation distribution function, $w(\xi = \cos\theta, \psi, \phi)$, can be described by the use of a series of generalized spherical harmonics:

$$w(\xi,\psi,\phi)=\sum_{l=0}^{\infty}\sum_{m=-l}^{l}\sum_{n=-l}^{l}W_{lmn}Z_{lmn}(\xi)e^{-im\psi}e^{-in\phi} \quad \text{with} \quad \int_0^{2\pi}\int_0^{2\pi}\int_{-1}^{1}w(\xi,\psi,\phi)d\xi d\psi d\phi=1 \tag{9}$$

where

$$W_{lmn}=\frac{1}{4\pi^2}\int_0^{2\pi}\int_0^{2\pi}\int_{-1}^{1}w(\xi,\psi,\phi)\times Z_{lmn}(\xi)e^{im\psi}e^{in\phi}d\xi d\psi d\phi \tag{10}$$

and

$$Z_{lmn}(\xi)=i^{n-m}\sqrt{\frac{2l+1}{2}}P_l^{mn}(\xi) \tag{11}$$

$$P_l^{mn}(\xi)=\frac{(-1)^{l-m}i^{n-m}}{2^l(l-m)!}\left[\frac{(l-m)!(l+n)!}{(l+m)!(l-n)!}\right]^{1/2}(1-\xi)^{-(n-m)/2}(1+\xi)^{-(n+m)/2}\frac{d^{l-n}}{d\xi^{l-n}}[(1-\xi)^{l-m}(1+\xi)^{l+m}]$$

In Equation 11 $Z_{lmn}(\xi)$ are generalized Legendre polynomials (Roe, 1965), and $P_l^{mn}(\xi)$ are the common Legendre polynomials (Bunge, 1982). For the determination of $\boldsymbol{T}^4$ and $\boldsymbol{T}^2$, $l\le 4$ and $l\le 2$, respectively. Also, due to circular reinforcement cross-section, and the orthotropy of the orientation distribution, the non-zero $W_{lmn}$ are all real and only several of the $W_{lmn}$ are required. The Appendix describes the details of the texture matrices, $\boldsymbol{T}^4$ and $\boldsymbol{T}^2$.

Based on (Dunn et al., 1996), for an orthotropic orientation distribution and spheroid fibers, the texture matrix, $\boldsymbol{T}^4$, is expressed as

$$\boldsymbol{T}^4=$$

$$\begin{bmatrix}
4(14/105+8a-16b+3c-4d+2e) & 0 & 4(7/105-2a+4b-6c+8d-4e) & 21/105-24a+48b+12c-16d+8e & 8(7/105-2a+4b-6c+8d-4e)\\
7/105-8a+4c-8e & 35/105+56a & 8(7/105-5a-c+2e) & 7/105-8a+4c-8e & 4(-7/105+8a-4c+8e)\\
7/105+4a+8b-16c+16d & 35/105-28a-56b & 4(14/105+5a+14b+8c) & 7/105+4a+8b-16c+16d & -4(7/105+4a+8b-16c+16d)\\
4(14/105+8a+16b+3c+4d+2e) & 0 & 4(7/105-2a-4b-6c-8d-4e) & 21/105-24a-48b+12c+16d+8e & 8(7/105-2a-4b-6c-8d-4e)\\
7/105+4a-8b-16c-16d & 35/105-28a+56b & 4(14/105+5a-14b+8c) & 7/105+4a-8b-16c-16d & -4(7/105+4a-8b-16c-16d)\\
8(7/105-8a+4c) & 0 & 4(7/105+4a-16c) & 21/105+48a+32c & 8(7/105+4a-16c)\\
49/2/105-10a+20b-16c-16d & -35/2/105+14a-28b & 2(-7/105-4a+8b+16c+16d) & 7/105+4a-8b-16c-16d & 2(21/105+6a-12b+32c+32d)\\
49/2/105-10a-20b-16c+16d & -35/2/105+14a+28b & 2(-7/105-4a-8b+16c-16d) & 7/105+4a+8b-16c+16d & 2(21/105+6a+12b+32c-32d)\\
49/2/105+20a+4c-8e & -35/2/105-28a & 2(-7/105+8a-4c+8e) & 7/105-8a+4c-8e & 2(21/105-12a-8c+16e)
\end{bmatrix} \tag{12}$$

where

$$a=\frac{\sqrt{10}\pi^2}{105}W_{200},\ b=\frac{\sqrt{15}\pi^2}{105}W_{220},\ c=\frac{\sqrt{2}\pi^2}{105}W_{400},\ d=\frac{\sqrt{5}\pi^2}{105}W_{420},\ e=\frac{\sqrt{35}\pi^2}{105}W_{440}. \tag{13}$$

Note that several of the coefficients of $\boldsymbol{T}^4$ deviate slightly from the expression in (Dunn et al., 1996).

For the thermal conductivity and the coefficients of thermal expansion of the composite, the texture matrix, $\boldsymbol{T}^2$, is expressed as (Dunn et al., 1999):

$$\boldsymbol{T}^2 = \begin{bmatrix} \frac{1}{3}+2f-4g & \frac{1}{3}+2f-4g & \frac{1}{3}-4f+8g \\ \frac{1}{3}+2f+4g & \frac{1}{3}+2f+4g & \frac{1}{3}-4f-8g \\ \frac{1}{3}-4f & \frac{1}{3}-4f & \frac{1}{3}+8f \end{bmatrix} \tag{14}$$

where

$$f = \frac{\sqrt{10}\pi^2}{15} W_{200} \qquad g = \frac{\sqrt{15}\pi^2}{15} W_{220} \tag{15}$$

In the present study, histograms are used to characterize the orientation distribution of the reinforcements. For an axisymmetric orientation distribution, the fraction of tows oriented between the Euler angle $\theta_{i-1}$ and $\theta_i$ is $R_i$, and $0<\theta_i<\pi/2$, where $\theta_i$ is the angle between the reinforcement axis and the normal to the deposition plane. For normalization purposes, a factor $\lambda$ relating the relative distribution density, $\rho$, and $R_i$ is introduced such that $\rho(\theta) = \lambda R_i$ for $\theta_{i-1} \le \theta \le \theta_i$. The normalization condition is $1 = \int_0^{\frac{\pi}{2}} \rho(\theta)\sin\theta d\theta$, from which integration provides the normalization factor in the form

$$\lambda = \left[ \sum_{i=1}^{N} R_i (\cos\theta_{i-1} - \cos\theta_i) \right]^{-1} \tag{16}$$

For the axisymmetric orientation distribution (with out-of-plane angles $\theta \neq 0^\circ$ and a random distribution of $\psi$) only the coefficients $W_{200}$ and $W_{400}$ are non-zero, and from Eq. 10 – and the histograms of the orientation distribution as obtained from the simulations described in the previous section - the values of $W_{200}$ and $W_{400}$ can then be obtained:

$$W_{200} = \frac{1}{4\pi^2}\frac{\sqrt{10}}{4}\lambda\sum_{i=1}^{N} R_i \left[ \frac{1}{4}(\cos\theta_i - \cos\theta_{i-1}) - \frac{1}{4}(\cos 3\theta_i - \cos 3\theta_{i-1}) \right] \tag{17}$$

$$W_{400} = \frac{1}{4\pi^2}\frac{3\sqrt{2}}{16}\lambda\sum_{i=1}^{N} R_i \left[ \frac{1}{8}(\cos\theta_i - \cos\theta_{i-1}) + \frac{5}{16}(\cos 3\theta_i - \cos 3\theta_{i-1}) - \frac{7}{16}(\cos 5\theta_i - \cos 5\theta_{i-1}) \right] \tag{18}$$

In the case of a combination of an axisymmetric orientation distribution and a transversely isotropic elastic material behavior on level II, $\boldsymbol{T}^4$ is a 5×5 matrix, which is only the function of the $W_{200}$ and $W_{400}$. $\boldsymbol{T}^2$ in the calculation of the thermal conductivity and the coefficient of thermal expansion, is a 3×3 matrix, which is only the function of the $W_{200}$.

In cases where an interaction of tow segments with a mold wall becomes important, the effect due to the in-plane orientation of the tows dominate over the effects caused by the out-of-plane orientation of the reinforcements. For this case $\theta = 90^\circ$ is assumed, and $\psi$ is non-random. In deriving the coefficients $W_{lmn}$ of the expansion of the orientation distribution function, the orthotropy of the orientation distribution with respect to the reference coordinate system $Ox_1x_2x_3$ (Fig. 3) is accounted for. In the case of a combination of an orthotropic orientation distribution and a transversely isotropic elastic material behavior on level II, $\boldsymbol{T}^4$ becomes a 9×5 matrix in dependence of $W_{200}, W_{220}, W_{400}, W_{420}$ and $W_{440}$. The matrix $\boldsymbol{T}^2$ in the calculation of the thermal conductivity and the coefficient of thermal expansion remains a 3×3 matrix in dependence of $W_{200}, W_{220}$.

Under the condition that $\xi = \cos\theta = \cos 90^\circ = 0$ the coefficients of the expansion are

$$\begin{aligned}
W_{200} &= -\frac{\sqrt{10}}{16\pi^2}, W_{400} = \frac{9}{32\sqrt{2}\pi^2} \\
W_{220} &= \frac{\sqrt{15}}{16\pi^2}\sum_{i=1}^{N} R_i \frac{1}{2}\left[(\sin 2\psi_i - \sin 2\psi_{i-1}) - i(\cos 2\psi_i - \cos 2\psi_{i-1})\right] \\
W_{420} &= \frac{1}{4\pi^2}\left(-\frac{3}{8}\sqrt{5}\right)\sum_{i=1}^{N} R_i \frac{1}{2}\left[(\sin 2\psi_i - \sin 2\psi_{i-1}) - i(\cos 2\psi_i - \cos 2\psi_{i-1})\right] \\
W_{440} &= \frac{1}{4\pi^2}\left(\frac{3}{16}\sqrt{35}\right)\sum_{i=1}^{N} R_i \frac{1}{4}\left[(\sin 4\psi_i - \sin 4\psi_{i-1}) - i(\cos 4\psi_i - \cos 4\psi_{i-1})\right]
\end{aligned} \tag{19}$$

In Equation 19, histograms are used to describe the in-plane reinforcement orientation distribution, $\rho(\psi) = R_i$ and for $\psi_{i-1} \le \psi \le \psi_i$.

**Results on thermomechanical properties.** In the present investigation, the framework outlined in the previous section is applied to a case where the reinforcement deposition process constitutes the preforming step for a carbon-carbon composite material (Siegmund et al., 2002).

**Table 1**: Matrix and fiber properties.

| Constituent | Fiber | Matrix |
|---|---|---|
| Elastic Properties | $E_{33}$=220 GPa | $E_M$=10.0 GPa |
| | $E_{11}$=13.8 GPa | $\nu_M$ =0.3 |
| | $G_{31}$=11.35 GPa | |
| | $G_{12}$=5.5 GPa | |
| | $\nu_{31}$=0.2 | |
| | $\nu_{12}$=0.25 | |
| Coefficient of Thermal Expansion [$\times 10^{-6}/^\circ$C] | | |
| Longitudinal | -1.5 | 3.0 |
| Transverse | 27.0 | 3.0 |
| Thermal Conductivity [W/(m°C)] | | |
| Longitudinal | 9.40 | 100 |
| Transverse | 0.67 | 100 |

In all cases considered, it is assumed that the reinforcement volume fraction in the final densified preform is equal to $V_r = 0.3$.

For the elastic behavior, the out-of plane stiffness, $\bar{E}_{33}/E_M$, is smaller than the in-plane stiffness, $\bar{E}_{11}/E_M = \bar{E}_{22}/E_M$. For the perfect in-plane random distribution ($\varphi = 0^\circ$) as depicted by the continuous lines, a strong increase in the in-plane elastic modulus, $\bar{E}_{11}/E_M$, and a small increase in the out-of plane modulus, $\bar{E}_{33}/E_M$, occurs with an increase in the aspect ratio of the reinforcements up to $L/d_r$=100. For larger aspect ratios, the modulus remains unchanged.

As the actual orientation distribution is considered in the calculation, the dependence of the elastic constants on aspect ratio changes only slightly, despite the fact that a large amount of fibers possesses $\varphi > 0^\circ$. For $\mu = 0.1$ and small out-of-plane angles, the elastic properties of the composite are practically identical to those of the ideal random in-plane case with $\varphi = 0^\circ$. For $\mu = 0.5$, for small aspect ratios the predicted properties are again identical to those of for the ideal random in-plane case. As $L/d_t$ increases, the in-plane elastic modulus initially increases further, but reaches a maximum value for $L/d_t$ =72, and subsequently decreases as $L/d_t$ further increases.

(a)

(b)

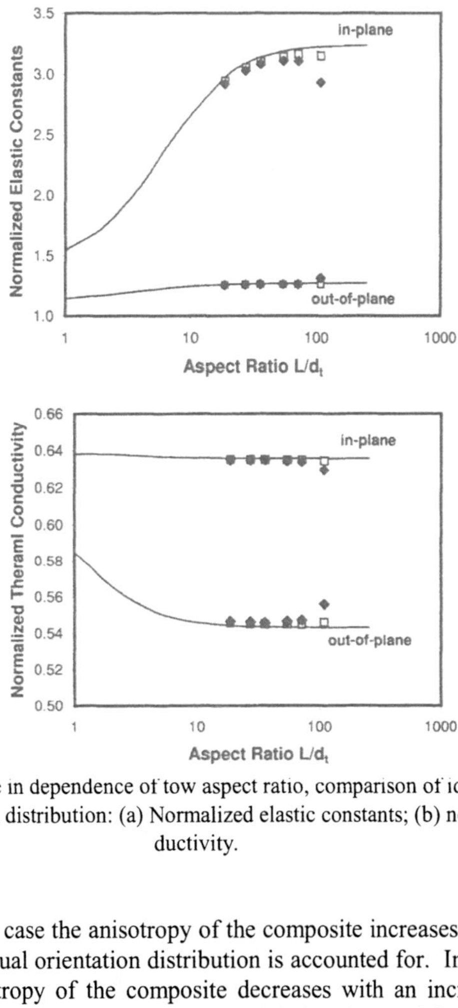


**Figure 8:** Predicted composite in dependence of tow aspect ratio, comparison of ideal random orientation to actual simulated orientation distribution: (a) Normalized elastic constants; (b) normalized thermal conductivity.

While for the ideal in-plane case the anisotropy of the composite increases with the aspect ratio, this is not the case if the actual orientation distribution is accounted for. In the case of $\mu = 0.5$, as $L/d_t$ increases the anisotropy of the composite decreases with an increase in aspect ratio. Similar trends in the influence of the tow segment length are observed for the predicted coefficients of thermal expansion and the thermal conductivity. Figure 8b shows the corresponding results for the predicted normalized thermal conductivities, $\bar{k}_1 / k_M = \bar{k}_2 / k_M$, $\bar{k}_3 / k_M$, respectively.

For thermal conductivities and assuming a perfect in-plane random distribution, the out-of-plane value, $\bar{k}_3 / k_M$, shows a stronger dependence on $L/d_t$ than the in-plane conductivity, $\bar{k}_1 / k_M = \bar{k}_2 / k_M$. Now, as the actual orientation distribution is accounted for, the in-plane conductivity decreases with increasing $L/d_t$, while the out-of-plane conductivity possesses minimum values for intermediate values of the aspect ratio.

As already observed for the statistical variation between the characteristics of individual simulations of the reinforcement orientation distribution, the predicted properties of the composite vary little from run to run. For the five simulation for $L$=2 $L^*$ and $\mu = 0.5$, the simulations predicted average values of $\bar{E}_{11} / E_M = \bar{E}_{22} / E_M$ =3.09 (standard deviation, $s$=0.019), $\bar{E}_{33} / E_M$ =1.26 ($s$=0.004), $\bar{k}_1 / k_M = \bar{k}_2 / k_M = 0.63$ ($s$=0.0001), and $\bar{k}_3 / k_M$=0.548 ($s$=0.001).

**Near-net shaped disk preform – in-plane distribution.** In the cases considering the in-plane orientation distribution of fiber tow the properties of the composite are orthotropic in the axi-symmetric coordinate system characterizing the ring. In the results shown in Fig. 9, $\bar{E}_{11} / E_M$, $\bar{\alpha}_1 / \alpha_M$, denotes radial, $\bar{E}_{22} / E_M$, $\bar{\alpha}_2 / \alpha_M$, tangential, and $\bar{E}_{33} / E_M$, $\bar{\alpha}_3 / \alpha_M$, through thickness properties, respectively. Property data in Figure 9 are obtained for the simulated orientation distributions of Fig. 6.

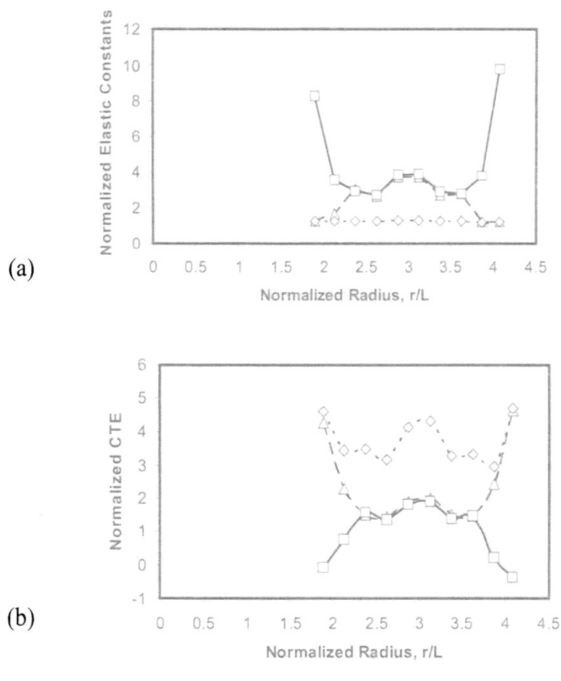


**Figure 9.** Predicted properties for the composite ring in dependence of the radial location

For the region close to the inner wall the composite possesses properties similar to those of a uniaxially oriented composite with transversely isotropic properties. In addition, the reinforcement content is high. The stiffness in tangential direction is high, while in the radial and through thickness direction a low stiffness is present. In the tangential direction, a negative coefficient of thermal expansion is present while for the radial and through thickness direction a large coefficient of as thermal expansion is found. For the region at the center of the disk, the composite behaves as a perfect in-plane random composite with transversely isotropic properties. Property variations along the radial direction are mostly related to the inhomogeneous mass distribution.

These changes in reinforcement content mainly affect the in-plane properties but leave the through thickness properties nearly without change. Subsequently, as the outer wall is approached, the composite changes from the in-plane transversely isotropic solid with $\bar{E}_{11}/E_M = \bar{E}_{22}/E_M$, $\bar{\alpha}_1/\alpha_M = \bar{\alpha}_2/\alpha_M$, $\bar{k}_1/k_M = \bar{k}_2/k_M$ to a transversely isotropic solid with $\bar{E}_{11}/E_M = \bar{E}_{33}/E_M$, $\bar{\alpha}_1/\alpha_M = \bar{\alpha}_3/\alpha_M$, $\bar{k}_1/k_M = \bar{k}_3/k_M$, which in the tangential direction again possesses the properties of a unidirectionally aligned composite. Table 2 summarizes the properties predicted using the actually measured orientation distributions for the cases 1 and 2 of Figure 1. The properties under consideration of the measured orientation distributions are compared to the perfectly random case. While the composite with random fiber orientation is transversely isotropic, the actual composites are orthotropic. Properties for case 1 and 2 differ considerably, especially for the elastic constant and thermal expansion.

**Table 2:** Predicted properties for composites using measured fiber orientation distribution data.

| | Random fiber ODF | | Actual fiber ODF | | |
|---|---|---|---|---|---|
| Elastic constants | $E_{11}/E_M = E_{22}/E_M$ | $E_{33}/E_M$ | $E_{11}/E_M$ | $E_{22}/E_M$ | $E_{33}/E_M$ |
| Case 1 | 2.9751 | 1.2588 | 2.7672 | 3.1934 | 1.2582 |
| Case 2 | | | 3.2004 | 2.8195 | 1.2583 |
| Thermal conductivity | $k_{11}/k_M = k_{22}/k_M$ | $k_{33}/k_M$ | $k_{11}/k_M$ | $k_{22}/k_M$ | $k_{33}/k_M$ |
| Case 1 | 0.6359 | 0.5443 | 0.6280 | 0.6437 | 0.5443 |
| Case 2 | | | 0.6429 | 0.6288 | 0.5443 |
| Thermal expansion | $\alpha_{11} = \alpha_{22}/\alpha_M$ | $\alpha_{33}/\alpha_M$ | $\alpha_{11}/\alpha_M$ | $\alpha_{22}/\alpha_M$ | $\alpha_{33}/\alpha_M$ |
| Case 1 | 1.6541 | 3.6554 | 1.8262 | 1.4819 | 3.6554 |
| Case 2 | | | 1.5009 | 1.8072 | 3.6554 |

### 3.2 Strength

**Theory.** The prediction of the composite strength is based on a concept of a "critical zone model" as proposed by Fukuda and Chou, and subsequently developed by Lauke and Fu (1999). In the model it is assumed that after matrix failure the load previously carried by the cracked region is transferred to the reinforcements (tows) that bridge the crack. Only tows that are located in the cracking zone can contribute to the crack bridging, and thus to the strength of the composite. The model for the composite strength discussed here again is a multi-scale model

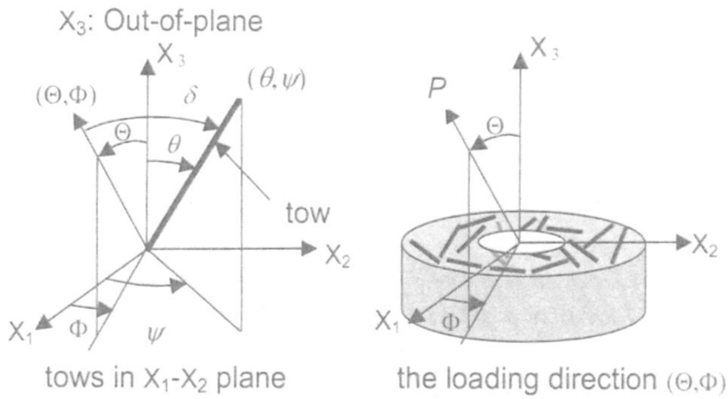


**Figure 10:** Schematic of strength modeling approach (a) General case, (b) In-plane orientation only.

and takes the tow character of the reinforcement into account. Figure 10 shows quantities needed in the modeling approach.

In a first step, the strength of the fiber tows, $\sigma_t$, is calculated. The distribution of the fiber strength is considered using the Weibull distribution function. The calculation of the tow strength in Level I is based on Rosen (1964). In this method, fibers are treated as having a statistical distribution of flaws or imperfections, which results in fiber failure at various stress levels. The Weibull distribution function is used to characterize the fiber strength distribution.

For the composite reinforced by the uniaxially oriented fibers when subjected to a uniaxial tensile load parallel to the fiber direction, the strength can be calculated using the following equation:

$$\sigma_t = V_f^I \alpha' \left( \frac{l}{l_i} \right)^{\frac{1}{\beta}} (\beta e)^{-\frac{1}{\beta}} \tag{20}$$

where $V_f^I$ is the fiber volume fraction in the tow. $\alpha'$ and $\beta$ are the Weibull parameters defining the fiber strength and determined by experimental tests of fiber strength vs. length. $l$ is the fiber length. $l_i$ is an ineffective fiber length determined from a shear lag model analysis as:

$$\frac{l_i}{d_f} = \frac{1}{2}\left[ \left( \frac{1-(V_f^I)^{1/2}}{(V_f^I)^{1/2}} \right) \left( \frac{E_f}{G_m} \right) \right]^{1/2} \times \cosh^{-1}\left[ \frac{1+(1-\eta)^2}{2(1-\eta)} \right] \tag{21}$$

The factor $\eta$ is the load-sharing ratio between the fiber stress $\sigma(l_i)$ and the undisturbed stress $\sigma_{f0}$, assumed as 0.9 here.

The model allows one to incorporate the orientation distribution of the composite by assuming that the matrix crack will form perpendicular to the direction of tensile load/or maximum principal stress. Then, the relative orientations of the tows to the loading/maximum principal stress direction need to be considered.

In Level II, the strength of the composite is predicted based on a concept of a modified rule of mixture approach, Fu and Lauke (1999). The composite strength, $\sigma_c$, under consideration of the actual reinforcement orientation distribution is calculated by:

$$\sigma_c = \chi V_{tow}^{II} \sigma_t + \left(1 - V_{tow}^{II}\right)\sigma_m \tag{22}$$

where $\sigma_m$ denotes the matrix strength. $V_{tow}^{II}$ is the tow volume fraction in the composite. The quantity $\chi$ is defined as the fiber efficiency factor. It can be calculated from considering the orientation of tows relative to the loading direction:

$$\chi = \int_{\psi_{min}}^{\psi_{max}} \int_{\theta_{min}}^{\theta_{max}} \int_{l_{min}}^{l_{c\delta}} f(l) g(\psi) h(\theta) \left(\frac{l}{l_{mean}}\right)\left(\frac{l}{2l_c}\right) e^{\mu\delta} \, dl d\theta d\psi + \int_{\psi_{min}}^{\psi_{max}} \int_{\theta_{min}}^{\theta_{max}} \int_{l_{c\delta}}^{l_{max}} f(l) g(\psi) h(\theta) \left(\frac{l}{l_{mean}}\right)(1 - A\tan\delta)\left[1 - \frac{l_c(1 - A\tan\delta)}{2le^{\mu\delta}}\right] dl d\theta d\psi \tag{23}$$

$g(\psi)$ and $h(\theta)$ are the distribution functions of the in-plane and out-of-plane tow orientation angles, respectively, while $f(l)$ denotes the tow length distribution function. The quantity $A$ is a material system specific constant. It can be approximately determined by $A \approx 5.5\tau_{my}/\sigma_t$. To contribute to the composite strength, tows must possess at least the critical tow length, $l_c$, which can be obtained by $l_c = \sigma_t d_t / 2\tau_i$. $\mu$ is the snubbing friction coefficient. For the present analysis, the tow length is assumed as constant, and larger than the critical tow length, $l_c$. For the tow orientation distribution, the discussion here is limited to the case of in-plane reinforcement orientation effects. Then Eq. 23 is simplified to

$$\chi = \int_{\psi_{min}}^{\psi_{max}} g(\psi)(1 - A\tan\delta)\left(1 - \frac{l_c(1 - A\tan\delta)}{2le^{\mu\delta}}\right) d\psi \tag{24}$$

In Equation 24, histograms are used to describe the in-plane reinforcement orientation distribution, $\rho(\psi) = aR_i$ and for $\psi_{i-1} \le \psi \le \psi_i$. The angle $\delta$ is measured between the tow axial direction $(\theta, \psi)$ and the direction of the applied load (or maximum principal stress) $(\Theta, \Phi)$. It is obtained by

$$\cos\delta = \cos\Theta\cos\theta + \sin\Theta\sin\theta\cos(\psi - \Phi) \tag{25}$$

**Results.** The model data employed in the strength prediction are assembled from published results, see Table 3. The data for the fiber strength distribution used in Weibull strength statistics were obtained from Fukuda (1994). In a first step the composite strength model was use to predict the strength of a unidirectionally reinforced composite material in dependence of the loading direction. Figure 11 depicts the normalized normal strength of this composite in dependence of the angle between load direction and reinforcement longitudinal axis for different reinforcement lengths. For the material parameters selected for the present study, the composite strengths remains above the matrix strength up to an angle of 70° between load and reinforcement. For higher angles, the composite strength essentially vanishes as the reinforcements then no longer are effective transferring load across the crack.

**Table 3:** Matrix and fiber properties

| Constituent | Fiber | Matrix |
|---|---|---|
| Diameter | $d_f$=7.1 μm | |
| Tensile strength | $\sigma_f$=2910 MPa | $\sigma_M$=300 MPa |
| Weibull Parameters | $\alpha'$=3.17GPa, $\beta$=5.1 | |
| Snubbing friction coefficient | $\mu$=0.1 | |
| Material constant $A$ | $A$=0.4 | |

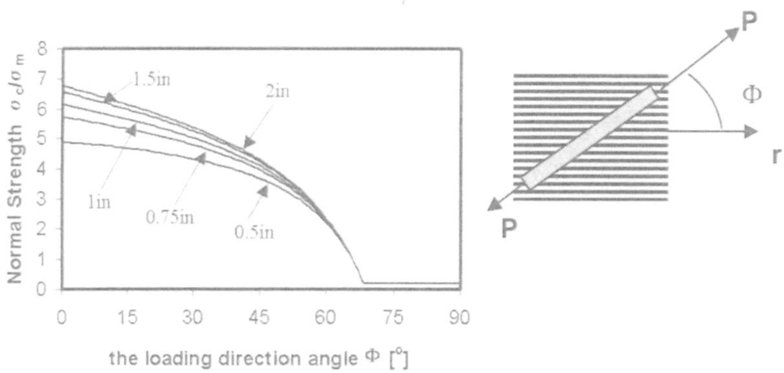


**Figure 11:** Strength vs. orientation for a composite with uniaxially aligned reinforcements.

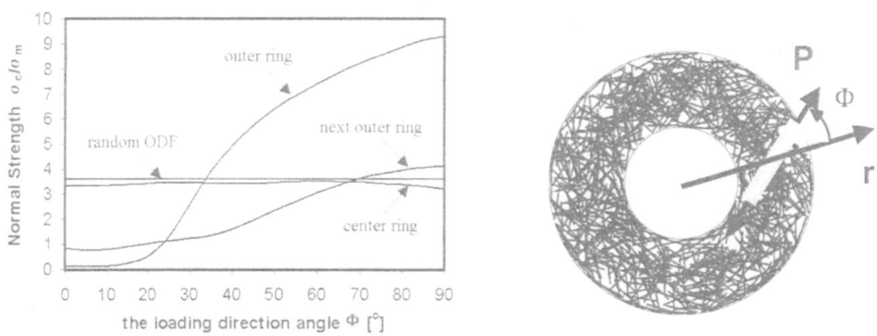


**Figure 12:** Predicted orientation dependent composite strength for three locations in the composite ring, comparison to a composite with ideal random reinforcement orientation.

In a next step, the strength prediction method was applied to study the strength characteristics in the composite ring of Fig. 6. The strength of the composite is predicted for three locations – at the outer wall, $r_o - 0.25L < r < r_o$, next to the outer wall, $r_o - 0.5L < r < r_o - 0.25L$, and at the center between inner and outer wall of the ring.

Strength predictions obtained under consideration of the simulated microstructure information are compared to that obtained for a composite with an ideal random orientation distribution, Figure 12. Close to the outer wall, the material is essentially uniaxially reinforced and possesses high strength under loading in tangential direction, but very low strength in radial direction. The strength anisotropy decreases dramatically already for a location slightly away from the wall. At the center between inner and outer wall, the composite essentially behaves like the one with a random orientation distribution.

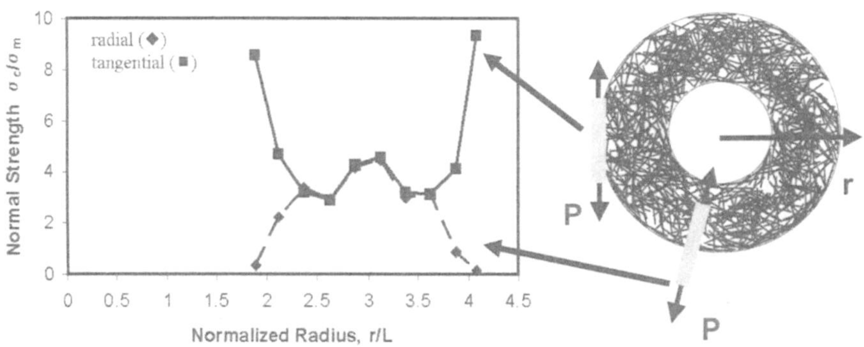


**Figure 13:** Predicted tangential and radial strength in dependence of the radial location.

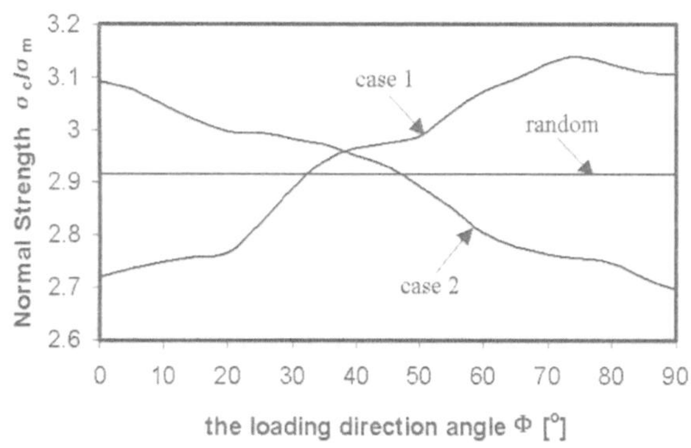


**Figure 14:** Predicted normalized strength for two experimentally determined reinforcement distribution functions.

Further details on the strength anisotropy are given in Figure 13. There the tangential and radial strength data are given in dependence of the radial location. The composite ring is predicted to be strongly anisotropic in strength at the outer and inner wall, but possesses isotropic strength in the area in between. While the overall tendencies are similar, the reinforcement orientation affects the composite strength considerably stronger than the continuum deformation and transport properties. The boundary layer effect will depend on mainly on the length of the reinforcements, but also on outer manufacturing parameters.

The strength prediction methodology was applied to the experimentally obtained reinforcement orientation distribution functions of Figure 2. The normalized strength data for case 1 and 2 are plotted in Figure 14, and compared to the case of the random composite. While no interaction of the deposition process with an external boundary was present, the model still predict that the composites produced by using deposition case 1 and 2 might have quite some difference in strength characteristics. Both materials are predicted to be of anisotropic strength, with the case 1 possessing low radial strength but higher tangential strength, while for case 2 the strength anisotropy character is opposite.

## Acknowledgement

The present work was performed with funding from the Indiana 21st Century Research and Technology Fund.

## References

R.S. Bay, C.L. Tucker III, Stereological measurements and error estimates for three-dimensional fiber orientation. *Polymer Engineering and Science* 2:240–253, 1992.

Y. Benveniste, A new approach to the application of Mori-Tanaka's theory in composite materials. *Mechanics of Materials* 6:147–157, 1987.

H. Bunge, Texture Analysis in Materials Science, Butterworths, London, 1982.

T. Cordell, T. Benson-Tolle, R. Rondeau, The programmable powdered preform process for aerospace – affordable performance through composites. *Science of Advanced Materials and Process Engineering Series* 45:254-265, 2000.

J.M. Corum, R.L. Battiste, M.B. Ruggles, W Ren, Durability-based design criteria for a chopped-glass-fiber automotive structural composite. *Composites Science and Technology* 61:1083-1095, 2001.

M.L. Dunn, H. Ledbetter, P.R. Heyliger, C.S. Choi, Elastic constants of textured short-fiber composites. *Journal of the Mechanics and Physics of Solids* 44:1509-1541, 1996.

M.L. Dunn, H. Ledbetter, Z. Li, Thermal expansion of morphologically textured short-fiber composites. *Metallurgical and Materials Transactions* A30:203-212, 1999.

H. Hatta, M. Taya, Equivalent inclusion method for steady state heat conduction in composites. *International Journal of Engineering Science* 7:1159-1172, 1987.

O. Jørgensen, A.E. Giannakopoulos, S. Suresh, Spherical indentation of composites laminates with controlled gradients in elastic constants. *International Journal of Solids and Structures* 35:5097-5113, 1998.

B. Lauke, S.Y. Fu, Strength anisotropy of misaligned short-fibre-reinforced polymers. *Composite Science and Technology* 59:699-708, 1999.

J. Liakus, B. Wang, R. Cipra, T. Siegmund, Processing–microstructure-property predictions for short fiber reinforced composite structures based on a spray deposition process. *Composite Structures* 61:363-374, 2003.

Y. Miyamoto, Functionally graded materials: design, processing and applications. Kluwer Academic Publishers, Boston, 1999.

J.C. Nadeau, M. Ferrari, Microstructural optimization of a functionally graded transversely isotropic layer. *Mechanics of Materials* 31:637-651, 1999.

R.J. Roe, Description of crystallite orientation in polycrystalline materials. III. General solution to pole figure inversion. *Journal of Applied Physics* 36:2024-2031, 1965.

B.W. Rosen, Tensile Failure of Fibrous Composites. *AIAA Journal* 2:1985-1991, 1964.

B.W. Rosen, Z. Hashin, Effective thermal expansion coefficients and specific heats of composite materials. *International Journal of Engineering Science* 8:157-173, 1970.

T. Siegmund, R. Cipra, J. Liakus, W. Strieder, F.R. Jacobs, A. Fatz, T. Cordell, C. Parker, F. Dillon, M. LaForest, A proposed novel approach to manufacturing low cost high temperature composite materials. *Transactions of the North American Manufacturing Research Institute of SME* 30:557-564, 2002.

# Discrete Dislocation Plasticity

Erik Van der Giessen

Dept. of Applied Physics, University of Groningen, Nijenborgh 4, 9747 AG Groningen, The Netherlands

## 1 Introduction

Plastic deformation of metals governs several length scales, Fig. 1. At the large length scale of the macroscopic world, plastic deformation is conveniently described by a phenomenological continuum theory. When zooming in, one will first start to observe that, in most engineering cases, the material is polycrystalline. At that scale, plastic deformation is a physical process that is inherently inhomogeneous and anisotropic. This is caused by the fact that each grain is anisotropic with a finite number of slip systems where glide can take place. When zooming in further, one will see that plastic deformation within each grain involves the collective motion of many dislocations. Zooming in on a single dislocation will reveal that it is an atomic line defect. It is the lattice distortion around this defect that contains sufficient energy that slip of close-packed atomic planes is possible by motion of dislocations, as any elementary materials textbook, such as Callister (2000), describes. This chapter is concerned with the bottom two length scales in Fig. 1, namely dislocations as discrete entities. The motion of individual dislocations and the creation of new ones, is what leads to plastic deformation. Crystal plasticity is a continuum description where the motion of many dislocations is averaged out in terms of an effective slip —this will not be discussed here; an excellent review is given by Asaro (1983).

## 2 Dislocation Concepts

A dislocation is a line defect in an atomic lattice (Hull and Bacon (2001); Hirth and Lothe (1968)). It is a line on an atomic plane that separates those regions of the plane that are intact from regions where the lattice has undergone slip, Fig. 2. The relative shift of atomic planes is in the direction denoted by the Burgers vector $\boldsymbol{b}$ and is essentially uniform in the region enclosed by the dislocation. The dislocated material can be constructed in a thought experiment from a perfect crystal by a cut-displace by $\boldsymbol{b}$-and-reweld procedure. Because of conservation of mass, a dislocation is a closed loop.

At point A in Fig. 2 the dislocation line is parallel to $\boldsymbol{b}$ —this is called the *screw* part of the dislocation. On the other hand, at B, the dislocation line is perpendicular to $\boldsymbol{b}$ and this is called the *edge* part. In between A and B, the character is mixed edge and screw. Edge dislocations are typically denoted by a $\perp$ symbol, and screws by an S.

Dislocations are observed in electron microscopy by virtue of the distortion of the lattice.

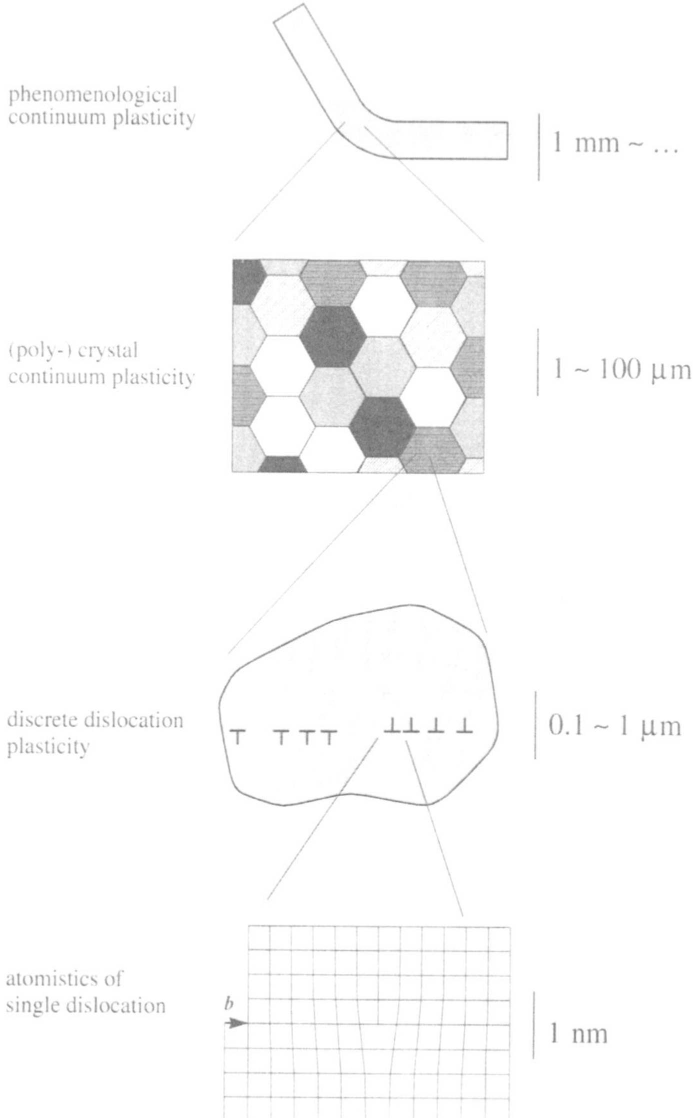


**Figure 1.** Various length scales in plasticity.

Plastic deformation is produced by the glide of dislocations in the slip plane (and/or by climb out of the plane), as illustrated in Fig. 3 [1]. A dislocation whose Burgers vector is a lattice translation vector is called a *perfect* or *unit dislocation*. The preferred glide plane in fcc crystals

[1] see also `http://www.enstimac.fr/recherche/mat/Formation/CRYSTAL/html/dislomo.html`

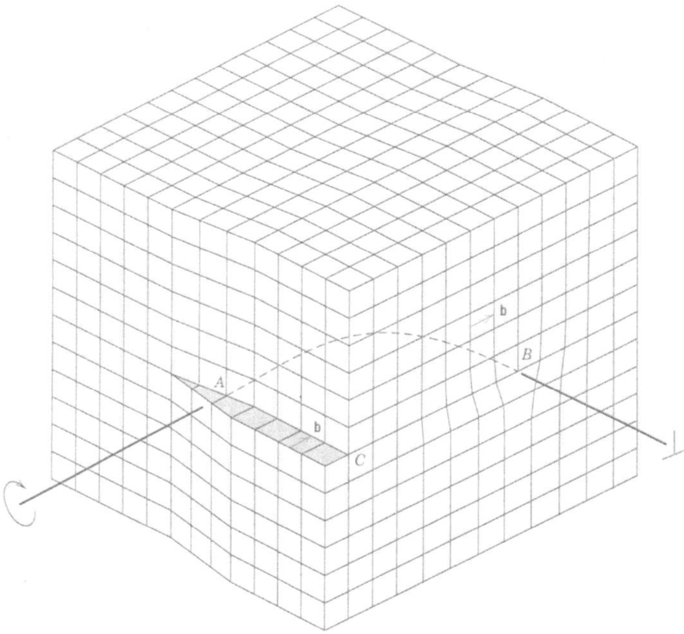


**Figure 2.** Schematic of a dislocation: inside the dislocation, the lattices on either side of the plane have shifted. From Callister (2000).

are the $\{111\}$ planes. The Burgers vector of perfect dislocations on such a plane are of the $\langle 110\rangle$ type and have $\boldsymbol{b} = \frac{1}{2}[110]$.

**Figure 3.** Illustration of the motion of an edge dislocation through the lattice (in analogy with that of a caterpillar), thus producing slip by an amount $b$. From Callister (2000).

## 3 Geometry of Dislocations and Slip

The geometry of a dislocation is governed by a number of variables:

- the slip plane, denoted with its unit normal vector $\boldsymbol{m}$;
- the dislocation line as a parameterized line on this plane and with a local tangent vector $\boldsymbol{t}$;
- the Burgers vector $\boldsymbol{b}$.

We shall here follow the convention of the right-handed rule for the circulation of the dislocation loop, i.e. the loop is traversed so that the corresponding vector according to the right-handed screw points in the direction of $\boldsymbol{m}$, see Fig. 4.

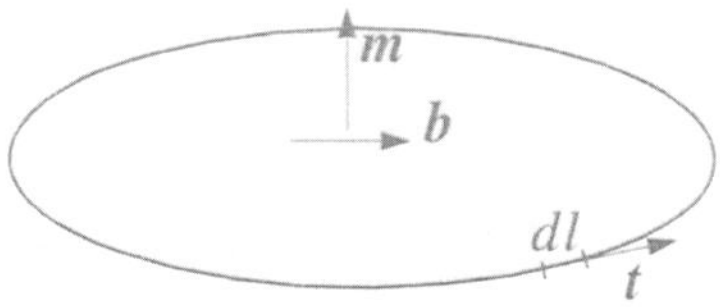


**Figure 4.** Definition of the geometry of a dislocation loop. The line direction corresponds to the slip plane unit normal vector $\boldsymbol{m}$ according to the right-handed screw rule.

There are a few special parts of a generic loop, namely

$$\text{edge:} \quad \boldsymbol{b}\cdot\boldsymbol{t} = 0; \tag{3.1}$$

$$\text{screw:} \quad \boldsymbol{b}\cdot\boldsymbol{t} = \pm b, \tag{3.2}$$

$b$ being the length of $\boldsymbol{b}$: $b = |\boldsymbol{b}|$. Edge and screw dislocations are the central notions in two-dimensional studies (see Sec. 4.2.

The Burgers vector[2] has been loosely defined in the above as the relative lattice displacement across the slip plane inside the dislocation loop. A more formal definition makes use of the concept of the *Burgers circuit*, as illustrated in Fig. 5. The key idea is that if one makes a closed contour in a perfect crystal, start and finish coincide; however, they do not if a dislocation is contained within the loop. Therefore, the procedure to define the Burgers vector in a dislocated crystal is to make a circuit that would be closed in a perfect crystal, the orientation of the circuit being right-handed relative to the line direction $\boldsymbol{t}$. Then, starting from S and tracing 1 thru 3, one finishes at F. The vector that closes the circuit, i.e. from F to S is defined to be the Burgers vector $\boldsymbol{b}$. Note that the direction of $\boldsymbol{b}$ depends on the line direction and on the chosen procedure. The one described above (and summarized in Fig. 5a) is referred to as the FS/*RH* convention Hirth and Lothe (1968): FS for closing the "Finish to Start" gap and *RH* for the "right-handed" sense relative to $\boldsymbol{t}$.

Also note that the precise value of $\boldsymbol{b}$ from this construction will depend somewhat on the shape and size of the circuit when it comes close to the core of the dislocation, since the atomic displacements there vary quite rapidly. Also, there is an effect of elastic strains and thermal vibrations. However, when the circuit is sufficiently large and lies in nearly perfect material, the closure vector will no longer depend on it.

There is an easily overseen subtlety in the drawing in Fig. 5b: it assumes that the line direction is known. The problem and the solution are shown in Fig. 6. First let us assume that we make a

[2] Introduced by the Dutch fluid mechanician J.M. Burgers from Delft University of Technology in a paper in the Proceedings of the Royal Academy of Arts and Sciences in 1939

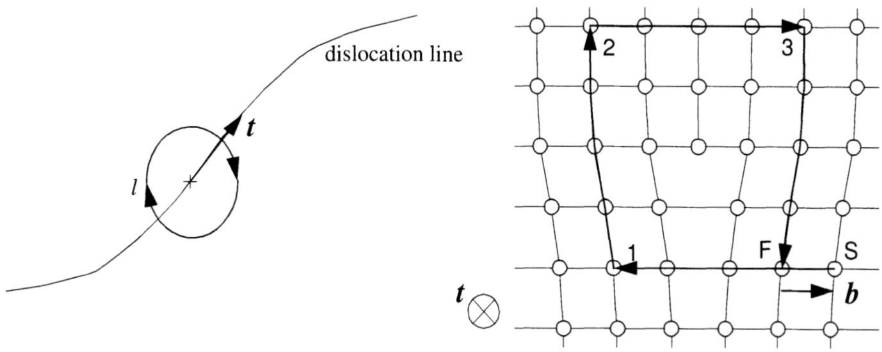


(a) (b)

**Figure 5.** Burgers circuit. (a) General definition of circuit around a dislocation line in the right-handed sense (the screw direction of the circuit is in the direction of the tangent ***t***). (b) Circuit around an edge dislocation (with ***t*** pointing into the paper): the closure vector from Finish to Start is the Burgers vector ***b***.

cross-section of the dislocation shown in Fig. 4 perpendicular to the glide plane and through the two opposite edge parts. Using the $\top$ and $\perp$ symbols for edges, we obtain the situation in Fig. 6a. The Burgers circuits around the two parts are shown as defined by the convention in Fig. 5a. From the atomic-scale picture of the dislocation in the right-hand side, Fig. 5b, we have seen above that the Burgers vector is ***b*** and pointing in the indicated direction; in a completely similar manner, we find the same result from the left-hand side dislocation. It is crucial to realize that this is correct, although the orientation of the $\top$ and $\perp$ symbols may perhaps suggest otherwise: the Burgers vector defines the relative shift of the atoms above the glide plane relative to those below it – both edges are parts of the same dislocation loop and therefore must have the same Burgers vector. The relative shift is always in between the two edges, so that the crystal with the cut is on different signs of the edges; but, the different orientations of the Burgers circuit ensure the same vector. However, if one only knows the two-dimensional atomic structure, the line direction is not known. Therefore, in two dimensions, we cannot decide what the circulation direction should be. Obviously there are two possibilities: one clockwise and the other counter-clockwise. Both of them are being used in the literature, which actually is quite confusing because the sign of the Burgers vector changes when changing circulation. The convention in two-dimensional views that we take here is that the Burgers circuit is always taken in the counter-clockwise direction, as illustrated in Fig. 6b. That is, we always think of the line directine as pointing out of the plane towards the reader. As a consequence, one finds the Burgers vector pointing in the right direction for the $\perp$ part, and the opposite result for the $\top$ edge. Though confusing this may seem at first, it can actually be used in a constructive manner by employing the notion of a *signed Burgers vector* in two dimensions, so that the $\perp$ edge has a positive Burgers vector and the $\top$ edge a negative one. Positive and negative now serve to remind one of the fact of the addition of an extra half-plane of atoms (from the positive side of the slip plane) or the removal, respectively.

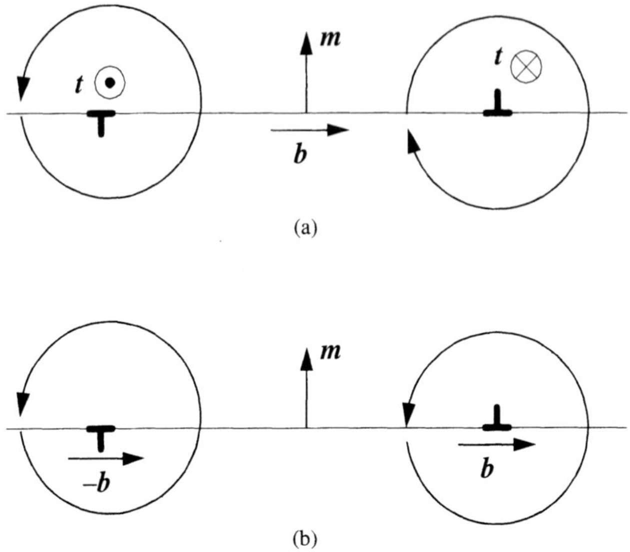


**Figure 6.** (a) Cross-section through the loop shown in Fig. 4 perpendicular to the two edge parts, with the local tangent vectors as indicated. (b) Convention for a purely two-dimensional view.

# 4 Elastic Models of Dislocations

## 4.1 General Idea

Although a dislocation is a lattice defect, it has proved very useful to describe it in the framework of continuum theory in which the atomic positions are averaged out. This reduces the total number of degrees of freedom enormously: from all atom positions to a mathematical, functional form of the geometric variables $\boldsymbol{m}$, $\boldsymbol{t}$ and $\boldsymbol{b}$. In a continuum framework the definition of the Burgers vector becomes

$$\boldsymbol{b} = \oint_C \frac{\partial \boldsymbol{u}}{\partial c} dc \tag{4.1}$$

where $C$ is a closed circuit around the dislocation, cf. Fig. 5a, traversed by local coordinate $c$ and $\boldsymbol{u}$ is the displacement field away from the perfect crystal. The real key to dislocations in a continuum description is that it involves a *displacement discontinuity* inside the dislocation loop. It is the expansion of dislocation loops that creates what we observe on a larger scale as *permanent*, that is, *plastic deformation*. Apart from the dislocation motion, the distortion of the lattice is entirely elastic. Thus, the picture of plasticity emerges of dislocation loops sweeping through an otherwise elastic continuum.

In such a continuum picture with elastic strains all around the dislocation and a discontinuity inside the loop, something special must happen right at the line for reasons of compatibility. Even from a schematic as in Fig. 2, it is not difficult to imagine that there are difficulties in using continuum ideas in the immediate neighborhood of the line. Therefore, one uses the continuum idea outside a certain radius $r_c$ around the line; inside this radius – the so-called *core region* – an atomistic description is really needed. The value of $r_c$ is a few times $b$.

Because of the core, dislocations are a source of internal stress and strain in the body. These stress and strain fields are internal since they exist even in the absence of an external load. The originate from the core itself where strains are needed to restore compatibility but they need not – and in fact, are not – restricted to this region, as can already be discerned from Figs. 2 and 5.

The current continuum theory of discrete dislocations employs linear elasticity. Clearly, it will break down inside the core region, where the strains will be too large for the linear approximation to hold. Away from the core by about 5 to $6b$, comparison with atomistics has shown that the linear elastic solution is very accurate. Hence, the view that will be developed from now is to describe the fields in a dislocated body by linear elasticity, excluding the core regions. The use of linear elasticity has the enormous advantage that many solutions for the governing equations are known and that one can use superposition. The latter will be exploited in full power in Sec. 5.

For completeness, we recall that the governing equations for linear elasticity are

$$\text{equilibrium} \quad : \sigma_{ij,j} = 0 \tag{4.2}$$

$$\text{elasticity} \quad : \sigma_{ij} = \mathcal{L}_{ijkl}\varepsilon_{kl} \tag{4.3}$$

$$\text{strains} \quad : \varepsilon_{ij} = \frac{1}{2}(u_{i,j} + u_{j,i}) \tag{4.4}$$

with $\mathcal{L}_{ijkl}$ the elastic moduli. Since the crystals we shall consider here are cubic, we should be using the cubic elastic moduli expressed in terms of the usual $C_{11}$, $C_{12}$ and $C_{44}$. However, for simplicity, we will assume isotropic elasticity, with the moduli expressed either in terms of Young's modulus $E$ and Poisson's ratio $\nu$,

$$\mathcal{L}_{ijkl} = \frac{E}{1+\nu}\left[\frac{1}{2}\left(\delta_{ik}\delta_{jl} + \delta_{il}\delta_{jk}\right) + \frac{\nu}{1-2\nu}\delta_{ij}\delta_{kl}\right] \tag{4.5}$$

or as

$$\mathcal{L}_{ijkl} = \mu\left(\delta_{ik}\delta_{jl} + \delta_{il}\delta_{jk}\right) + (k - \frac{2}{3}\mu)\delta_{ij}\delta_{kl}\,. \tag{4.6}$$

in terms of the shear modulus $\mu = E/2(1+\nu)$ and the bulk (or compression) modulus $k = E/3(1-2\nu)$.

### 4.2 Two-dimensional Dislocations

There are two useful, idealized dislocations: pure screw dislocations and pure edge dislocations. Both are straight lines with the Burgers vectors satisfying (3.1). They admit rather simple elasticity solution, and are very useful as idealizations for real curved dislocations.

For the analysis of a *screw dislocation*, we choose base vectors so that $\boldsymbol{e}_3$ is parallel to the line direction, i.e. $\boldsymbol{e}_3 = \boldsymbol{t}$, so that all fields are independent of the $x_3$ coordinate, see Fig. 7a. Thus, it becomes a two-dimensional, so-called anti-plane shear problem:

$$u_1 = u_2 = 0\,, \quad u_3 = u_3(x_1, x_2)\,.$$

Combination of the governing equations (4.2)–(4.4) in terms of $u_3$ then reduces to the Laplace equation (see Ex. 8.1)

$$\nabla^2 u_3 = 0\,. \tag{4.7}$$

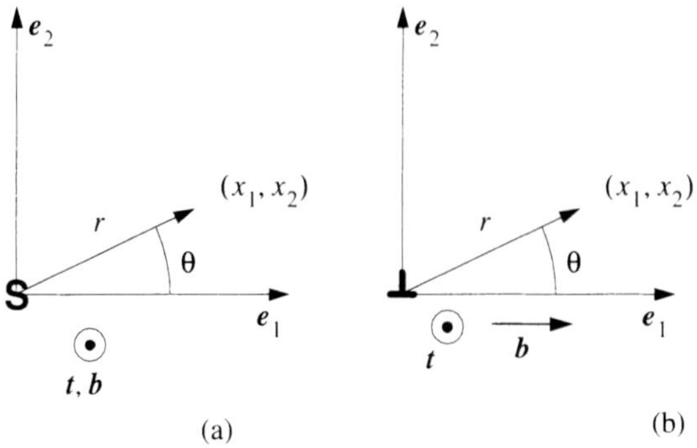


**Figure 7.** Definitions of (a) screw and (b) edge dislocation configurations discussed in the text.

In addition to satisfying this equation, the displacement field must satisfy the condition (4.1) that there be a displacement jump across the slip plane of $\pm b$. To make things specific, let us assume that the slip plane coincides with the $x_1$–$x_3$ plane, i.e. $\boldsymbol{m} = \boldsymbol{e}_2$, and that $\boldsymbol{b} \cdot \boldsymbol{t} = b$. Then,

$$u_3(x_1, 0^-) - u_3(x_1, 0^+) = b \tag{4.8}$$

This reminds one of a helical ramp (as in a parking garage), with the displacement increasing linearly with the winding angle $\theta = \arctan(x_2/x_1)$. Therefore we try

$$u_3 = \frac{b}{2\pi} \arctan\left(\frac{x_2}{x_1}\right) . \tag{4.9}$$

Substitution confirms that this function satisfies the Laplace equation (4.7), while, because of the multi-valuedness of the arctan function, it also satisfies (4.8). The solution is plotted in Fig. 8. The non-vanishing components of the stress tensor are

$$\sigma_{13} = -\frac{\mu b}{2\pi} \frac{x_2}{x_1{}^2 + x_2{}^2}, \quad \sigma_{23} = \frac{\mu b}{2\pi} \frac{x_1}{x_1{}^2 + x_2{}^2}$$

and are clearly singular at the dislocation origin. When expressed in terms of polar coordinates $(r, \theta, z)$, the only non-vanishing stress components is

$$\sigma_{\theta z} = \frac{\mu b}{2\pi r}$$

which reveals an important property of dislocations: the internal stress field caused by the dislocation decays with distance $r$ as $1/r$. Hence, dislocations have a *long-range effect*. The associated elastic energy (per unit of dislocation line length) in a cylinder of radius $R$ around the dislocation is given by (see Ex. 8.2)

$$W = \frac{\mu b^2}{4\pi} \ln\left(\frac{R}{r_c}\right) \tag{4.10}$$

where $r_c$ is the core radius. Note that the energy would diverge as $r_c \to 0$. Comparison with atomistic estimates reveals that $r_c \approx b/4$ gives a fairly good approximation of the energy of screw dislocations.

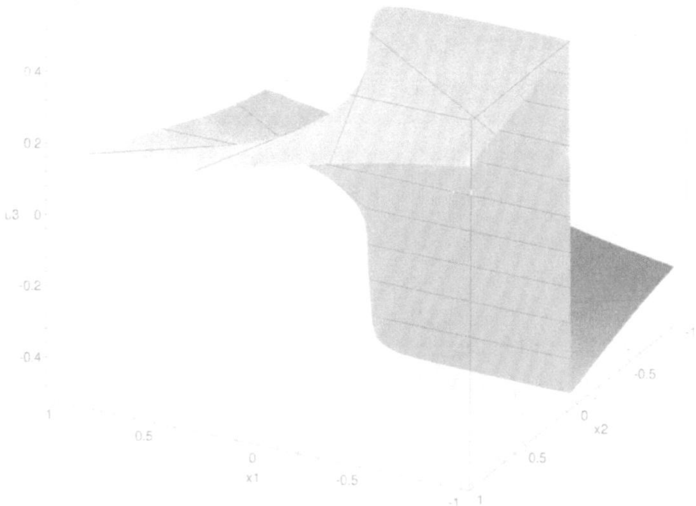


**Figure 8.** Displacement field $u_3(x_1, x_2)$, normalized by $b$, of a screw dislocation with $\boldsymbol{b} = b\boldsymbol{e}_3$ at $(x_1, x_2) = (0, 0)$.

An *edge dislocation* poses a plane strain problem. If the dislocation line is arranged in the same way as the screw above, i.e. $\boldsymbol{t} = \boldsymbol{e}_3$, the Burgers vector lies in the $x_1$–$x_2$ plane. Specifically, we shall assume again that the $x_1$–$x_3$ plane is the slip plane and that $\boldsymbol{b}$ points in the positive $x_1$ direction, see Fig. 7b. This problem is then conveniently solved by application of the Airy stress function approach in a manner that is similar to that leading to the asymptotic singular field near a sharp crack tip. Leaving the details to Hirth and Lothe (1968), the solution reads

$$\sigma_{11} = -\frac{\mu b}{2\pi(1-\nu)} \frac{x_2(3x_1^2 + x_2^2)}{(x_1^2 + x_2^2)^2} \tag{4.11}$$

$$\sigma_{22} = \frac{\mu b}{2\pi(1-\nu)} \frac{x_2(x_1^2 - x_2^2)}{(x_1^2 + x_2^2)^2} \tag{4.12}$$

$$\sigma_{12} = \frac{\mu b}{2\pi(1-\nu)} \frac{x_1(x_1^2 - x_2^2)}{(x_1^2 + x_2^2)^2} \tag{4.13}$$

for the in-plane stress components and $\sigma_{33} = \nu(\sigma_{11} + \sigma_{22})$. As for the screw dislocation, the stress field is singular at the dislocation and decays with $r$ as $1/r$ (see Ex. 8.3). The displacement field is given by

$$u_1 = \frac{b}{2\pi(1-\nu)}\left[\frac{1}{2}\frac{x_1 x_2}{x_1{}^2+x_2{}^2} - (1-\nu)\arctan\left(\frac{x_1}{x_2}\right)\right] \tag{4.14}$$

$$u_2 = \frac{b}{2\pi(1-\nu)}\left[\frac{1}{2}\frac{x_2{}^2}{x_1{}^2+x_2{}^2} - \frac{1}{4}(1-2\nu)\ln\frac{x_1{}^2+x_2{}^2}{b^2}\right] \tag{4.15}$$

The $u_1$ field, i.e. parallel to the slip plane, is shown in Fig. 9.

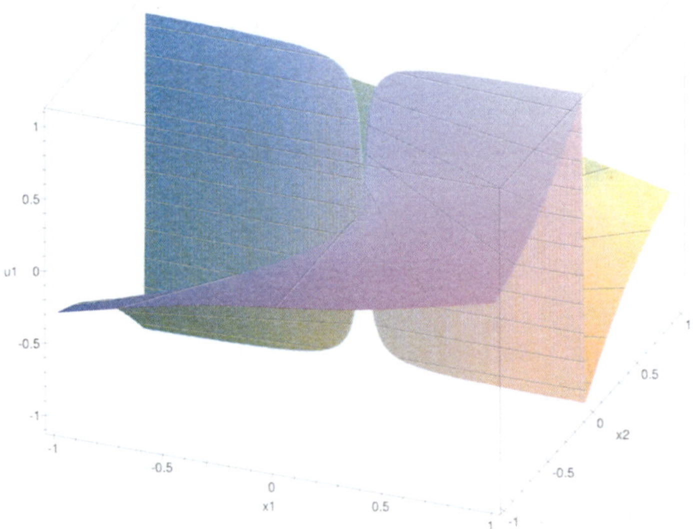


**Figure 9.** Displacement field $u_1(x_1, x_2)$, normalized by $b/2\pi(1-\nu)$, of an edge dislocation with $\boldsymbol{b} = b\boldsymbol{e}_1$ at $(x_1, x_2) = (0, 0)$.

Note that the solutions of the screw and edge dislocations assume the body to be infinitely large – no boundary conditions have been incorporated. Since both of the solutions are singular, it is to be expected that a correction due to boundary conditions has no significant effect close to the dislocation. This will be discussed further in Sec. 5.

### 4.3 Dislocation structures

Plastic deformation generally involves many dislocations. Since they all have long-range interactions, they are able to form structures. One way to classify dislocation structures is the *net Burgers vector* over a certain area. This is most clearly seen for straight dislocations, as illustrated in Fig. 10 for edge dislocations. In one extreme case, illustrated in Fig. 10a, there are just as many positive dislocations (i.e. dislocations with their Burgers vector in a chosen direction) as negative ones; the net-Burgers vector —in this case the sum of all Burgers vectors— then adds up to zero. In the other case, Fig. 10b, there is a surplus of negative dislocations ($\top$) so that there is a non-zero net-Burgers vector.

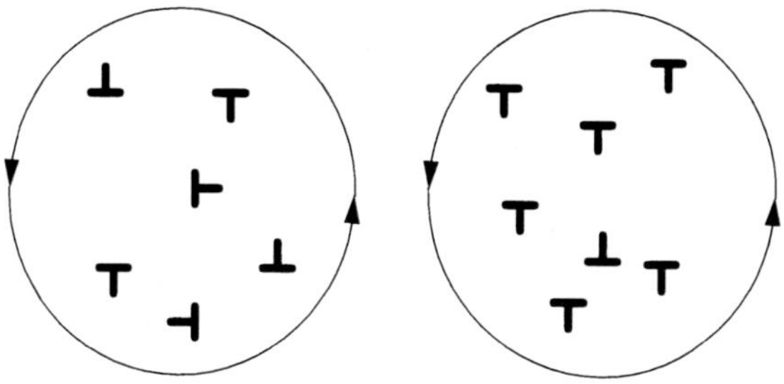

(a) (b)

**Figure 10.** Two distinctly different dislocation distributions: (a) one where the net Burgers vector in the drawn circuit vanishes; (b) a distribution with a non-vanishing net-Burgers vector.

A physically relevant distribution of edge dislocations of the same sign is the *wall* of dislocations shown in Fig. 11. When this wall is infinitely long, with the same spacing $h$ between

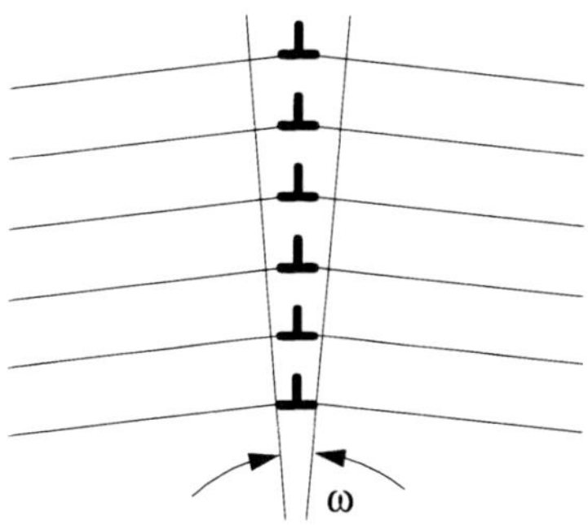


**Figure 11.** A wall of dislocations produces a relative rotation of the slip planes on either side by an angle $\omega$.

dislocations, the collective effect is a rotation of the slip planes on either side of the wall. Since each dislocation gives a relative displacement of $b$ above versus below the dislocation, this results in a rotation of $b/h$. For an infinite wall, this rotation is the same everywhere and therefore equal to the misorientation $\omega = b/h$. Because of this, a dislocation wall serves as a representation of a *low-angle (tilt) grain boundary*. Walls of dislocations have anther interesting property, which can be illustrated by considering the total shear stress field. This can be obtained by summing the contributions $\sigma_{12}$ according to (4.13) of all dislocations:

$$\sigma_{12} = \frac{\mu b}{2\pi(1-\nu)} \sum_{n=-\infty}^{\infty} \frac{x_1({x_1}^2 - (x_2 - nh)^2)}{({x_1}^2 + (x_2 - nh)^2)^2}$$

Making use of tools from analysis, the sum can be performed analytically, resulting in (see Hirth and Lothe (1968))

$$\sigma_{12} = \frac{\mu b}{2\pi(1-\nu)} \frac{2\pi^2\xi(\cosh 2\pi\xi \cos 2\pi\eta - 1)}{(\cosh 2\pi\xi - \cos 2\pi\eta)^2}$$

where $\xi$ and $\eta$ are reduced variables: $\xi = x_1/h$, $\eta = x_2/h$. The important thing to notice here is that the shear stress drops with the distance $\xi$ from the wall as $1/\cosh(\text{distance})$ rather than as $1/\text{distance}$ as it does for a single dislocation. The reason is that the positive and negative regions of individual dislocation fields partly cancel in a wall.

A more or less random distribution of edge dislocations of the same Burgers vector in a rectangular crystal, as shown in Fig. 12 gives rise to bending. When there are $N$ parallel

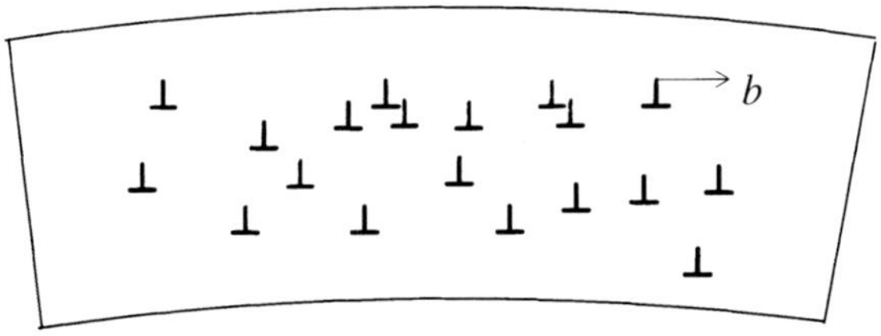


**Figure 12.** A distribution of positive edge dislocations gives rise to bending.

dislocations with Burgers vector length $b$, the net-Burgers vector has a length $Nb$. Similar to the argument above for a wall, this *super dislocation* produces a misorientation of one edge of the crystal relative to the other of $\theta = Nb/h$, when $h$ is the height of the crystal. As a consequence, the crystal bends, with a curvature equal to

$$\varkappa = \frac{\theta}{L} = \frac{Nb}{hL} = b\rho_G$$

where $\rho_G = N/(hL)$ is the density of dislocations. Conversely, if one wants to plastically bend a crystal to a curvature $\varkappa$ (i.e. radius of curvature $1/\varkappa$), one needs to have a density of dislocations with the same Burgers vector $b$ of

$$\rho_G = \frac{\varkappa}{b} \tag{4.16}$$

In addition to these there may be other dislocations in the crystal but there has to be a minimum of this density for the given curvature to be accommodated by dislocations. These were called *geometrically necessary dislocations* by Nye (1953) and Ashby (1970).

## 5 Boundary Value Problems

Even for straight two-dimensional dislocations, as discussed in the previous section, the solutions do not incorporate any boundaries of the crystal. The solutions are, strictly speaking, for dislocations in infinite space. It is not hard to imagine that in three dimensions, the situation is certainly not better. Interactions with the boundaries of course do exist, and their are commonly Hirth and Lothe (1968) referred to as *image effects*. Rather clever image constructions have been developed but they remain limited to particular configurations.

Several years ago, Van der Giessen and Needleman (1995) proposed a versatile approach based on superposition. The idea is to make use of the known solutions in infinite space and to superpose an 'image' solution to correct for the boundary conditions. To this end, the displace-

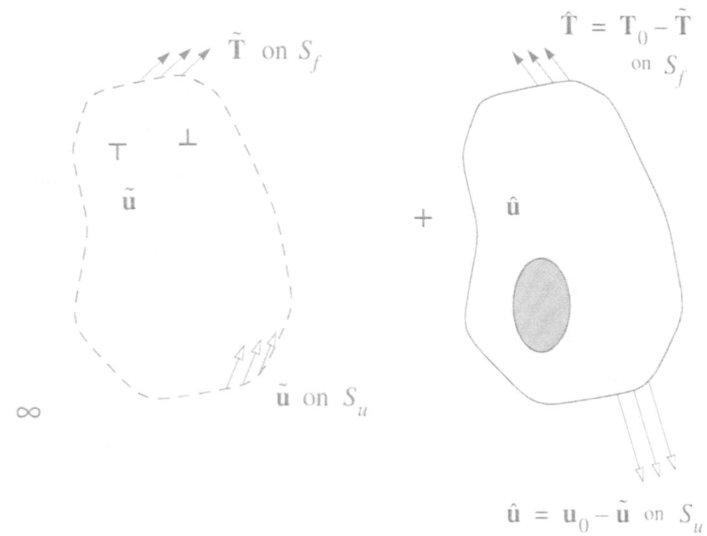


**Figure 13.** Decomposition of the problem for the dislocated body into the problem of interacting dislocations in an infinite solid (˜ fields) and the complementary problem for the body without dislocations (ˆ fields).

ment, strain and stress fields are decomposed as

$$\boldsymbol{u} = \tilde{\boldsymbol{u}} + \hat{\boldsymbol{u}}, \quad \boldsymbol{\varepsilon} = \tilde{\boldsymbol{\varepsilon}} + \hat{\boldsymbol{\varepsilon}}, \quad \boldsymbol{\sigma} = \tilde{\boldsymbol{\sigma}} + \hat{\boldsymbol{\sigma}}. \tag{5.1}$$

The (˜) fields are the superposition of the singular fields of the individual dislocations, as discussed above, in infinite space. Identifying the fields for dislocation $I$ by a superscript $(I)$, the (˜) stress field, for example, is obtained as

$$\tilde{\boldsymbol{\sigma}} = \sum_I \boldsymbol{\sigma}^{(I)}$$

The (˜) fields will in general not meet the boundary conditions in terms of tractions on part $S_f$ of the boundary nor the prescribed displacements on part $S_u$. Instead they will give rise to displacements $\tilde{\boldsymbol{u}}$ on $S_u$ and tractions $\tilde{\boldsymbol{T}} = \tilde{\boldsymbol{\sigma}} \cdot \boldsymbol{n}$ on $S_f$ (with normal $\boldsymbol{n}$). The actual boundary conditions, $\boldsymbol{u}^0$ on $S_u$ and $\boldsymbol{T}^0$ on $S_f$, are imposed through the (ˆ) fields, in such a way that the sum of the (˜) and the (ˆ) fields in (5.1) gives the solution that satisfies all boundary conditions. Since the (˜) fields satisfy the governing equations (4.2)–(4.4), the (ˆ) fields also have to satisfy the elasticity equations, i.e.

$$\text{equilibrium:} \quad \operatorname{div}\hat{\boldsymbol{\sigma}} = 0$$
$$\text{elasticity:} \quad \hat{\boldsymbol{\sigma}} = \mathcal{L} : \hat{\boldsymbol{\varepsilon}}$$
$$\text{strains:} \quad \hat{\boldsymbol{\varepsilon}} = \frac{1}{2}[\operatorname{grad}\hat{\boldsymbol{u}} + (\operatorname{grad}\hat{\boldsymbol{u}})^T]$$

supplemented with the boundary conditions

$$\hat{\boldsymbol{T}} = \boldsymbol{T}^0 - \tilde{\boldsymbol{T}} \text{ on } S_f \qquad \hat{\boldsymbol{u}} = \boldsymbol{u}^0 - \tilde{\boldsymbol{u}} \text{ on } S_u$$

It is important to note that the solution of the $(\hat{\ })$ problem does not involve any dislocations. Therefore, the $(\hat{\ })$ fields (often called 'image' fields) are smooth and the boundary value problem for them can conveniently be solved using a finite element method.

A crucial concept for the evolution of dislocations, to be discussed in the next section, is the *Peach–Koehler* force. It is defined as the configurational force associated to motion of the dislocations: the work of these forces as the dislocations move is the change in *potential energy* $\Pi$. Delaying the formulation of the potential energy of a dislocated body for a moment, let us first introduce the quantities to describe the motion of each dislocation. As in Fig. 4, let $l$ be the coordinate around the loop $\mathcal{L}^{(I)}$. Infinitesimal changes in dislocation geometry are then described by variations $\delta \boldsymbol{s}^{(I)}(l)$ of the dislocation position. The Peach–Koehle force is now formally defined as the configurational force following from

$$\delta\Pi = -\sum_I \oint_{\mathcal{L}^{(I)}} \boldsymbol{f}^{(I)} \cdot \delta\boldsymbol{s}^{(I)} dl$$

It is seen that $\boldsymbol{f}^{(I)}$ has the dimension of force per unit (dislocation line) length and that it generally changes along the loop.

The Peach–Koehle force can be more explicit, in this superposition approach, by calculating the potential energy: the elastic energy in the body $V$ minus the work of the tractions on the boundary $\partial V = S_f \cap S_u$. Since we have seen that the energy of the linear elastic field of a dislocations is singular at the core, we will exclude the core regions as illustrated for an arbitrary dislocation loop $I$ in Fig. 14. First we introduce a small core region $C^{(I)}$ around each dislocation loop $i$, which is formed by a torus of radius $r_0$ along the dislocation line, cut open internally by the slip plane and bounded by two surfaces $S_+^{(I)}$ and $S_-^{(I)}$ (see Fig. 14). The total surface of the core region, including $S_+^{(I)}$ and $S_-^{(I)}$, is denoted by $\partial C^{(I)}$. Then, the volume of the body excluding the core region $C^{(I)}$ is $\hat{V}^{(I)} = V \setminus C^{(I)}$. The body volume excluding all core regions $\tilde{C} = \cap_{(I)} C^{(I)}$ is $\hat{V} = V \setminus \tilde{C}$. The potential energy $\Pi$ for the dislocated body is then found as the potential energy of the volume $\hat{V}$ excluding all core volumes,

$$\Pi = \frac{1}{2}\int_{\hat{V}} \boldsymbol{\sigma} : \boldsymbol{\varepsilon} dV - \frac{1}{2}\int_{\partial\tilde{C}} (-\boldsymbol{T}) \cdot \boldsymbol{u} dS - \int_{Sf} \boldsymbol{T}^0 \cdot \boldsymbol{u} dS,$$

with the second term in the right-hand size accounting for the core energies through the work of tractions $-\boldsymbol{T}$ on the interal surfaces $\partial C^{(I)}$ of $\hat{V}$, see Fig. 14)b. All integrands in the above expression are functions of the dislocation positions $\boldsymbol{s}^{(I)}$, so that one can perform a variation of

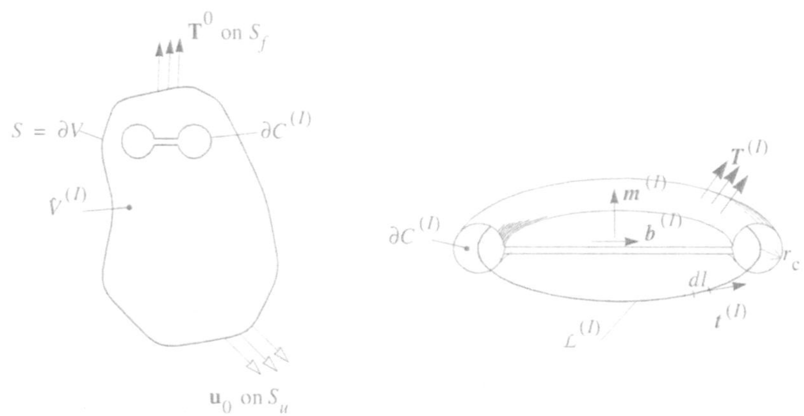


(a) (b)

**Figure 14.** (a) Body $V$ with the dislocations removed. (b) Open view of the small core region $C^i$ around dislocation loop $I$ with Burgers vector $\boldsymbol{b}^{(I)}$. The slip plane normal is $\boldsymbol{m}^{(I)}$ and $\boldsymbol{T}$ is the surface traction.

$\Pi$ with respect to $\boldsymbol{s}^{(I)}$. Leaving the details to Van der Giessen and Needleman (1995), the final result is that the Peach–Koehler force can be written in the present approach as

$$\boldsymbol{f}^{(I)} = \boldsymbol{t}^{(I)} \times \left[ \left( \hat{\boldsymbol{\sigma}} + \sum_{J \neq I} \boldsymbol{\sigma}^{(J)} \right) \cdot \boldsymbol{b}^{(I)} \right] .$$

It is important to note that the Peach–Koehler force on dislocation $I$ does not depend on its own stress field but only on the fields generated by the others. The fact that $\boldsymbol{\sigma}^{(I)}$ is singular at $I$ therefore does not pose any problem at all. The deeper reason is that the Peach–Koehler force is a *configurational force*: as dislocation $I$ moves, it only sees the changes from its point of view. Of particular interest is the component $f^{(I)}$ of $\boldsymbol{f}^{(I)}$ in the glide plane and in the direction $\boldsymbol{t}^{(I)} \times \boldsymbol{m}^{(I)}$ normal to the dislocation:

$$f^{(I)} = \boldsymbol{m}^{(I)} \cdot \left( \hat{\boldsymbol{\sigma}} + \sum_{J \neq I} \boldsymbol{\sigma}^{(J)} \right) \cdot \boldsymbol{b}^{(I)} . \tag{5.2}$$

## 6 Dislocation Dynamics

So far, we have discussed the state of the material in the presence of dislocations, but they have to move in order to produce plastic deformation. In this section, we give a brief summary of various physical phenomena that govern the motion of dislocations. For brevity, we will focus on those aspects that are connected to *dislocation glide*. Climb, i.e. motion perpendicular to the plane, also occurs under certain circumstances but will not be treated, nor will cross slip (the slip out of the original glide plane).

The driving force for glide is the Peach–Koehler force component given by (5.2). During glide, however, the dislocation may be subjected to various sources of resistance against motion. Denoting their collective force by $f^{(I)}_{resist}$, one can write the (Newtonian) equation of motion for a dislocation as

$$m^{(I)}\dot{v}^{(I)} = f^{(I)} - f^{(I)}_{resist}$$

where $v^{(I)}$ is the velocity of the dislocation and $m^{(I)}$ its effective mass. For most applications, the velocity changes of dislocations occur on such short time scales that they are not relevant for most areas of plastic deformation. At the same time, we have treated the fields of dislocations when in a quasi-static framework, so that dynamic effects related with stress waves have been ignored. It is therefore useful and consistent to neglect the inertia of dislocations as well, and let their motion be governed by the force balance

$$f^{(I)} = f^{(I)}_{resist} \,. \tag{6.1}$$

## 6.1 Peierls–Nabarro stress

As a dislocation moves (see Fig. 3 for an edge dislocation) existing atomic bonds have to be broken and new ones formed. Because of the periodic nature of the lattice, the energy landscape that the dislocation moves across is periodic. This acts as a *friction stress* against the initiation of dislocation motion, corresponding to a resistance force of $b\tau_f$ in the direction opposing the dislocation velocity $v^{(I)}$. This stress is referred to as the *Peierls–Nabarro* stress in honour of the two scientists who studied this first in the 1940's. The value of $\tau_f$ depends sensitively on the crystal structure. At 0 K, the values for fcc and bcc crystals typically are $10^{-6}\mu$ and $10^{-3}\mu$, respectively, while for intermetallics its value can be as large as $0.1\mu$.

## 6.2 Drag

As a dislocation glides, it experiences *drag* (just as you do while you are biking or swimming). Dislocation drag originates from: (i) phonon drag; (ii) electron drag; (iii) impurity effects. Of these, phonon drag is dominating in many materials. As a dislocation moves, it generates phonons (elastic waves in the lattice); during motion it feels more phonons in the direction of motion than in other directions (phonon wind) which results in a net force $Bv^{(I)}$. Other sources of phonons include thermal phonons exciting vibrations in the dislocation line, and those due to elastic waves when the dislocation accelerates or decelerates. It is possible to make estimates of each of the contributions, but in practice $B$ is measured experimentally. Values of $B$ show quite some scatter; a typical value for aluminum is on the order of $B = 10^{-4}$ Pa s. When drag is the only source of resistance, the balance (6.1) states

$$f^{(I)} = Bv^{(I)}$$

so that the velocity, for a given Peach-Koehler force, can be calculated from

$$v^{(I)} = f^{(I)}/B \tag{6.2}$$

Such a linear relationship between velocity and driving force has been confirmed experimentally. Data for various materials over various temperatures and stress levels suggest a relation of the type

$$v^{(I)} \propto (f^{(I)})^m e^{-E/kT}$$

but for relatively low stress levels, the exponent $m$ is close to unity.

### 6.3 Junctions

When dislocations move, they can meet other dislocations. Depending on the slip plane orientation and Burgers vectors, segments of the two dislocations can be attracted to one another or be repelled (see Exer. 8.4 for edge dislocations). When they repel each other, they frustrate or even block the motion of the other dislocation; this is observed as hardening of the material. Attraction can lead to the formation of *junctions* which can, again depending on orientation etc., serve as pinning points for parts of the dislocations involved.

### 6.4 Annihilation

Attracting of dislocations can also lead to *annihilation*. This is easily seen for straight edge or screw dislocations of opposite sign coming close together: the defect is eliminated by coalescence of the dislocations once they are sufficiently close together [3].

### 6.5 Frank–Read sources

So far, we have just assumed that dislocations were present, but have not addressed the question were they come from. One source is the generation of new dislocations through the so-called Frank–Read mechanism illustrated in Fig. 15a. The source of this mechanism is a dislocation segment that is being pinned between two hard points, commonly junctions formed between dislocations on other slip systems. This segment bows out under the influence of a Peach-Koehler force, in the figure originating from a shear stress $\tau$. As it continues to bow out, the self-interaction can become so strong that the dislocation configuration becomes unstable and curved segments in the wake (i.e. opposite to the shear stress) are mutually attracted. When these segments approach to within a sufficiently small distance, these parts start to annihilate. This is the beginning of a closed loop surrounding the initial pinned segment, which continues under influence of the self-stress. Ultimately, a smooth new dislocation loop has emerged as well as a copy of the initial segment. This copy can operate as a source again, etc., so as to generate more dislocations.

Figure 15 shows a two-dimensional version of this mechanism, which one can imagine as a cross-section of the situation in Fig. 15a and subsequent projection onto the plane of observation normal to the slip plane. The pinned segment now appears as a point, and the Frank-Read mechanism has been translated by Van der Giessen and Needleman (1995) into two dimensions to generate edge dislocations as follows. When the shear stress on the source is sufficiently high for a sufficiently long time, an edge dipole is generated. The strength of the source is, in principle, determined by the three-dimensional dislocation configuration (initial segment length, Burgers vector, etc.); in the two-dimensional model it becomes a parameter. The same holds for the nucleation time. The polarity of the dipole is determined by the direction of the shear stress,

[3] See `http://www.enstimac.fr/recherche/mat/Formation/CRYSTAL/html/disloani.html` for an animation

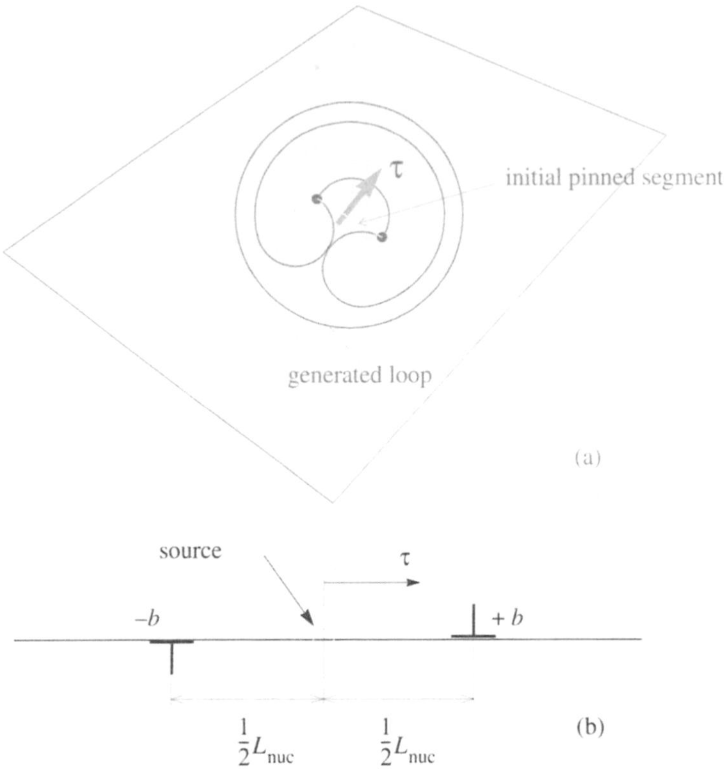


**Figure 15.** (a) Frank–Read mechanism of generating a new loop from an initial pinned segment. (b) Two-dimensional simplification (from Van der Giessen and Needleman (1995)) as a mechanism for nucleation of edge dislocations.

or more correctly, the Peach-Koehler force; the sign changes (i.e. the right-hand side dislocation becomes negative etc.) compared to Fig. 15b when the force changes direction. The width of the dipole, $L_{nuc}$, is the two-dimensional cross-section of the loop at the moment that the initial pinned segment has been re-formed and the source is ready to operate again. The value of $L_{nuc}$, again, can be determined from three-dimensional simulations. However, a different and practical criterion has been proposed by Van der Giessen and Needleman (1995) based on the observation that the dislocations in a dipole feel strong attractive forces that are inversely proportional to their distance. If their distance is $L_{nuc}$, the left-hand side dislocation in Fig. 15b feels a force to the right of

$$F = \frac{\mu b}{2\pi(1-\nu)} \frac{b}{L_{nuc}} \tag{6.3}$$

(see Exer. 8.7). If the shear stress at that moment is $\tau$, the dislocation feels a force of $\tau b$ in the opposite direction. When the latter is smaller than the attractive force, the dislocations will move towards each other and annihilate so that effectively no dislocation has been generated. Hence, there is a minimum distance beyond such collapse under an applied stress $\tau$ will not take place. Assuming that this applied stress is equal to the source strength $\tau_{nuc}$, the critical dipole width is given by Van der Giessen and Needleman (1995) as

$$L_{nuc} = \frac{\mu b}{2\pi(1-\nu)\tau_{nuc}}.$$

# 7 Applications

For space reasons, I will only give a short summary of simulations that have been carried out in two dimensions using the framework described previously. The reader is referred to the cited references for further details. It should be noted that several of the discrete dislocation results have been used subsequently as "reference" results for strain-gradient continuum theories; an overview of these can be found in Van der Giessen and Needleman (2003).

## 7.1 Shear of a composite

One of the first applications was the analysis of a model composite under shear parallel to its single slip system, Fig. 16. Both having a volume fraction of 20% of reinforcing particles,

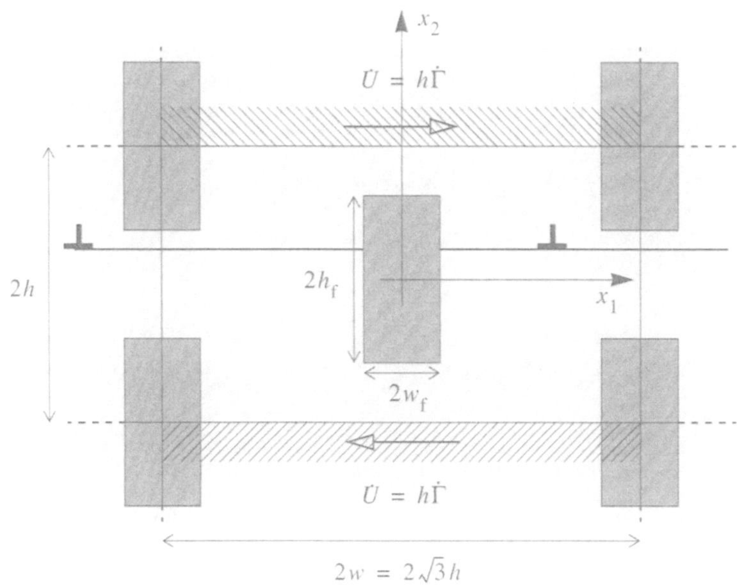


**Figure 16.** Simple shear of a model composite material with elastic reinforcements in a hexagonal pattern. All slip planes are parallel to the shearing direction. Only the $2h \times 2w$ unit cell is analyzed.

two morphologies have been considered by Cleveringa *et al.* (1997): one, called (i), with square particles ($h_f = w_f$) and another, called (iii), with rectangular particles ($h_f = 2w_f$). Morphology (i) leaves channels of unreinforced material in between the particles, and dislocation motion localizes in these channels. By contrast, dislocations in (iii) pile-up against the vertical interfaces between matrix and particles, thus giving rise to the geometrically necessary dislocations in the sense of Ashby (1970). As a consequence, morphology (iii) exhibits size dependence —with smaller particles giving harder material— while (i) does not.

### 7.2 Bending

Figure 12 obviously is a cartoon. Cleveringa *et al.* (1999) carried out a discrete dislocation analysis of bending of single crystals with three slip systems. A typical result for a crystal with slip planes at $\pm 60^\circ$ and $0^\circ$ from the longitudinal axis is shown in Fig. 17. A few noteworhty

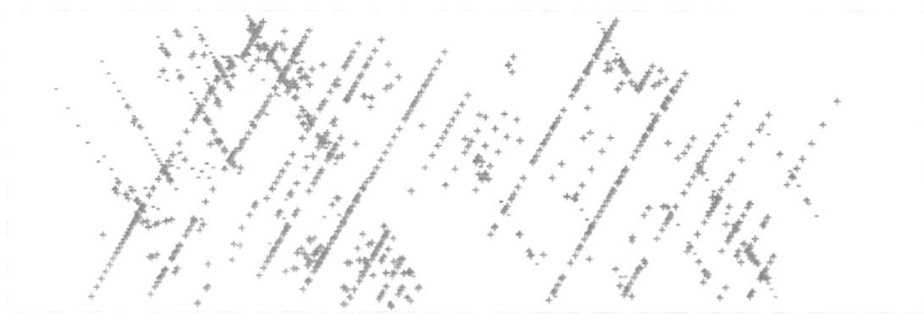

(a)

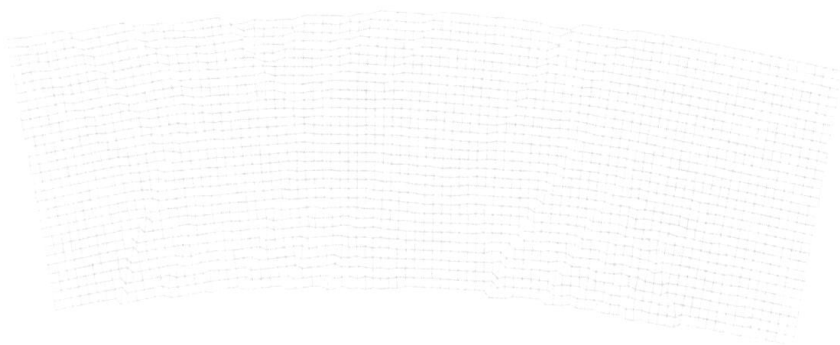

(b)

**Figure 17.** Dislocation distribution in a single crystal specimen under pure bending. From Cleveringa *et al.* (1999).

observations from the simulations are: (i) the inclined slip planes have a higher resolved shear stress than the horizontal one, so that the vast majority of dislocations have their Burgers vector in the $\pm 60^\circ$; collectively, however, their net Burgers vector is in the direction shown in Fig. 12; (ii) the density of dislocations is often somewhat higher than the density of geometrically necessary dislcoations according to (4.16)—these are the statistically stored dislocations; (iii) smaller crystals are stronger in bending than larger ones (while their tensile response is identical).

### 7.3 Shear of a constrained layer

Shu *et al.* (2001) have analyzed a seemingly very simple problem, but which was found to contain a number of subtle interesting features. The problem is to shear an infinitely long, single crystalline strip of height $H$ between rigid walls. The walls act as impenetrable barriers to dislocation motion. As a consequence of this, pile-ups evolve at the walls and plastic deformation in the strip becomes highly inhomogeneous, eventhough the initially elastic state is fully homogeneous. The inhomogeneity of plastic deformation is illustrated by the profile of local slip $\gamma$ across the strip height, see Fig. 18. A boundary layer appears to evolve inside of which little plastic

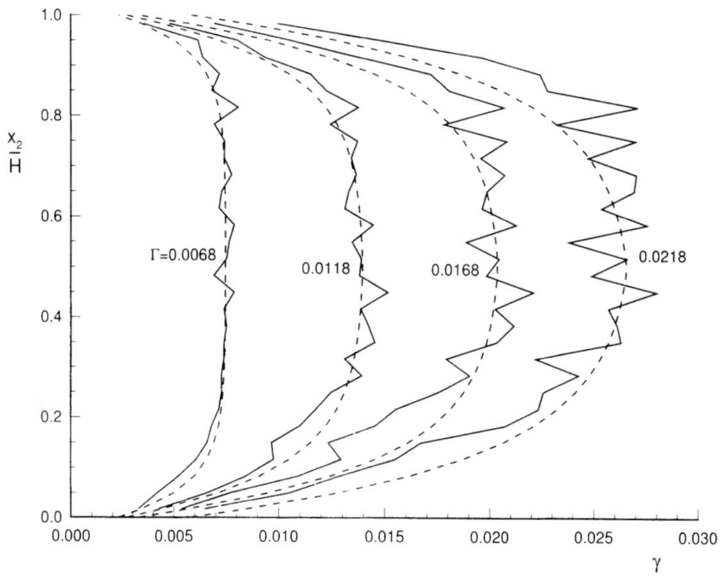


**Figure 18.** Shear strain profiles at various values of the applied shear $\Gamma$ for a single-crystalline strip of height $H = 1\mu$m with two symmetric slip systems. The dashed lines are fitted exponential strain profiles. From Shu *et al.* (2001).

slip occurs. The thickness of this layer was found to be rather insensitive to the strip height $H$, this leading to a size-dependent overall response.

### 7.4 Fracture

In cases where fracture takes place with limited plasticity around the crack tip, continuum crystal plasticity may no longer apply. It is the stress concentrations at the discrete dslocation scale that determine whether or not the crack propagates. Therefore, Cleveringa *et al.* (2000) carried out a small-scale yielding simulation of crack growth with discrete dislocations in the neighbourhood of the crack tip. As in the simulations mentioned previously, there are no dislocations initially, and sources and obstacles are distributed randomly. There is no special mechanism for dislocation nucleation from the crack tip, i.e. we are focusing on metallic crystals, where dislocation sources in the bulk are unavoidable.

While conventional continuum crystal plasticity theory predicts that stress levels at the crack tip are limited to being on the order of a few times the flow strength (roughly $10^{-3}E$), the discrete dislocation simulations yield stress levels many times higher. The distribution of opening stresses near the crack tip, after it has propagated by almost $0.5\mu$m, is shown in Fig. 19. These

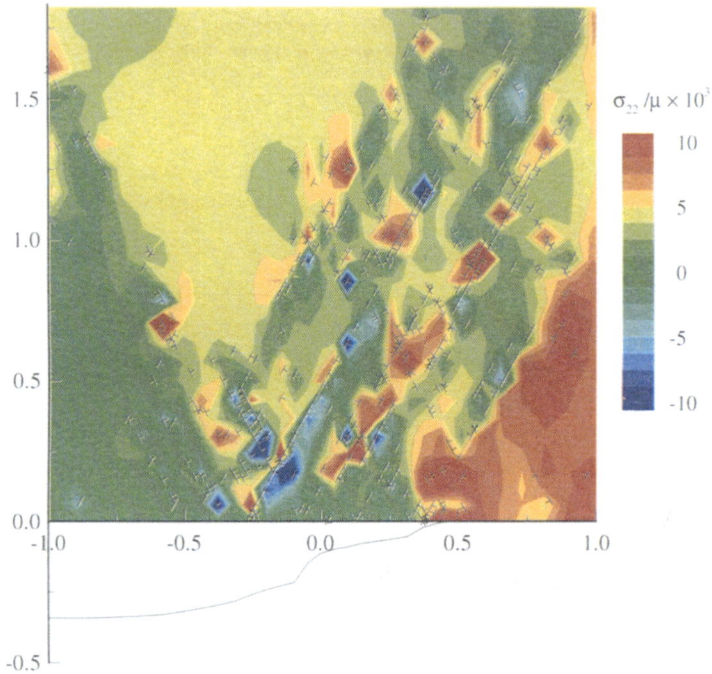


**Figure 19.** Distribution of the opening stress $\sigma_{22}$, i.e. in the vertical direction, near the tip of a crack in a crystal with two slip systems (slip planes oriented at $\pm - 60°$ with respect to the horizontal crack plane). The curve below the horizontal axis gives the opening profile (multiplied by a factor of 10). From Cleveringa *et al.* (2000).

calculations, Cleveringa *et al.* (2000), have taught us that dislocations play a dual role in the fracture process. On the one hand, dislocation activity gives rise to values of the crack driving force at initiation that are significantly higher than for an elastic solid with the same cohesive properties. On the other hand, it is the local stress concentration associated with discrete dislocations in the vicinity of the crack tip that leads to stress levels of the magnitude of the cohesive strength, causing the crack to propagate.

Subsequent more detailed studies revealed by that outside a region of about $1\mu$m away from the tip, the average stress fields agree rather well with continuum slip solutions, having stress levels on the order of the yield strength, Van der Giessen *et al.* (2001).

### 7.5 Thin Films

Thin films are being used for coating purposes as well as for functional purposes in electronic devices. High stresses develop in the film during cooling and re-heating due to the difference in thermal expansion coefficient between the film and its substrate. Even though stresses in metalic films get relaxed by dislocation motion, the sress levels in thin films can raise to several times those in bulk —the higher, the thinner the film. In an attempt to understand the origin of the size effect in thin films, Nicola *et al.* (2003) have devised a plane strain model where edge dislocations move on a set of three slip systems in a single crystalline film on an infinitely large substrate.

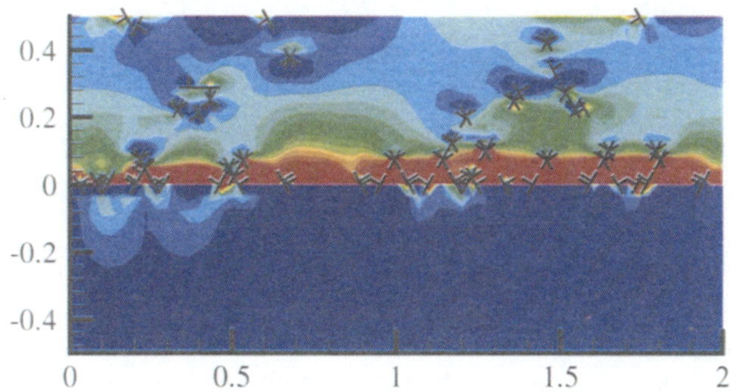


**Figure 20.** Distribution of stress $\sigma_{11}$, i.e. parallel to the interface, in a (2$\mu$m-wide unit cell of a) single crystal film of thickness 0.5$\mu$m with three slip systems (slip planes oriented at $\pm - 60°$ and 0° with respect to the horizontal interface). From Nicola *et al.* (2003).

Figure 20 shows the distribution of the in-plane stress, normalized by the stress $\sigma_n$ if there were no dislocations, for a 0.5$\mu$m thick film. The salient features are: (i) there is a boundary layer of relatively high stresses next to the interface, while (ii) the rest of the film has relaxed substantially. The width of the boundary layer, which arises from dislocation pile-ups at the interface, does not scale with the film thickness. This, along with the fact that the relaxation in the bulk is less effective for thinner films, gives that the average film stress scales with the film thickness $h$ roughly between $h^{-1/2}$ and $h^{-1}$.

## 8 Exercises

**Exercise 8.1.** Derive the Laplace equation (4.7) for the anti-plane shear problem of a screw dislocation from the general elasticity equations (4.2)–(4.4).

**Exercise 8.2.** Show that the elastic energy (per unit of dislocation line length) in a cylinder of radius $R$ around a straight screw dislocation is given by (4.10). Calculate the average shear stress in the same cylinder.

**Exercise 8.3.** Show that the stress field of an edge dislocation decays with the distance $r$ as $1/r$.

**Exercise 8.4.** Consider a slip plane in two dimensions. Show that two edge dislocations of the same sign repel each other, while two opposite ones attract one another.

**Exercise 8.5.** Show that the Burgers circuit (4.1) applied to the displacement field (4.14)–(4.15) indeed recovers the Burgers vector of the edge dislocation in Fig. 7b. What would the result be if the line direction, and therefore the circulation direction, were reversed?

**Exercise 8.6.** Modify the velocity law (6.2) of a dislocation to incorporate the Peierls-Nabarro stress.

**Exercise 8.7.** Show that the attractive force between dislocation in the dipole of Fig. 15 is given by (6.3).

## Bibliography

R.J. Asaro. Micromechanics of crystals and polycrystals. *Adv. Appl. Mech.*, 23:1–115, 1983.

M.F. Ashby. The deformation of plastically non-homogeneous materials. *Phil. Mag.*, 21:399–424, 1970.

W.D. Callister, *Materials Science and Engineering: An Introduction*. John Wiley & Sons, New York, NY, 2000.

H.H.M. Cleveringa, E. Van der Giessen and A. Needleman. Comparison of discrete dislocation and continuum plasticity predictions for a composite material. *Acta Mat.*, 45:3163–3179, 1997.

H.H.M. Cleveringa, E. Van der Giessen and A. Needleman. A discrete dislocation analysis of bending. *Int. J. Plast.*, 15:837–868, 1999.

H.H.M. Cleveringa, E. Van der Giessen and A. Needleman. A discrete dislocation analysis of mode I crack growth. *J. Mech. Phys. Solids*, 48:1133–1157, 2000.

V.S. Deshpande, A. Needleman and E. Van der Giessen. Discrete dislocation modeling of fatigue crack propagation. *Acta Mat.*, 50:831–846, 2002.

V.S. Deshpande, A. Needleman and E. Van der Giessen. Discrete dislocation plasticity modeling of short cracks in single crystals. *Acta Mat.*, 51:1–15, 2003.

J.P. Hirth and J. Lothe. *Theory of Dislocations*. McGraw–Hill, New York, NY, 1968.

D. Hull and D.J. Bacon. *Introduction to Dislocations*. Butterworh–Heinemann, Oxford, UK, 2001.

L. Nicola, E. Van der Giessen and A. Needleman. Discrete dislocation analysis of size effects in thin films. *J. Appl. Phys.*, 93:5920–5928, 2003.

J.F. Nye. Some geometrical relations in dislocated crystals. *Acta. Metall.*, 1:153–162, 1953.

J.Y. Shu, N.A. Fleck, E. Van der Giessen and A. Needleman. Boundary layers in constrained plastic flow: comparison of nonlocal and discrete dislocation plasticity. *J. Mech. Phys. Solids*, 49:1361–1395, 2001.

E. Van der Giessen and A. Needleman. Discrete dislocation plasticity: a simple planar model. *Model. Simul. Mater. Sci. Eng.*, 3:689–735, 1995.

E. Van der Giessen, V.S. Deshpande, H.H.M. Cleveringa and A. Needleman. Discrete dislocation plasticity and crack tip fields in single crystals. *J. Mech. Phys. Solids*, 49:2133–2153, 2001.

E. Van der Giessen and A. Needleman. GNDs in nonlocal plasticity theories: lessons from discrete dislocation simulations. *Scripta Mat.*, 48:127–132, 2003.

# Creep Rupture in Polycrystalline Materials

Erik Van der Giessen

Dept. of Applied Physics, University of Groningen, Nijenborgh 4, 9747 AG Groningen, The Netherlands

## 1 Introduction

To gain efficiency in technological processes, such as energy transformations (power plants, jet engines) and (petro)chemical reactions, the temperature should be as high as possible. Therefore, advanced, high-temperature materials are used. Their safe operation in such high-temperature environments, however, is severely limited by creep fracture, which can occur suddenly after years of stationary loading. Creep fracture can lead to high costs (down time, repair) and even loss of human life.

Another impetus for studying creep fracture stems from the need to assess the remaining lifetime of components and structures in service. Many high-temperature installations have been in service for 30 years or so and have already reached or will soon reach their original design life. These lifetime estimates are based on rather conservative assumptions, so that urgent questions arise concerning the remaining time during which these installations can still be operated safely.

The creep response of a material is commonly classified in three regimes: primary, secondary and tertiary creep, as illustrated in Fig. 1. Initially, after the instantaneous elastic response, the creep rate is relatively high due to the fact that creep of a polycrystalline material is highly non-uniform because of the variability in grain orientations. During the primary creep stage, the creep rate then reduces as creep becomes gradually more uniform. Then follows the so-called secondary creep stage, during which the creep rate is more or less constant in time. However, the creep rate is very sensitive to the stress level; in practical applications, this dependence is often written in the form of a power law,

$$\dot{\varepsilon} = B\sigma^n , \tag{1.1}$$

known as the Norton law, with $n$ the creep exponent (for metals in between 5 and 10) and $B$ a temperature-dependent pre-factor. At some point, however, one observes that the creep rate increases again, which is a signal that creep failure is on its way. The secondary creep regime is best known. Its characteristics for metals are that the creep rate is very sensitive to the stress level and to temperature.

Extensive experimental investigations have revealed that creep fracture in polycrystalline materials starts with the nucleation of sub-micron cavities at the grain boundaries (see Fig. 2). The cavities grow until they coalesce and form grain boundary microcracks. Final intergranular fracture occurs when the facet-sized microcracks link-up (see Fig. 3).

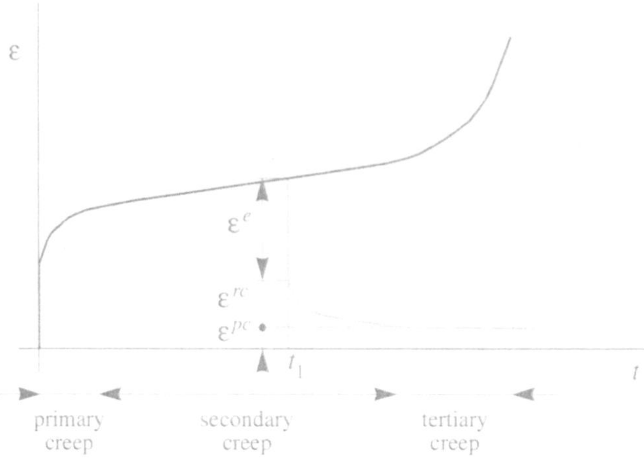


**Figure 1.** Typical creep curve of a metal under constant stress or load.

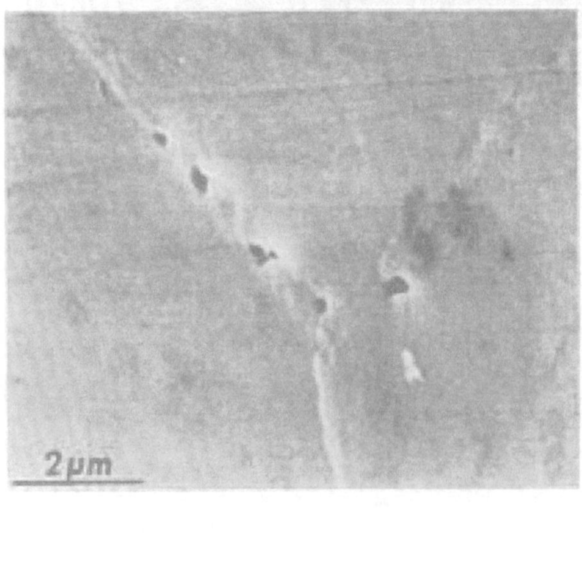


**Figure 2.** Micrograph of a distribution of cavities along grain boundaries in a stainless steel. From Argon *et al.* (1980).

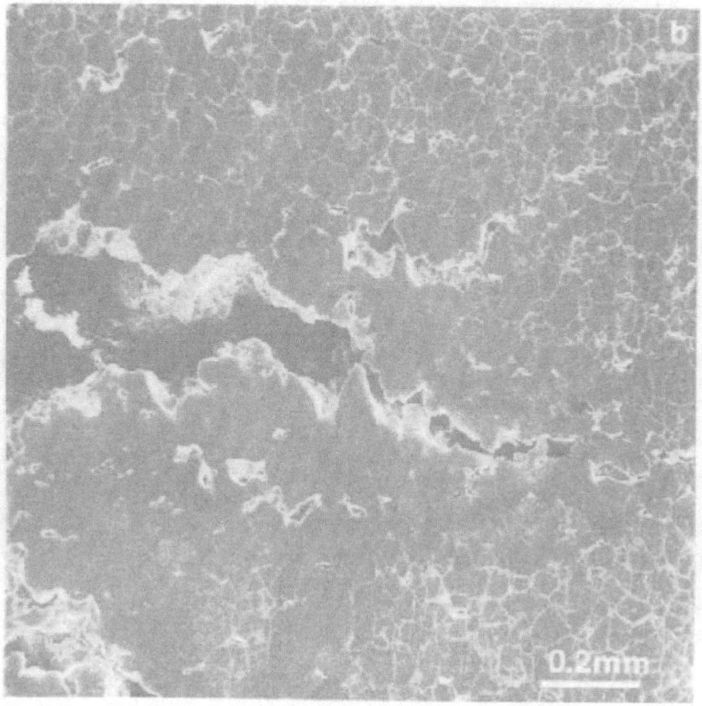


**Figure 3.** Micrograph showing intergranular creep fracture by the linking-up of microcracks. From Ozmat *et al.* (1991).

Clearly, high-temperature failure of polycrystalline metals (see Riedel (1987) for an overview) spans a wide range of length scales, as illustrated schematically in Fig. 4. At the largest, i.e. macroscopic, scale we consider a component or test specimen containing a crack (Fig. 4a), while the smallest relevant scale is that of the key failure mechanism, i.e. the nucleation and growth of small cavities along the grain boundaries (Fig. 4e). The intermediate scales determine how these elemental mechanisms lead to growth of the macroscopic crack. At the second smallest length scale (Fig. 4d), we observe the individual grains in the material and the distribution of cavitation damage along its grain boundaries. This is also the scale where two other key mechanisms are operating, namely creep of the grains themselves and sliding of adjacent grains relative to each other. Coalescence of the cavities after sufficient growth leads to microcracks along the grain facets, and at the next larger length scale we are concerned with the distribution of these microcracks near the tip of the crack (Fig. 4c). Growth of the crack at this scale occurs by the linking-up of facet microcracks with the main crack. Zooming out further (Fig. 4b) brings us to the size scale at which we are no longer able to distinguish the grain microstructure of the material; what remains is that we can identify a zone in the neighborhood of the crack tip in which damage occurs. This damage zone is surrounded by a zone of material that is not damaged and which only creeps (Fig. 4b).

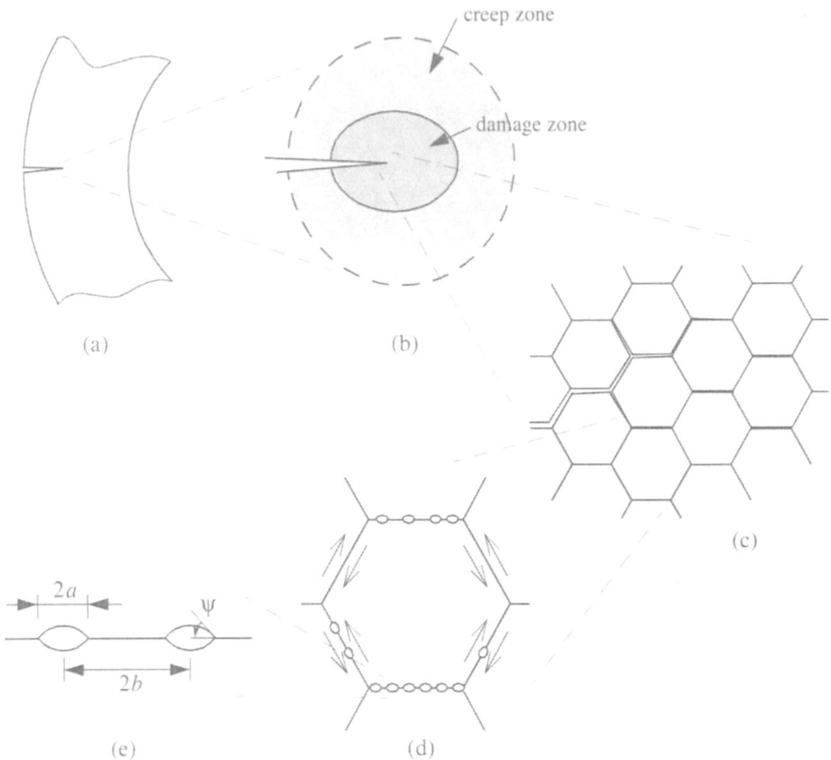


**Figure 4.** Size scales involved in creep fracture: (a) macroscopic crack; (b) crack-tip neighborhood with creep and damage zones; (c) mesoscopic near-tip region inside damage zone; (d) individual grains and grain boundary damage; (e) microscopic grain boundary cavities.

This chapter will present some techniques that have shown to be crucial in bridging some of the important length scales. The first sections discuss the treatment of cavity growth (on the cavity scale $\approx 10^{-7}$ m, see Fig. 4e) and grain boundary sliding (on the grain scale $\approx 10^{-5}$ m, see Fig. 4d ). Subsequently, methods will be discussed to bridge the length scales between the cavity scale (Fig. 4e) to the grain scale (Fig. 4d), from the grain scale (Fig. 4d) to the multiple-grain scale (Fig. 3 and Fig. 4c), and finally to the crack scale (Fig. 4b). The final scale transition is outside the scope of this chapter.

## 2 Cavity growth

Grain boundary cavities are often observed to have a spherical-caps shape and are specified by their radius $a$, half-spacing $b$ and cavity tip angle $\psi$ (see Fig. 5). The cavity volume can be expressed as $V = 4/3\pi a^3 h(\psi)$, where $h$ is the cavity shape parameter, defined by $h(\psi) = [(1+\cos\psi)^{-1} - \frac{1}{2}\cos\psi]/\sin\psi$. The cavities grow by the diffusional flow of atoms from their surface into the grain boundary layer. If we assume surface diffusion to be much faster than grain boundary diffusion, the cavities retain their quasi-equilibrium spherical-caps shape. Not only do cavities grow by diffusion, they also grow by creep of the surrounding grain material.

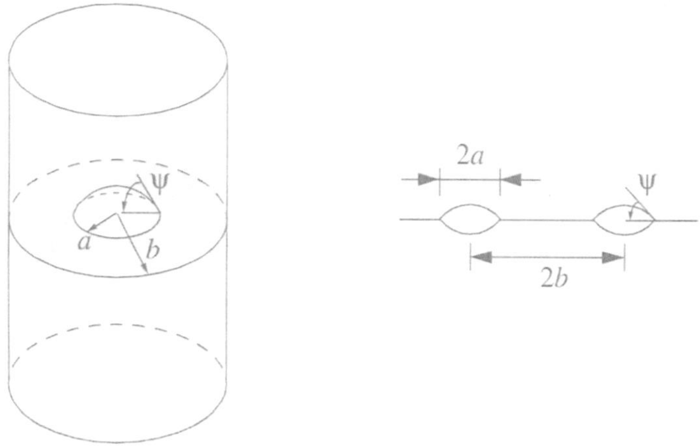


**Figure 5.** (a) Geometry of a quasi-equilibrium spherical-caps shaped cavity with radius $a$, half spacing $b$ and cavity tip angle $\psi$. (b) Representation of cavities on a grain boundary in a planar polycrystal model.

Hull and Rimmer (1959) were the first to analyze purely diffusional cavity growth by assuming the grains to be rigid. This will be discussed in the next section. Then, we summarize the results of modeling cavity growth by creep only. Finally, we consider the full problem of coupled growth as treated by Needleman and Rice (1980), and summarize the key findings.

### 2.1 Diffusional cavity growth

Hull and Rimmer (1959) considered an axisymmetric unit cell of radius $b$ —the average spacing between cavities— containing a grain boundary void of radius $a$, and subject to a remote stress normal to the boundary, see Fig. 6. The cavity has a tip angle $\psi$ which is determined by equilibrium of the surface energy $\gamma_s$ and grain boundary energy $\gamma_b$ through $\cos\psi = \gamma_b/2\gamma_s$.

The key approximation by Hull and Rimmer (1959) was that grains on either side of the boundary are considered to be undeformable. They then argued that the void could grow by (i) material from the inside of the void being transported in the direction of the triple point, which is then (ii) uniformly plated out over the grain boundary surface ("uniformity" is a consequence of the grains being rigid). This plated-out material thus thickens the grain boundary (and separates the grains) by an amount $\delta$. Conservation of mass must ensure that $\delta$ correlates with the amount of volume being "scooped out" from the void's inner surface. The rate at which the process takes place is determined by the rate of diffusion along the void surface and that along the grain boundary. At the typical temperatures where creep rupture is dominated by void growth, surface diffusion is generally faster than grain boundary diffusion, Hull and Rimmer (1959) actually

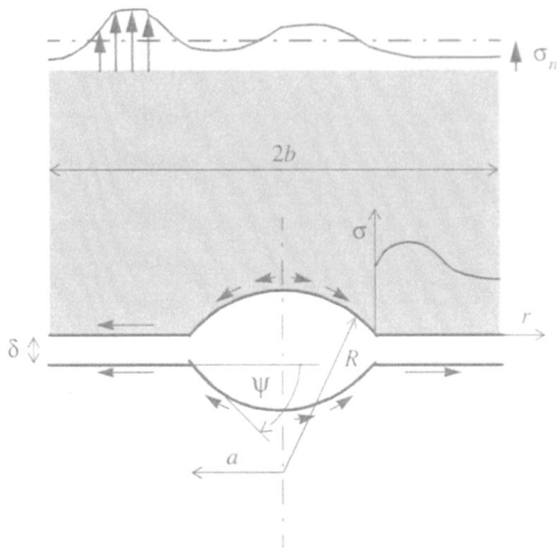


**Figure 6.** Axisymmetric unit cell model of a void on the boundary between two rigid grains, which grows by surface and grain boundary diffusion under the influence of a remote stress normal to the boundary.

assumed that surface diffusion is so much faster than grain boundary diffusion that the shape of the void remains the same spherical-cap shape and that only the radius changes. Thus, the problem reduces to calculating the instantaneous growth rate at an arbitrary void radius $a$. With $\psi$ being the constant tip angle, the void consists of two caps of a sphere with radius $R = a/\sin\psi$.

With the above assumptions, the governing equations are

$$\text{conservation of matter:} \quad \frac{1}{r}\frac{\partial}{\partial r}(r\Omega J_b) + \dot{\delta} = 0\,, \tag{2.1}$$

$$\text{diffusion law:} \quad \Omega J_b = -\frac{D_b\delta_b}{\Omega kT}\frac{\partial\mu}{\partial r} \tag{2.2}$$

along $a < r < b$, where the chemical potential of atoms along the grain boundary, $\mu$, is given by

$$\mu = \mu_0 - \Omega\sigma$$

in terms of the grain boundary stress $\sigma(r)$. Here, $\Omega J_b$ is the volumetric rate of matter diffusion ($\Omega$ is atomic volume), $D_b$ the grain boundary diffusion coefficient and $\delta_b$ the boundary thickness. The boundary conditions for the equations (2.1) are $\Omega J_b = 0$ at $r = b$ (because of symmetry) and $\sigma = \sigma_s$ at $r = a$, with the sintering stress $\sigma_s$ given by

$$\sigma_s = 2\gamma_s/R\,.$$

After elimination of $\mu$ and $J_b$, a single differential equation for the boundary stress results:

$$\frac{\mathcal{D}}{r}\frac{\partial}{\partial r}\left(r\frac{\partial\sigma}{\partial r}\right) + \dot{\delta} = 0\,, \quad \mathcal{D} \equiv \frac{D_b\delta_b\Omega}{kT}$$

with boundary conditions

$$\sigma = \sigma_s \text{ at } r = a; \quad \partial\sigma/\partial r = 0 \text{ at } r = b.$$

The solution reads

$$\sigma = \sigma_s + \frac{1}{2}b^2\frac{\dot{\delta}}{\mathcal{D}}\left\{\ln\left(\frac{r}{a}\right) - \frac{f}{2}\left[\left(\frac{r}{a}\right)^2 - 1\right]\right\} \tag{2.3}$$

with

$$f = \left(\frac{a}{b}\right)^2$$

the area fraction of cavities. In (2.3), $\dot{\delta}$ is assumed to be known. Since,

$$\dot{\delta} = \frac{\dot{V}}{\pi b^2} \tag{2.4}$$

because of, again, conservation of mass, eq. (2.3) can be solved for the volumetric cavity growth rate $\dot{V}$. Noting, furthermore, that from overall equilibrium,

$$\sigma_n = \frac{1}{\pi b^2}\int_a^b 2\pi r\sigma(r)dr,$$

one finds

$$\dot{V} = 4\pi\mathcal{D}\frac{\sigma_n - (1-f)\sigma_s}{\ln(1/f) - \frac{1}{2}(3-f)(1-f)}. \tag{2.5}$$

As pointed out by Needleman and Rice (1980), the derivation and the result of Hull and Rimmer (1959) contains an error; the result (2.5) is correct.

### 2.2 Cavity growth by creep

The modeling of growth of voids in plastically deforming materials received a major impulse by the paper by Budiansky *et al.* (1982). The key result for our purpose here is that of a spherical void with radius $a$ in an infinite creeping solid under an axisymmetric remote stress state characterized by a principal stress $S$ and a transverse stress $T$. The constitutive model is that of an isotropic creeping solid with the creep strain rate tensor being given by

$$\dot{\varepsilon}_{ij} = \frac{3}{2}\frac{\sigma'_{ij}}{\sigma_e}\dot{\varepsilon}_e \tag{2.6}$$

with the Von Mises equivalent stress and strain rate,

$$\sigma_e \equiv \sqrt{\frac{3}{2}\sigma'_{ij}\sigma'_{ij}}, \quad \dot{\varepsilon}_e \equiv \sqrt{\frac{2}{3}\dot{\varepsilon}_{ij}\dot{\varepsilon}_{ij}}, \tag{2.7}$$

respectively, being connected through the Norton power law (1.1). The remote equivalent Von Mises stress and the mean stress in this case are

$$\sigma_e = |S - T| \quad , \quad \sigma_m = \frac{1}{3}(S + 2T).$$

The analysis involves finding an approximate Rayleigh–Ritz solution to a minimum principle appropriate for the infinite domain under consideration, and uses an asymptotic expansion of the relations for high triaxialities $\sigma_m/\sigma_e$ (subject to $S > T$). The volumetric growth rate is then found as

$$\frac{\dot{V}}{\dot{\varepsilon}_e V} = \frac{3}{2}\left[\alpha_n\left|\frac{\sigma_m}{\sigma_e}\right| + \beta_n\right]^n, \tag{2.8}$$

where the constants $\alpha_n$ and $\beta_n$ are given by $\alpha_n = 3/2n$ and $\beta_n = (n-1)(n+0.4319)/n^2$.

### 2.3 Coupled cavity growth

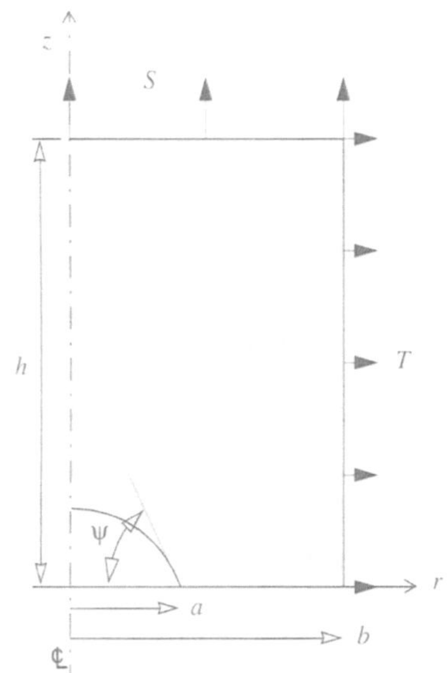


**Figure 7.** Axisymmetric unit cell model of a void subject to an average stress $S$ normal to the grain boundary ($z = 0$) and an axisymmetric transverse stress $T$.

As cavities grow by diffusion, as discussed above, creep accelerates the plating of atoms from the cavity surface into the grain boundary layer. Needleman and Rice (1980) analyzed an axisymmetric cavity model, illustrated in Fig. 7, and investigated the combined influence of creep and diffusion. For this purpose, they showed that this problem can be described by minimizing the functional $F(v_i)$:

$$\begin{aligned}\frac{F(v_i)}{4\pi} = {} & \frac{n}{n+1}\sigma_0\dot{\varepsilon}_0\int_0^h\int_{r_0(z)}^b\left(\frac{\dot{\varepsilon}_e}{\dot{\varepsilon}_0}\right)^{(n+1)/n} r\,dr\,dz - S\int_0^b r v_z(r,h)\,dr \\ & + \frac{1}{\mathcal{D}}\int_a^b\frac{1}{r}\left[\int_r^b r' v_z(r',0)\,dr'\right]^2 + \sigma_s\int_a^b r v_z(r,0)\,dr,\end{aligned} \tag{2.9}$$

subject to the boundary condition

$$v_r = b\dot{\varepsilon}_{rr}^{\infty} = -\frac{1}{2}b\,\mathrm{sign}(S-T)\dot{\varepsilon}_0 \quad \text{at } r=b\,. \tag{2.10}$$

In (2.9) the strain rates are related to the velocity field through

$$\dot{\varepsilon}_{ij} = \frac{1}{2}(v_{i,j}+v_{j,i}),$$

and $\dot{\varepsilon}_e$ is related to these creep strain rates according to (2.7)b. The last two contributions in (2.9) to $F$ account for grain boundary diffusion and are equivalent to the set of equations (2.1) upon noting that $\dot{\delta}$ in (2.9) is equivalent in this formulation to $2v_z$ and $S=\sigma_n$. The unit cell boundary conditions (2.10) express that the radial contraction of the cell on $z=0$ is constrained by the contraction of the cylinder of radius $b$ far from the grain boundary.

The resulting equations were solved by Needleman and Rice (1980) through a detailed finite element calculation. They summarized their numerical results into an expression for the volumetric cavity growth rate $\dot{V}$. This expression was later modified by Tvergaard (1984), resulting in

$$\dot{V} = \dot{V}_1 + \dot{V}_2. \tag{2.11}$$

Here, $\dot{V}_1$ is the contribution of diffusion, specified through

$$\dot{V}_1 = 4\pi\mathcal{D}\frac{S-(1-f)\sigma_s}{\ln(1/f)-\frac{1}{2}(3-f)(1-f)}, \tag{2.12}$$

with

$$f = \max\left[\left(\frac{a}{b}\right)^2, \left(\frac{a}{a+1.5L}\right)^2\right], \tag{2.13}$$

and the contribution of creep, $\dot{V}_2$, is given by

$$\dot{V}_2 = \begin{cases} \pm 2\pi\dot{\varepsilon}_e^C a^3 h(\psi)\left[\alpha_n\left|\dfrac{\sigma_m}{\sigma_e}\right| + \beta_n\right]^n, & \text{for } \pm\dfrac{\sigma_m}{\sigma_e} > 1; \\[2ex] 2\pi\dot{\varepsilon}_e^C a^3 h(\psi)\left[\alpha_n + \beta_n\right]^n \dfrac{\sigma_m}{\sigma_e}, & \text{for } \left|\dfrac{\sigma_m}{\sigma_e}\right| \le 1. \end{cases} \tag{2.14}$$

The high-triaxiality part of this expression follows directly from application of (2.8) to the void in Fig. 7.

The coupling between diffusive and creep contributions to void growth enters in (2.13) through the stress and temperature dependent length scale

$$L = \left[\mathcal{D}\sigma_e/\dot{\varepsilon}_e^C\right]^{1/3}, \tag{2.15}$$

that originates from re-normalization of (2.9) by the effective stress $\sigma_e$, Needleman and Rice (1980). For small values of $a/L$ (say, $a/L < 0.1$) cavity growth is dominated by diffusion, while for larger values of $a/L$ creep growth becomes more and more important. The range of validity of expression (2.11) was specified to be $a/L \le 10$. Furthermore in (2.12), $\mathcal{D}$ is the grain boundary diffusion parameter and $\sigma_s$ is the sintering stress which will be neglected. The effective stress $\sigma_e$, the mean stress $\sigma_m$ and the normal stress $\sigma_n$ are considered to be stresses remote from each cavity (on the scale of individual cavities).

### 2.4 Cavitation-enhanced creep

As the cavities on a grain boundary grow in volume, the grains on either side move relative to each other. Thus, cavitation gives a contribution to the overall strain and this is part of the explanation of the tertiary creep regime.

The strain contribution of cavitation can be easily calculated for a distribution of cavities on a grain boundary as illustrated in Fig. 5, where the average spacing between cavities is $2b$. Then one can consider a single cavity of radius $a$ in a cylinder with radius $b$. As this cavity grows, the cylinder plus void elongates by an amount $\delta_c$. Conservation of mass (per unit time) requires that

$$\dot{\delta}_c = \frac{\dot{V}}{\pi b^2}, \tag{2.16}$$

cf. eq. (2.4). Substitution of the volumetric growth relations (2.11) gives an expression for $\dot{\delta}_c$ as a function of the normal stress on the grain boundary, $\sigma_n$, the effective Von Mises stress, $\sigma_e$, and the mean stress $\sigma_m$ in the neighborhood of the cavity, i.e.

$$\dot{\delta}_c = \dot{\delta}_c(\sigma_n, \sigma_e, \sigma_m), \tag{2.17}$$

with the function depending on the current cavity volume fraction $f$, and the material through $\mathcal{D}$, $B$ and $n$.

Notice that in (2.16) it is implicitly assumed that $b$ does not change. However, if new cavities get nucleated, $b$ will change and (2.16) needs to be extended. A simple manner to do so is to employ a nucleation law in the form $\dot{b} = \dot{b}(\text{local state})$ and to differentiate the full expression $\delta_c = V/\pi b^2$. This leads to

$$\dot{\delta}_c = \frac{\dot{V}}{\pi b^2} - \frac{2V}{\pi b^2}\frac{\dot{b}}{b} \tag{2.18}$$

with the first term on the right-hand side equal to that in (2.16) and the second term due to nucleation. This approach neglects the fact that freshly nucleated cavities will be smaller than the current family of cavities; however, the time it takes for the small voids to catch up in size is often negligible. Possible nucleation laws will be discussed subsequently in Sec. 4.

## 3 Grain boundary sliding

An important mechanism at the grain scale is grain boundary sliding. At typical creep conditions the resistance of the grain boundaries becomes lower than the resistance of the grains, so that the grain boundary shear stresses relax. This causes a profound stress-redistribution which affects the creep failure process considerably.

According to Ashby (1972), grain boundaries can be modeled as thin layers that slide in a Newtonian viscous manner, governed by

$$\dot{\gamma} = \frac{\tau}{\eta_B}, \tag{3.1}$$

where $\dot{\gamma}$ is the relative sliding velocity of adjacent grains due to a shear stress $\tau$, divided by the thickness of the boundary, $w$. The grain boundary viscosity $\eta_B$ can be related to the grain boundary diffusivity, while irregularities in the grain boundary, such as ledges or second-phase particles, can increase the boundary viscosity. Here, $\eta_B/w$ is regarded as a separate material parameter, which can be expressed in terms of the strain-rate parameter

$$\dot{\varepsilon}_B = \left( \frac{w}{d} \frac{B^{-1/n}}{\eta_B} \right)^{n/(n-1)}, \tag{3.2}$$

with $d$ a length parameter related to the grain size, $d \approx 2\sqrt{3}R$, with $2R$ the facet length. Ghahremani (1980) studied a regular hexagonal microstructure and identified a unit-cell similar to Fig. 8. Effectively, grain boundary sliding can be characterized by the ratio $\dot{E}_e^C/\dot{\varepsilon}_B$, which denotes the relative resistance of the grain material and grain boundary layer. Here, $\dot{E}_e^C$ is the macroscopic Norton creep rate, obtained by substituting the macroscopic effective stress $\Sigma_e$ in (1.1). Free sliding ($\eta_B = 0$) corresponds to $\dot{E}_e^C/\dot{\varepsilon}_B = 0$, while no sliding ($\eta_B \to \infty$) is equivalent to $\dot{E}_e^C/\dot{\varepsilon}_B \to \infty$.

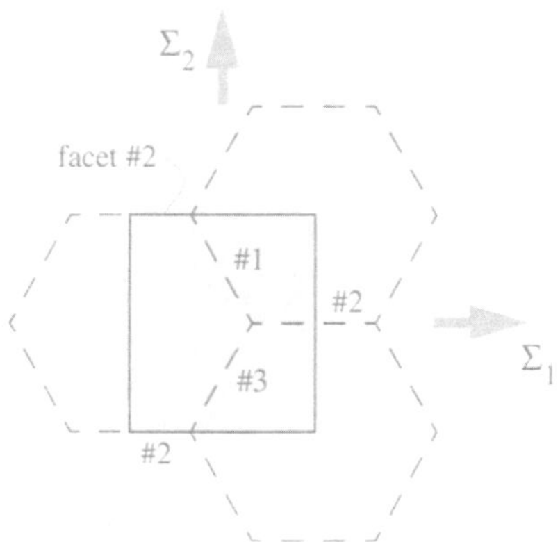


**Figure 8.** Unit-cell from a regular hexagonal structure, specifying three families of facets.

As revealed by detailed micromechanical investigations, e.g. Ghahremani (1980), grain boundary sliding enhances the creep rate of an aggregate of grains. In fact, under sliding conditions, the stress inside the grain is nonuniform, and as a consequence, the overall creep strain rate of the aggregate is significantly increased. This effect is usually incorporated at the macroscale through a stress enhancement factor $f^*$, such that the overall creep law is

$$\dot{E}_e^C = B(f^* \Sigma_e)^n. \tag{3.3}$$

For a certain applied stress level, the value of $f^*$ depends on $n$ and $\dot{E}_e^C/\dot{\varepsilon}_B$. It varies smoothly from $f^* = 1$ without grain boundary sliding ($\dot{E}_e^C/\dot{\varepsilon}_B \to \infty$) to a finite value ($f^* \approx 1.2$ for $n = 5$) under free sliding ($\dot{E}_e^C/\dot{\varepsilon}_B = 0$) in a regular hexagonal polycrystal, see Fig. 9. This dependence can be expressed in a fully equivalent manner in terms of the ratio between $\Sigma_e$ and the stress measure $\sigma_B = (\dot{\varepsilon}_B/B)^{1/n}$, i.e. $f^* = f^*(n, \sigma_B/\Sigma_e)$.

The above results have been obtained by Ghahremani (1980) via a rather ingenious renormalization of the governing equations. The governing equations for creep in a unit cell of

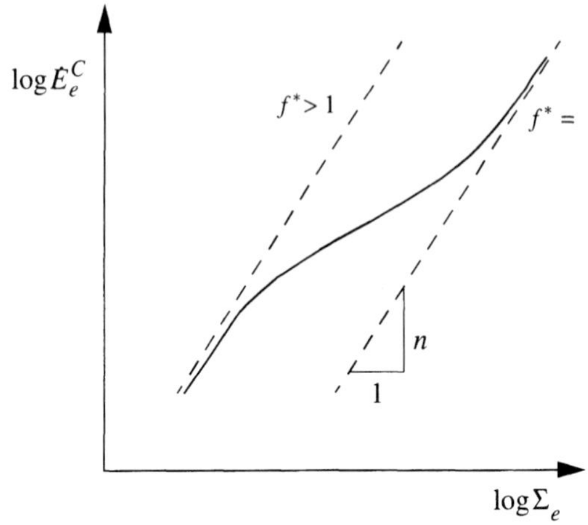


**Figure 9.** Transition from power-law creep without grain boundary sliding to creep with free sliding, $f^* > 1$.

sliding grains as in Fig. 8 read[1]:

$$\text{equilibrium: } \frac{\partial \sigma_{ij}}{\partial x_j} = 0 \tag{3.4}$$

$$\text{kinematics: } \dot{\varepsilon}_{ij} = \frac{1}{2}\left(\frac{\partial v_i}{\partial x_j} + \frac{\partial v_j}{\partial x_i}\right) \tag{3.5}$$

$$\text{Norton creep law: } \sigma'_{ij} = \frac{2}{3}\frac{\sigma_e}{\dot{\varepsilon}_e} = \frac{2}{3}\frac{\sigma_0}{\dot{\varepsilon}_0}\left(\frac{\dot{\varepsilon}_e}{\dot{\varepsilon}_0}\right)^{\frac{1-n}{n}} \dot{\varepsilon}_{ij} \tag{3.6}$$

$$\text{sliding condition: } \tau = \frac{\eta_B}{w}[\![v]\!] \tag{3.7}$$

where $[\![v]\!]$ denotes the jump of velocity from one side of the grain boundary to the other. The sliding condition is identical to (3.1), and Norton's creep law is the limit of the viscoplastic constitutive law used in Chap. 3 in the absence of elasticity.

In preparation of the renormalization, it is useful to list the various types of quantities that govern the problem:

$$\begin{array}{llll} \text{lengths:} & x_i, & w, & d \\ \text{strain rates:} & \dot{\varepsilon}_{ij}, & \dfrac{v_i}{d}, & \dfrac{[\![v]\!]}{w} \\ \text{stresses:} & \sigma_{ij}, & \tau & \\ \text{viscosities:} & \eta_B, & \eta_C = \dfrac{\sigma_0}{\dot{\varepsilon}_0}\left(\dfrac{\dot{\varepsilon}_e}{\dot{\varepsilon}_0}\right)^{\frac{1-n}{n}} & \end{array} \tag{3.8}$$

The variable $d$ is a measure of the grain size as defined before. The $\eta_C$ is the stress-dependent, equivalent viscosity caused by creep. The solution of the equations obviously depends only on

[1] Small letters refer to microscopic quantities, capitals to overall quantities

the non-dimensional ratios of the various quantities listed in (3.8). As for the ratio of viscosities, Ghahremani showed that it is convenient to consider the following non-dimensional variable:

$$\frac{d}{w}\frac{\eta_B}{\eta_C} = \left(\frac{\dot{\varepsilon}_B}{\dot{\varepsilon}_0}\right)^{\frac{1-n}{n}}$$

where $\dot{\varepsilon}_B$ is introduced as a "creep equivalent grain boundary strain rate" which can be written as

$$\dot{\varepsilon}_B = \dot{\varepsilon}_0 \left(\frac{\eta_B d\dot{\varepsilon}_0}{w\sigma_0}\right)^{\frac{n}{1-n}} .$$

Inserting this back into the Norton creep law, we obtain the associated stress

$$\sigma_B = \sigma_0 \left(\frac{\dot{\varepsilon}_B}{\dot{\varepsilon}_0}\right)^{\frac{1}{n}} ,$$

so that

$$\frac{\sigma_B}{\dot{\varepsilon}_B} = \frac{\eta_B d}{w} .$$

Now, we proceed by normalizing all lengths in the problem by $d$, all strain rates by $\dot{\varepsilon}_B$ and the stress by $\sigma_B$. Denoting the normalized quantities with a ˆ, the governing equations become

$$\text{equilibrium: } \frac{\partial \hat{\sigma}_{ij}}{\partial \hat{x}_j} = 0 \tag{3.9}$$

$$\text{kinematics: } \hat{\dot{\varepsilon}}_{ij} = \frac{1}{2}\left(\frac{\partial \hat{v}_i}{\partial \hat{x}_j} + \frac{\partial \hat{v}_j}{\partial \hat{x}_i}\right) \tag{3.10}$$

$$\text{Norton creep law: } \hat{\sigma}'_{ij} = \frac{2}{3}\left(\hat{\dot{\varepsilon}}_e\right)^{\frac{1-n}{n}} \hat{\dot{\varepsilon}}_{ij} \tag{3.11}$$

$$\text{sliding condition: } \hat{\tau} = [\![\hat{v}]\!] \tag{3.12}$$

The interesting thing to note about these equations is that they are independent of the material properties. The solutions of these equations are therefore universal, independent of the ratio between creep properties and GB viscosity. These material properties only re-enter once the solution is transformed back to physical quantities by elimination of the normalization.

Without actually solving the equations, you can make some interesting observations about the overall response. To this end consider the macroscopic deviatoric stress,

$$\hat{\Sigma}'_{ij} = \frac{1}{V}\int \frac{\sigma'_{ij}}{\sigma_B} dV = \frac{1}{V/d^3}\int \hat{\sigma}'_{ij}\, d\hat{V} = \frac{2}{3}\frac{1}{V/d^3}\int \left(\hat{\dot{\varepsilon}}_e\right)^{\frac{1-n}{n}} \hat{\dot{\varepsilon}}_{ij}\, d\hat{V}$$

when the system is subjected to a prescribed macroscopic strain rate $\hat{\dot{E}}_{ij}$:

$$\hat{\Sigma}'_{ij} = \frac{2}{3}\hat{\dot{E}}_e^{C\,\frac{1}{n}} \underbrace{\frac{1}{V/d^3}\int \left(\frac{\hat{\dot{\varepsilon}}_e}{\hat{\dot{E}}_e^C}\right)^{\frac{1-n}{n}} \left(\frac{\hat{\dot{\varepsilon}}_{ij}}{\hat{\dot{E}}_e^C}\right) d\hat{V}}_{\chi_{ij}} . \tag{3.13}$$

When we know the term $\chi_{ij}$, we know the macroscopic stress. In general, this requires the solution. However, it is known that an array of hexagonal (isotropic) grains responds in an isotropic manner. Thus, $\chi_{ij}$ must be proportional to $\delta_{ij}$. With this, we can calculate the normalized macroscopic Von Mises stress from (3.13) as

$$\hat{\Sigma}_e = \sqrt{\frac{3}{2}\hat{\Sigma}'_{ij}\hat{\Sigma}'_{ij}} = \frac{1}{g}\hat{\dot{E}}_e^{C\,\frac{1}{n}}$$

where the coefficient $g$ is the eventual proportionality factor. In terms of dimensionalized quantities, the final result is

$$\frac{\dot{E}_e^C}{\dot{\varepsilon}_B} = \left(g\frac{\Sigma_e}{\sigma_B}\right)^n ,$$

which is equivalent to (3.3) (after noting that $B = \dot{\varepsilon}_0/\sigma_0^n$ and rewriting $g$ in terms of $f^*$).

Grain boundary sliding also affects the normal stresses on the facets. In the absence of sliding, the average facet normal stress on facet $k$ ($k = 1, \dots, 3$, see Fig. 8), ${}^{nsl}\sigma_n^{(k)}$, is simply the resolved normal stress according to

$${}^{nsl}\sigma_n^{(k)} = n_i^{(k)}\Sigma_{ij}^{(k)}n_j^{(k)} , \tag{3.14}$$

where $n_i^{(k)}$ is the unit normal of facet $k$. For completely free sliding (in the absence of any damage), the average normal stress follows directly from equilibrium of a single grain, Rice (1981),

$${}^{fsl}\sigma_n^{(k)} = 2n_i^{(k)}\Sigma_{ij}^{(k)}n_j^{(k)} - \Sigma_m^{(k)} . \tag{3.15}$$

Under a uniaxial state of stress, i.e. $\Sigma_1 = 0$ in Fig. 8, the average normal stress on facets in family #2 is ${}^{fsl}\sigma_n^{(2)} = (3/2)\Sigma_2$ (see Exer. 7.2).

## 4 Cavity Nucleation

The nucleation of cavities at grain boundaries actually is among the least understood aspects of creep fracture. There is consensus about the fact that cavities nucleate continuously during creep, but the physical mechanism is not well understood. There are two prevailing theories at the moment:

1. Cavity nucleation is the consequence of the *condensation of vacancies* on stressed interfaces. This is a physically convincing explanation, but suffers from two problems: (i) classical nucleation theory leads to a critical cavity size of $2\gamma_s/\sigma_n$, which typically amounts to 2nm and therefore is too small to be experimentally validated; (ii) for it to operate at experimentally observed rates, the grain boundary stresses should be at least an order of magnitude larger than expected.
2. Cavity nucleation is due to *athermal decohesions* at grain boundaries originating from stress concentrations due to slip bands inside the creeping grains intersection with grain boundary particles. This explanation is due to Dyson (1983), who also formulated that the nucleation rate (number of new cavities per unit area) be described by

$$\dot{N} = F_n\left(\frac{\sigma_n}{\Sigma_0}\right)^2 \dot{\varepsilon}_e^C \tag{4.1}$$

i.e. proportional to the creep rate in the neighboring grains. Here, $F_n$ is a material parameter that specifies the nucleation activity, while $\Sigma_0$ is just a normalizing parameter.

By lack of an accurate *and* convincing explanation, the examples in the sequel have used (4.1). The connection with the modeling in the previous section is made by observing that

$$\frac{\dot{b}}{b} = -\frac{1}{2}\frac{\dot{N}}{N}, \tag{4.2}$$

which is to be substituted into (2.18).

# 5 Cavitation in Polycrystals

The void growth expressions (2.11)–(2.14) together with the evolution law for the cavity spacing (2.18) contain the information to calculate the rates of changes of all cavity dimensions. Assuming constant stress, as is typical for creep conditions, and assuming that the near-facet stress state is equal to the applied macroscopic stress, one can readily integrate these equations in time. When the ratio $a/b$ approaches unity, coalescence of cavities occurs. The moment that this occurs may be identified in a first crude approach as the failure time. However, this procedure gives times to failure that are much shorter than found experimentally. Dyson (1976) pointed out what the prime reason for this is. Cavitation does not take place homogeneously throughout the polycrystalline material. Hence, for a substantial part of the lifetime, cavitation grain boundaries are more or less isolated from each other. But, as cavitation proceeds and the separation between the two adjacent grains, $\delta_c$ from (2.18), increases, this needs to be accommodated by the surrounding material. If this is not cavitating, but just creeping, an internal stress re-distribution must take place in order to match both processes. At typical creep conditions of relatively low stress levels, cavitation tends to be faster than creep so that stress must be shedded away from the cavitating grain boundary and towards the surrounding grains. Effectively, the stress on cavitating facets is lower than the applied stress (and higher in the surrounding grains): Dyson (1976) called this *constrained cavitation.*

## 5.1 Polycrystal Model

Constrained cavitation is accounted for in a natural manner by actually modeling the polycrystalline structure of the material. Van der Giessen and Tvergaard (1994) and later Onck and Van der Giessen (1997) did so for a planar material model illustrated in Fig. 10. It assumes a doubly-periodic stacking of unit cells, each of which contains $m_1 \times m_2$ regular hexagonal grains, which itself is assumed to be symmetric about the $x_1$ and about the $x_2$ direction. The material is subjected to (macroscopic) remote stresses $\Sigma_1$ and $\Sigma_2$.

For simplicity and to focus on cavitation, the grains are assumed to be isotropic. Total strain rates are decomposed as usual into elastic parts $d^e_{ij}$ and creep parts $d^c_{ij}$, with the latter being determined (in a finite strain context) by (2.6). Elasticity is described by

$$\overset{\nabla}{\sigma}_{ij} = \mathcal{R}_{ijkl} d^e_{kl} = \mathcal{R}_{ijkl}(d_{kl} - d^c_{kl})$$

with $\overset{\nabla}{\sigma}_{ij}$ the Jaumann rate of Cauchy stress and $\mathcal{R}_{ijkl}$ the isotropic elastic stiffness tensor in terms of Young's modulus $E$ and Poisson's ration $\nu$.

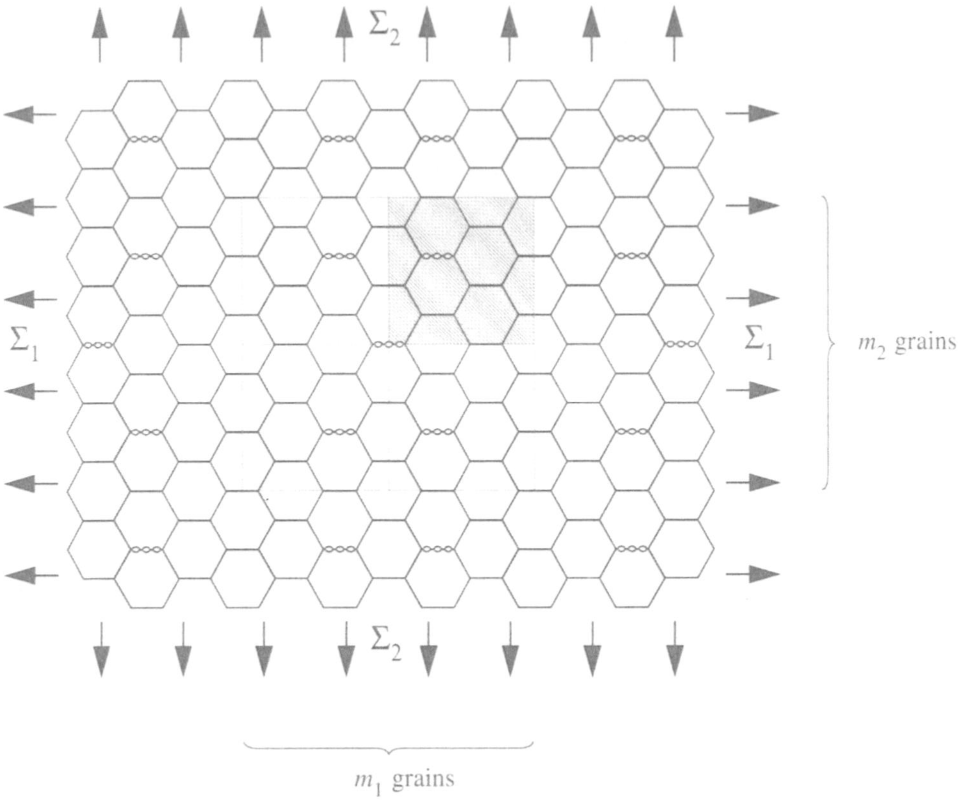


**Figure 10.** Planar polycrystal unit-cell model comprising $m_1 \times m_2$ grains (in this example, $(m_1, m_2) = (6,5)$). Only a quarter of the unit cell is analyzed because of symmetry. From Onck and Van der Giessen (1997).

The grain boundaries are represented through a cohesive zone model, which takes into account separation of grains because of cavitation and grain boundary sliding. As illustrated in Fig. 11, the modeling of cavitation in itself involves a scale transition, namely from the description of individual cavities to that of a distribution of cavities on a grain boundary. Sliding is modeled in a linearly viscous manner, cf. (3.7). For numerical convenience, fictitious springs are introduced with stiffness $k_n$ in normal direction and $k_s$ in tangential direction, so that the constitutive equations for the interface become

$$\dot{\sigma}_n = k_n(\dot{\delta} - \dot{\delta}_c), \quad \dot{\tau} = k_s(\dot{u} - \dot{u}_s)$$

with $\dot{\delta}$ and $\dot{u}$ being the total opening rates, and

$$\dot{\delta}_c = \frac{\dot{V}}{\pi b^2} - \frac{2V}{\pi b^2}\frac{\dot{b}}{b}, \quad \dot{u}_s = \frac{\tau}{\eta_B/w}.$$

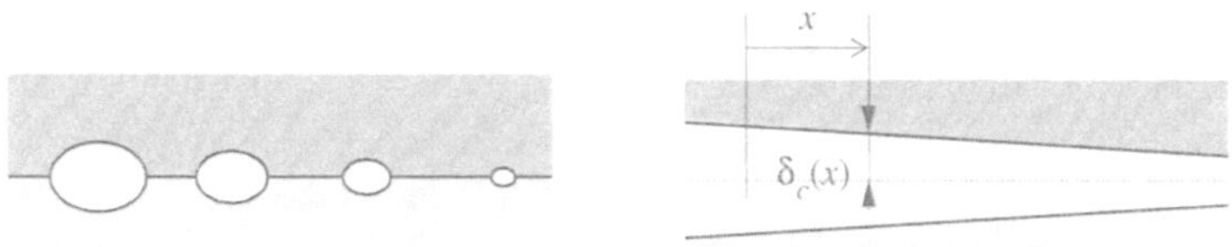


**Figure 11.** Illustration of the procedure used to numerically handle cavitating grain boundary facets. (a) Individual cavities with varying radius on a grain boundary; (b) 'smeared out' model in terms of a layer of thickness $\delta_c$. From Van der Giessen and Tvergaard (1994).

All grains are meshed by quadrilateral finite elements, with a refinement near the grain boundaries, and linear interface elements are used to connect the grains through their grain boundaries, see Van der Giessen and Tvergaard (1994).

The effect of constrained cavitation is demonstrated in Fig. 12, where different cell sizes are studied in which cavitation only takes place on the central grain boundary facet. The initial cavitation state is taken to be uniform over the facet and is specified by the initial radius $a_I = 0.1b_I$ and by the initial spacing $b_I = 0.1R_0$, $2R_0$ being the initial facet width. The grain boundary diffusion parameter $\mathcal{D}$ is specified in terms of the length scale $L$ measured relative to the initial cavity radius $a_I$ by the initial value $(a/L)_I = 0.025$. Here, $L_I$ corresponds to substituting the macroscopic effective stress $\Sigma_e$ and the associated creep rate $\dot{E}_e^C$ according to (1.1) into (2.15). The creep exponent, here and in the sequel, is $n = 5$. The applied macroscopic stress state is take to be such that $\Sigma_e = 0.5 \times 10^{-3}E$, while $\Sigma_1 = 0.5\Sigma_2$. Time $t$ is normalized by the reference time $t_R = \Sigma_e/(E\dot{E}_e^C)$. One observes from Fig. 12 that in the early stages of growth there is not much effect of the cell size, hence not a strong creep constraint. But for longer times, when $a/b$ approaches 1, the effect of constraint becomes very significant, as evidenced by the dependence on cell size $(m_1, m_2)$. Roughly speaking, one can say that the time to coalescence decreases when the density of cavitating grain boundaries, $1/(m_1 m_2)$, increases (thus confirming Dyson (1976)), but there also is an effect of the cell's aspect ratio $m_2/m_1$, i.e. the relative orientation of interacting facets.

## 5.2 Rupture Development

As creep rupture evolves, there is a period during which one can envision more or less isolated cavitating facets, but at some moment neighboring facets have to start cavitating in order for rupture to end in failure. An example of this process, simulated first by Onck and Van der Giessen (1997), is shown in Fig. 13. The differences with Fig. 12 are that the central grain boundary facet is taken to have already failed completely and that cavitation is also possible on all other facets. The initial cavities there are assumed to be very small, $a_I = 0.1b_I$, and nucleation is taken to be governed by (4.1) with $F_n = 100(\pi R_I^2)$ (the scaling parameter is arbitrarily set equal to $\Sigma_e$).

For a long time from the initial state, damage remains very small on all other facets. Only after about 80% of the lifetime $t_f$ does cavity coalescence lead to the formation of a microcrack on a neighboring facet, see Fig. 13a. Also on many other facets that are normal to the macroscopic tensile stress $\Sigma_2$ damage has develops, except for the ones above the central cracked facet; they are shielded. The damage distribution on other transverse facets is quite non-uniform with significant elevations near the grain triple points. This is due to the stress concentrations arising there from the free sliding condition, $\tau = 0$, on the inclined facets. The enhanced damage on the facet in the right-hand top corner is an artifact of the cell analysis and expresses the interaction with neighboring cells. From this state on, the rupture process proceeds very quickly, Fig. 13b. The nearest-neighbor facets fails completely, so that the stress on the remaining transverse facets increases even further. Coalescence soon leads to a string of failed transverse facets from the central one to the cell edge, after which the material falls apart by the grains sliding off.

## 6 Creep Crack Growth

The previous sections have dealt with the scales from Fig. 4e up to Fig. 4c. The next scale transition is one where the creep fracture mechanisms interact with the crack tip. As illustrated in Fig. 4b, this requires that we distinguish the region in which creep damage takes place *—the process zone—* from the surrounding material where just creep takes place (which, in turn, may be contained within a region of pure elastic deformation). In principle, this scale can be reached

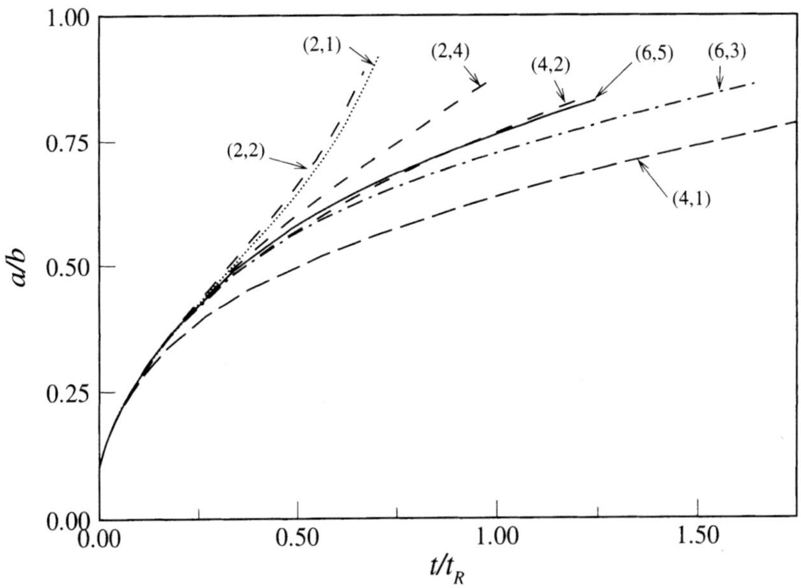


**Figure 12.** Development of damage, $a/b$, at the center of the central facet in different $(m1, m2)$ unit cells for an applied stress state with $\Sigma_1 = 0.5\Sigma_2$. The material is assumed to have free grain boundary sliding ($\eta_B = 0$), diffusion properties specified through $(a/L)_I = 0.025$ (see text), no nucleation of new cavities and the initial values of cavity radius and half-spacing are $a_I = 0.1b_I$ and $b_I = 0.1R_0$. From Van der Giessen and Tvergaard (1994).

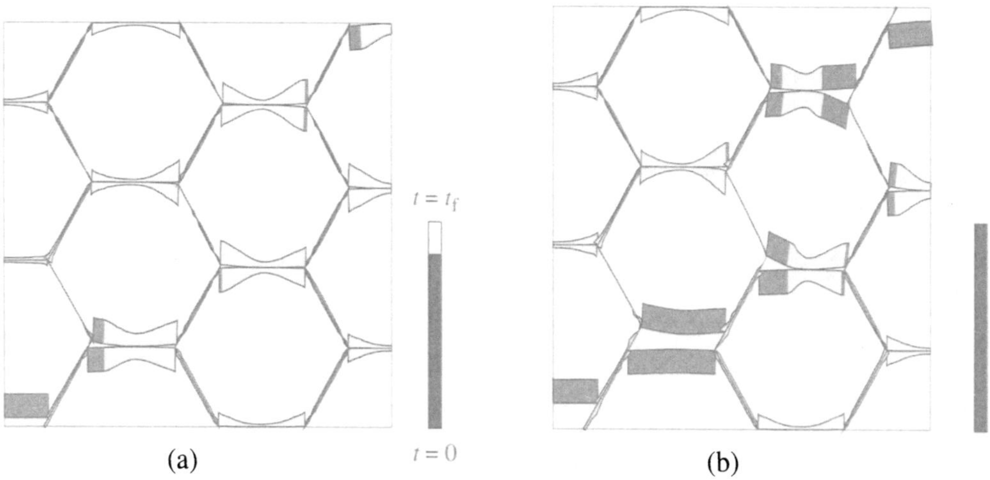


**Figure 13.** The state of damage at two instants $t/t_f$ in a quarter unit-cell under uniaxial tension ($\Sigma_1$) starting from a facet microcrack at the center of the unit cell (left-hand bottom corner in plot). Values of $a/b$ are plotted along, and on either side of the grain boundary facets. Microcracked regions where $a/b = 0.7$ are indicated by the darker gray scale. (a) $t/t_f = 0.83$; (b) $t/t_f \approx 1$. From Onck and Van der Giessen (1997).

with the multi-grain description of the previous section, but the number of grains contained inside the process zone maybe be prohibitively large for a computation with a fine resolution of the fields inside each grain. Therefore, Onck and Van der Giessen (1997) developed so-called *grain elements* which involves a coarser description that allows to analyze larger aggregates of grains.

### 6.1 Grain Elements

A grain element essentially is a special six-node finite element to model a single grain. It is defined so that displacements vary linearly along grain boundary facets; thus, the associated grain boundary elements are linear elements. A grain element provides a relatively crude description of the elastic deformations and creep inside the grain. In order to deal properly with incompressibility associated with creep deformations, the super element is built up of two quadrilateral elements, each of which consists of four-crossed linear triangles, as shown in Fig. 14. Other arrangements can easily lead to locking.

### 6.2 Small-scale Damage Analysis of Creep Crack Growth

Onck and Van der Giessen (1999) carried out a multi-grain analysis of mode-I creep crack growth on the basis of the model described above. It is based on the assumption of small-scale damage, i.e. cavitation damage is assumed to be confined to a region close to the crack tip,

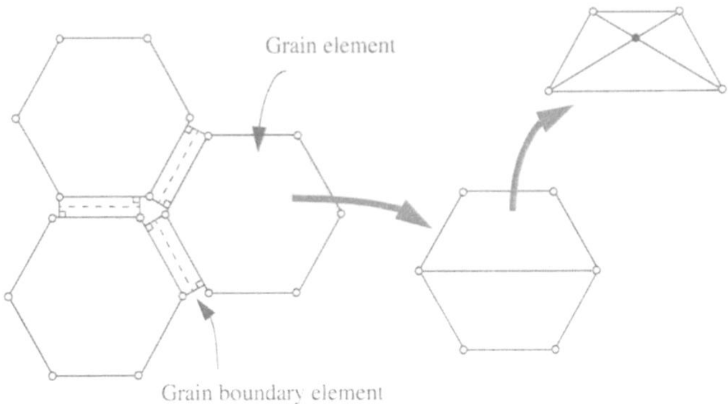


**Figure 14.** Grain-elements representation of a polycrystalline aggregate. Each grain element consists of two quadrilaterals, built up of four constant strain triangles. The grains are connected by grain boundary elements. From Onck and Van der Giessen (1997).

which in turn is surrounded by an HRR field[2]. The latter is the near-tip field of a purely creeping continuum —when the process zone is sufficiently small, this field at remote distances is not perturbed by the damage. Assuming that all elastic transients have relaxed before cavitation damage develops, the amplitude of the HRR field is uniquely determined by the steady-state value $C^*$, i.e.

$$\sigma^{ij} = \left[\frac{C^*}{BI_n r}\right]^{1/(n+1)} \tilde{\sigma}^{ij}(\theta, n),$$

with $r$ and $\theta$ polar coordinates centered at the initial crack tip, $B$ the creep parameter, eq. (1.1), and $I_n$ an $n$-dependent dimensionless constant. The novelty in their analysis was to explicitly represent the grain microstructure near the crack tip in the process zone, while the damage-free creeping material was represented by a continuum, as illustrated in Fig. 15. The un-perturbed HRR field is prescribed at the outer circular boundary at almost 5000 grain sizes away from the tip.

For the example result shown in Fig. 16 it is first pointed out that the simulations are done for material parameters specified similar as above, but in terms of a specific reference stress. This stress, $\Sigma$, is defined as the amplitude of the HRR stress field at a distance $R_I$ —the initial half-facet width—in front of the crack tip, i.e. $\Sigma = [C^*/BI_n R_I]^{1/(n+1)}$. Time is normalized by the characteristic time scale for creep, $t_R = 1/\dot{E}_e^C$, with $\dot{E}_e^C = B\Sigma^n$. The diffusion parameter $\mathcal{D}$ is specified through the reference (diffusion) length $L_R = (\mathcal{D}\Sigma/\dot{E}_e^C)^{1/3}$ defined in accordance with (2.15); the value $L_R/R_I = 0.032$ is paramount to $a_I/b_I = 0.1$ in the rupture study of the previous subsection. $\Sigma$ is also used as the normalization stress $\Sigma_0$ in (4.1). The value of the nucleation activity parameter $F_n$ is randomly assigned to all grain facets, according to a normal distribution. Because of the regular hexagonal structure, the behavior with zero boundary viscosity is pathological and therefore not shown. Figure 16 shows the result for a particular

[2]Named after the three authors, Hutchinson, Rice and Rosengren, who developed the asymptotic near-tip fields for a sharp crack in a plastic solid.

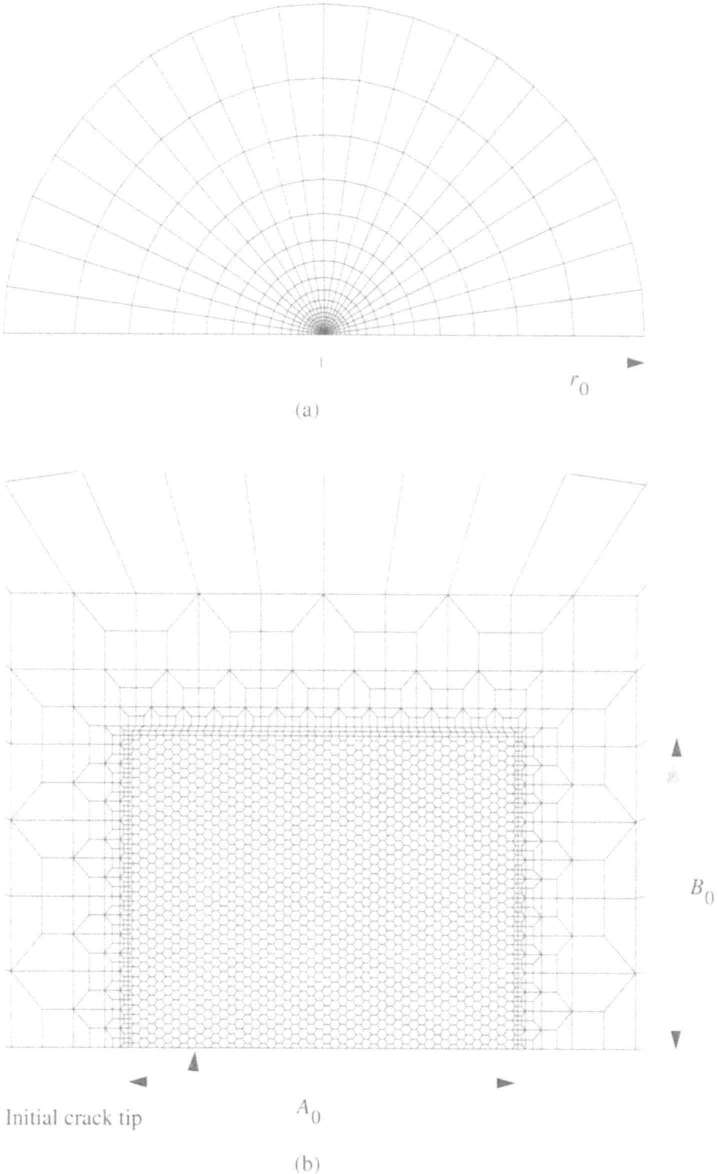


**Figure 15.** Finite element mesh used for small-scale damage analysis. (a) Circular domain with radius $r_0$. (b) Crack tip region, showing the process window with dimensions $A_0$, $B_0$, consisting of grain elements and grain boundary elements. Cavitation damage is allowed to occur along all grain boundaries in the process window, containing $48 \times 33$ grains. From Onck and Van der Giessen (1999).

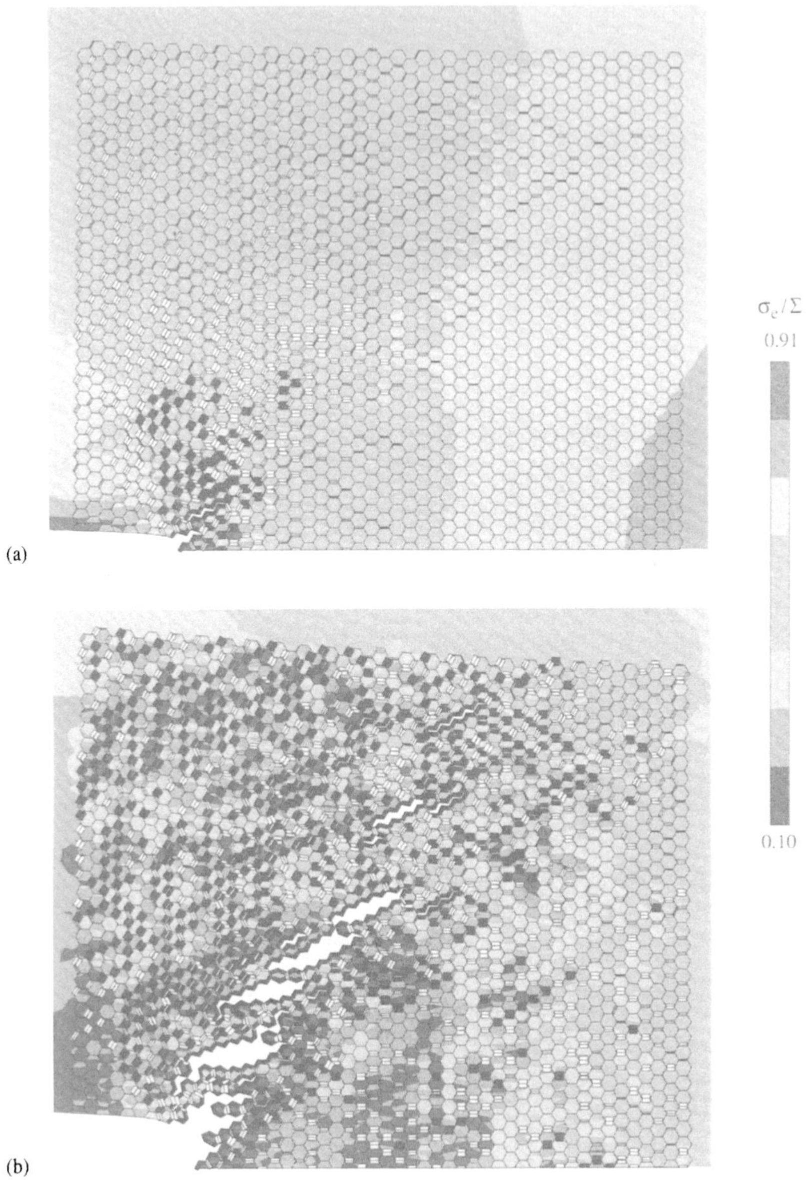


**Figure 16.** Grain boundary cavitation state and local effective stress distribution at two instances during crack extension for $L_R/R_I = 0.032$ with a random distribution of $F_n$. (a) $t/t_R = 1.10$; (b) $t/t_R = 2.73$. From Onck and Van der Giessen (1999).

value of $\eta_B/w$, chosen such that $\dot{E}_e^C/\dot{\varepsilon}_B = 10$ at a distance $R_I$ from the crack tip (corresponding to $f^* = 1.01$).

Relatively early in the process, Fig. 16a, some local damage has developed in a region of about 10 grains around the tip, which apparently has not yet destroyed the HRR stress field. A number of facet cracks has formed (noticeable by the black marks), and due to the random nature of the nucleation activity, cavitation damage has also started at larger distances from the tip. In Fig. 16b, crack extension is well underway. The key feature is that the main crack grows by the linking up of short cracks, spanning some eight grains or so. Foot-printed by the regular hexagonal structure, the short cracks are oriented at 30° from the symmetry plane. The linking-up process renders the main crack to grow at about 55°.

## 7 Exercises

**Exercise 7.1.** Show that in plane strain, pure creep in the $x_1$–$x_2$ plane, $\sigma_{33} = \frac{1}{2}(\sigma_{11} + \sigma_{22})$.

**Exercise 7.2.** Show from (3.15) that the average normal stress on facets in family #2 under a uniaxial state of remote stress, i.e. $\Sigma_1 = 0$ in Fig. 8, is ${}^{fsl}\sigma_n^{(2)} = (3/2)\Sigma_2$, i.e. an amplification by 50% due to free sliding. Also calculate the average facet stresses on facets #1 and #3 when $\Sigma_2 = 0$.

**Exercise 7.3.** Integrate the cavity growth relations (2.5) by assuming that the near facet stress state is equal to the macroscopic stress state considered in Fig. 12. The result neglects cavity growth by creep; extend the calculations by using the full growth relations (2.11) for a diffusivity specified through $a_I/L = 0.025$. Compare this *unconstrained cavitation* result with the ones in Fig. 12.

## Bibliography

A.S. Argon, I.-W. Chen and C.W. Lau. Intergranular cavitation in creep: theory and experiments. In R.M. Pelloux and N.S. Stoloff, editors, *Creep-Fatigue-Environment Interactions*, pages 46–83. New York, NY, 1980.

M.F. Ashby. Boundary defects and atomistic aspects of boundary sliding and diffusional creep. *Surface Sci.*, 31:498–542, 1972.

B. Budiansky, J.W. Hutchinson and S. Slutsky, S.. Void growth and collapse in viscous solids. In H.G. Hopkins and M.J. Sewell, editors, *Mechanics of Solids: The Rodney Hill 60th Anniversary Volume*, pages 607–652, Pergamon Press, Oxford, 1982.

B.F. Dyson. Constraints on diffusional cavity growth rates. *Metal Sci.*, 10:349–353, 1976.

B.F. Dyson. Continuous cavity nucleation and creep fracture. *Scr. Metall.*, 17:31–37, 1983.

F. Ghahremani. Effect of grain boundary sliding on steady creep of polycrystals. *Int. J. Solids Struct.*, 16:847–862, 1980.

D. Hull and D.E. Rimmer, The growth of grain boundary voids under stress. *Phil. Mag.*, 4:673–687, 1959.

A. Needleman and J.R. Rice. Plastic creep flow effects in the diffusive cavitation of grain boundaries. *Acta Metall.*, 28:1315–1332, 1980.

P.R. Onck and E. van der Giessen. Microstructurally-based modelling of intergranular creep fracture using grain elements. *Mech. Mater.*, 26:109–126, 1997.

P.R. Onck and E. van der Giessen. Growth of an initially sharp crack by grain boundary cavitation. *J. Mech. Phys. Solids*, 47:99–139, 1999.

B. Ozmat, A.S. Argon and D.M. Parks. Growth modes of cracks in creeping Type 304 stainless steel. *Mech. Mater.*, 11:1–17, 1991.

J.R. Rice. Constraints on the diffusive cavitation of isolated grain boundary facets in creeping polycrystals. *Acta Metall.*, 29:675–681, 1981.

H. Riedel. *Fracture at high temperatures*. Springer–Verlag, Berlin, 1987.

V. Tvergaard. On the creep constrained diffusive cavitation of grain boundary facets. *J. Mech. Phys. Solids*, 32:373–393, 1984.

E. Van der Giessen and V. Tvergaard. Interaction of cavitating grain boundary facets in creeping polycrystals. *Mech. Mater.*, 17:47–69, 1994.

FSC
www.fsc.org
MIX
Papier | Fördert gute Waldnutzung
FSC® C083411

Kolibri 360 GmbH
Ferdinand-Jühlke-Straße 7
99095 Erfurt, Deutschland
produktsicherheit@kolibri360.de